D0698958

Control
of
Human
Movement

Mark L. Latash, PhD
Rush-Presbyterian–St. Luke's Medical Center
Chicago, Illinois

Human Kinetics Publishers

Library of Congress Cataloging-in-Publication Data

Latash, Mark L., 1953-
    Control of human movement / Mark L. Latash.
        p.   cm.
    Includes bibliographical references and index.
    ISBN 0-87322-455-8
    1. Motor learning.   2. Human locomotion.   I. Title.
    QP301.L36   1993
    152.3--dc20   92-36991
                                                            CIP

ISBN: 0-87322-455-8

Copyright © 1993 by Mark L. Latash

All rights reserved. Except for use in a review, the reproduction or utilization of this work in any form or by any electronic, mechanical, or other means, now known or hereafter invented, including xerography, photocopying, and recording, and in any information storage and retrieval system, is forbidden without the written permission of the publisher.

Acquisitions Editor: Rick Frey, PhD; **Developmental Editor**: Larret Galasyn-Wright; **Assistant Editors**: Laura Bofinger and John Wentworth; **Copyeditor**: Chris DeVito; **Proofreader**: Pam Johnson; **Indexer**: Theresa Schaefer; **Production Director**: Ernie Noa; **Typesetter and Text Layout**: Yvonne Winsor; **Text Design**: Keith Blomberg; **Interior Art**: Denise Lowry; **Cover Design**: Jack Davis; **Printer**: Braun-Brumfield

Printed in the United States of America   10 9 8 7 6 5 4 3 2 1

Human Kinetics Publishers
Box 5076, Champaign, IL 61825-5076
1-800-747-4457

Canada: Human Kinetics Publishers, P.O. Box 2503, Windsor, ON N8Y 4S2
1-800-465-7301 (in Canada only)

Europe: Human Kinetics Publishers (Europe) Ltd., P.O. Box IW14,
Leeds LS16 6TR, England
0532-781708

Australia: Human Kinetics Publishers, P.O. Box 80, Kingswood 5062,
South Australia
618-374-0433

New Zealand: Human Kinetics Publishers, P.O. Box 105-231, Auckland 1
(09) 309-2259

OLSON LIBRARY
NORTHERN MICHIGAN UNIVERSITY
MARQUETTE, MICHIGAN 49855

To my dear parents—
the best parents on
this planet—
Sara and Lev Latash

Contents

Preface

Don't try to embrace the unembraceable.
                                        Koz'ma Prutkov

The approach to motor control that I advocate here originates from the classic works by Nicholai Bernstein and was intensively developed during the last three decades. The last book devoted to this approach was written by Bernstein a quarter of a century ago. Around the same time, Ragnar Granit wrote his classic book *The Basis of Motor Control*, which is still one of the most exhaustive and commonly used sources of information in the field. More recently, several monographs on motor control have been published, including outstanding books by Jeannerod (1988) and Schmidt (1988). However, Schmidt's book is devoted mostly to the behavioral aspect of motor control, whereas the book by Jeannerod puts greater emphasis on the role of perception. Both these aspects are extremely important, but there was still a detectable lack of an update that would follow Bernstein's tradition. My primary aspiration is to fill this gap while attempting to answer three questions:

- What are the variables used by the central nervous system to control voluntary movements?
- How do these variables interact with muscle reflexes?
- How are these variables translated into peripheral motor patterns?

Although intended as both a reference book and a text for graduate students, this book's complex and controversial nature may well prove difficult for unprepared students. I wholeheartedly tried to include all the necessary information to make the book self-contained and readable, however, a strong background in physiology, physics, mathematics, and control theory is certainly desirable. The book is also targeted at potential readers working at the border of motor control and related applied areas, such as kinesiology, clinical neurophysiology, and physical therapy. Chapters 6 and 9 are devoted to certain more applied aspects of the general ideas I advocate.

This book was not designed merely as a review. The main goal was to present an approach to motor control studies based upon control of the equilibrium states of the neuromotor system. The "key" concept here is probably *interaction*. According to this approach, peripheral patterns of voluntary movements are not prescribed by a smart controlling system but rather emerge as a result of interaction between the body and the environment and between different structures within the body. This approach is illustrated throughout the book in chapters addressing such issues as single-joint and multijoint movements, the emergence of electromyographic patterns, the phenomena of motor learning and variability, and some of the pathological states of the motor system.

Three chapters stand somewhat aside from the others. Chapter 2 is devoted to the problem of measuring joint compliance. This may seem relatively unimportant and technical but it is crucial for understanding the experimental foundations of the equilibrium-point hypothesis. Chapter 8 describes models of motor control based on optimization criteria, which may seem to be in contradiction with the rest of the book. However, it sharpens (I hope!) the general picture. The first draft of chapter 10—on the similarities between language and motor function—was written about 10 years ago when I was an unemployed refusenik in Russia and had a lot of free time for groundless speculations. This may still be reflected in the chapter's somewhat relaxed style of presentation.

Writing a reference book examining the works of colleagues is probably the best way to turn friends into enemies and provoke "enemies" to even more belligerence. This may be especially true of this book because it was originally conceived as biased. Besides that, the controversial nature of the book means that on some points it could well be *wrong*, although I honestly do not know where.

Acknowledgments

My scientific career has been very lucky; even eight years of virtual unemployment in Russia failed to spoil this belief. Throughout my life, I have been blessed with the caring scientific guidance of my father, Dr. Lev Latash. I have had outstanding friends-advisors who have profoundly influenced my thinking: Victor Gurfinkel, Anatol Feldman, and Gerald Gottlieb. I have had the opportunity to work in both basic and clinical research environments studying patients, rats, athletes, frogs, and colleagues; I have been part of archaeological expeditions excavating ancient Greek colonies and the capital of the Golden Hordes; I have even worked as a cook. Finally, I have met only very nice people during my research activities. I would particularly like to thank those with whom I have collaborated on exciting projects: Gil Almeida, Michael Berkinblit, Daniel Corcos, Olga Fukson, Simon Gutman, Mark Lipshits, and Konstantin Popov, and those with whom I have discussed ideas and results: Gyan Agarwal, Greg Anson, Simon Bouisset, Dick Burgess, Pat Crago, Roger Enoka, Tamar Flash, Stan Gielen, Peter Greene, Zia Hasan, Ken Holt, Jim Houk, Scott Kelso, Emily Keshner, Jeff Kroin, Ning Lan, Karl Newell, Zev Rymer, Gregor Schoner, Mark Shik, Doug Stuart, Michael Turvey, Charles Walter, and Charles Worringham. Very special thanks are due to Richard Nichols and Les Carlton for a thorough, productive critique, to Lloyd Partridge for his generous help and support during my emigration, and to Richard Penn and Jeff Nicholas for their financial support during different phases of work on this book.

Chapter 1

What Muscle Parameters Are Controlled by the Nervous System?

To study voluntary control of a muscle or a joint, one needs to introduce hypothetical variables that are likely to be used by the control structure as a language for communication with the "lower" neural structures and the muscles. Our present knowledge of the central structures and connections of muscles and their receptors is very limited and forces a metaphorical approach based on ideal rather than on directly observable variables. The black box approach—that is, attempts to understand a complex system by analyzing input-output relations—has led to a number of mass-spring models of motor control. The equilibrium point hypothesis (λ-model) I advocate introduces a monoparametric control for a muscle and a two-parametric control for a joint.

The question of what parameters are controlled by the nervous system has been investigated extensively and discussed at practically every stage of the history of motor control studies. Many monographs, reviews, and conferences have been devoted to the problem (Granit, 1970; Greene, 1972; Arbib, 1980; Houk & Rymer, 1981; Stein, 1982; Berkinblit et al., 1986a). But even the formulation of the original problem does not have uniform understanding. This is exemplified by a series of review articles and open peer commentaries published in the journal *Behavioral and Brain Sciences* (Stein, 1982; Berkinblit et al., 1986a; Gottlieb et al., 1989a).

One important source of misunderstanding lies in the word *controlled*. From everyday experience and experimental studies, we know that human beings and animals can be trained to "control" virtually any

1

variable characterizing voluntary movement: joint angle (position), joint torque (force), movement speed, accuracy, and even one as exotic as the second derivative of torque changes (Ghez & Gordon, 1987). All these variables describe voluntary movements at the level of performance. We are going to try to answer this question: Can we find a variable at another level of the motor control hierarchy whose time changes would be "controlled" (supplied by the higher levels) for all the variety of motor tasks?

Thus, *controlled variable* will be understood as a signal supplied by one "upper" level of the motor control system to another (or to the executing apparatus) independently of the current conditions of task execution that frequently depend on external, and sometimes unpredictable, factors. Independence of the peripheral information about the external conditions does not mean that the upper hypothetical level does not receive this information or is unable to use it. For example, this information can be used for planning or correcting chosen patterns of the controlled variables. It is important, however, that if a certain time profile of a controlled variable is chosen, it is independent of the peripheral information until the upper level decides to change it. Signals from one level to another may not have well-defined neuronal counterparts and therefore are close to an analytical fiction, but I hope to demonstrate that they are a very helpful analytical fiction. In this framework, it appears that none of the variables cited earlier can be controlled, because all of them depend on current conditions of motor task execution.

For example, if an experimenter asks a subject to occupy a constant static position in a joint and applies an unexpected change in the external load, the position will change (cf. Feldman, 1966a,b; Vincken et al., 1983). So when an experimenter asks a subject to occupy a position, the subject's central motor command does not encode the position by itself, but rather encodes something different that implies this position in the given external conditions (load). The question is, What is this something?

A systematic search for the "independently controlled" variables (variables supplied by the higher levels of the motor control system, independent of peripheral conditions) began in the second half of the twentieth century, obviously inspired by the overwhelming success of cybernetics. However, the first works in this field were written much earlier (Bernstein, 1926, 1935), and their author, Nicholai Bernstein, introduced the cybernetic approach 20 years before the classic works by Norbert Wiener (1948). The depth of Bernstein's understanding of the motor control problems has hardly been surpassed since, and many contemporary authors acknowledge that their works are founded on the ideas originally expressed by Bernstein (Brookhart, 1979; Feldman, 1979, 1986; Talbott, 1979; Brooks, 1981; Kelso et al., 1980; Kugler & Turvey, 1987; Jeannerod, 1988; Schmidt, 1988; Turvey, 1990b).

Bernstein was the first to address the motor control system as a "black box" with a virtually unknown internal structure that must control an effector apparatus of multiple links and degrees of freedom. His studies were based on the physical approach to such systems, namely investigation of "input-output" relations. The input was modulated by giving different instructions to the subjects, by changing peripheral conditions of task execution, or both; the output was monitored as kinematics of the movements. Cinematographic analysis of voluntary movements led Bernstein to very profound conclusions concerning general properties of control systems that would be capable of demonstrating such behaviors. His conclusions included the following:

1. The control system should represent a hierarchy of several levels.
2. There should be feedback loops connecting the "lower" levels with the "higher" ones used for tuning the "descending" commands.
3. Inevitable time delays in the feedback loops require combining feedback and predictive, open-loop modes of control.
4. The number of degrees of freedom in a motor system is always excessive, and the process of control can be regarded as overcoming the ambiguity caused by redundant degrees of freedom (the Bernstein Problem).

For the purposes of our discussion, we will not analyze all the levels of Bernstein's original scheme of motor control (Bernstein, 1947, 1967); instead, we will accept the black box approach and introduce only three levels in the motor control system that are useful for describing how the system functions and for interpreting a body of experimental observations. Figure 1.1 shows a scheme of voluntary motor control that is organized on a functional basis. So, when we speak about upper or lower levels of motor control, this will not necessarily mean anatomical location along the neural axis. The hierarchical scheme shown in Figure 1.1 should not be considered a reflection of actual organization of the central nervous system controlling voluntary movements. It is rather a metaphorical, simplified model that appears helpful for discussing some of the properties of the "actual" system with a virtually unknown internal structure.

The upper level in the scheme (level 1, "decision making" in Figure 1.1) may tentatively be associated with production of a "voluntary central motor command," although "voluntary" should not be confused with "conscious." It is tempting to describe this level as "supraspinal," but there are a lot of data demonstrating complex coordinated behavior in decerebrate or spinal animals (Shik & Orlovsky, 1976; Forssberg, 1979a,b; Fukson et al., 1980; Stein, 1983; Berkinblit et al., 1986a), which prevent even such minor "localizationism." On the other hand, can this behavior be considered voluntary?

Let us move away from this linguistic argument. Knowledge of functional structures and connections at this level of the black box is limited. So, we will hypothesize that there is a smart and experienced homunculus

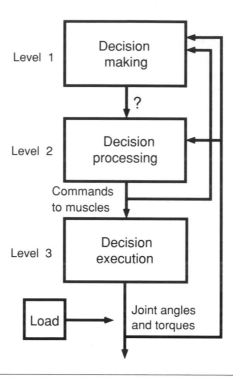

Figure 1.1 This tentative scheme of voluntary motor control includes three major levels: decision making, decision processing, and decision execution. Note feedback loops connecting all three levels. We will be mostly interested in the functioning of the decision-processing level (level 2).

sitting at the upper level who gets proprioceptive information from the afferent sources, combines it with information from other sources (visual, auditory, and others), and generates relevant descending signals to the "intermediate" level based on the task (will) and memory. The main question of this chapter can now be reformulated: How can one describe relevant motor command when the only reliable measure, and therefore the only reliable language for the description, is given by the output characteristics?

We have already started using a number of loosely defined terms, in particular *motor program*. Following Bernstein, let us use this term to denote a prototype of a planned movement expressed in an abstract time with a hypothetical central "language" whose features encode specific characteristics of the planned movement. This definition has been elaborated by Schmidt (1980a,b, 1988) in the form of a concept of generalized motor program. Note, however, that in some of the contemporary dynamic models of motor control (Saltzmann & Kelso, 1987; Kelso & Schoner, 1988; Schoner & Kelso, 1988), the notion of a motor program seemingly disappears and is replaced by dynamic transitions between the system's equilibrium states. Let us stop now before going too far into the notion of motor program and related problems; they will be discussed in detail in chapter 6.

The lower level of the scheme (level 3, "decision execution" in Figure 1.1), combines two important functions: executing the movement and informing the upper control levels on the course of the movement and changes in the external conditions. These two functions look quite different, but some studies have suggested that they are very closely interrelated and are likely to use the same receptors, afferent pathways, and central connections (Von Holst, 1954; McCloskey, 1978; Feldman & Latash, 1982a,b; Brodie & Ross, 1984).

The intermediate level of the scheme in Figure 1.1 (level 2, "decision processing") is the black box in whose function we are particularly interested. The upper decision-making level is a smart and well-informed homunculus, so for the time being we abstain from any attempt at understanding how it comes to its wise decisions. The lower decision-executing level is much more accessible for direct measurements and looks, in our context, more "simple" (although it is far from fully understood; see Partridge & Benton, 1981; Agarwal & Gottlieb, 1982; Hoyle, 1983; Enoka, 1988) than the decision-making and the decision-processing levels. The intermediate level, which we will not even try to localize, takes all the responsibility for correctly interpreting the descending command signals and processing them so that the decision-executing level performs at appropriate values of joint angles and torques.

In one of the chapters in his monograph, which appeared originally as an article in 1935, Bernstein (1967) warned against drawing too close a parallel between coordination and localization. From his view, the models of motor control should not be directly based on anatomical findings because the presence of an anatomical connection by itself says nothing about its possible functional importance, and the absence of direct anatomical linkage does not prove functional independence. However, information on anatomical connections, although far from exhaustive, can help in restricting possible models to compatibility with known facts. So, let us now try to "open" the three boxes in Figure 1.1 and look inside. Figure 1.2, which should be taken with a grain of salt, shows some of the known internal structures of the three levels involved in performing a voluntary movement.

The decision-executing level 3 incorporates muscles, joints, tendons, skin, and subcutaneous structures with all the receptors sensitive to stimuli of different modalities. It also includes the postsynaptic structures of α-motoneuron and γ-motoneuron pools. As such, this level represents a system that can execute the input commands (the total presynaptic input to the α-motoneurons) and inform the upper levels about the progress of movement execution.

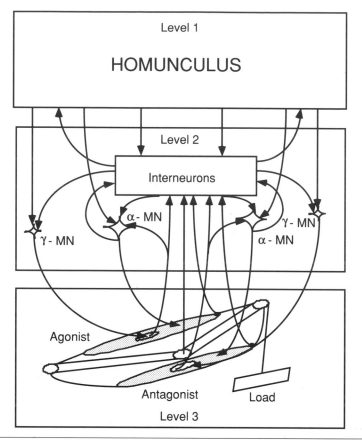

Figure 1.2   A tentative scheme of voluntary motor control with "open boxes" (cf. Figure 1.1). Note multiple feedback loops between the levels. The scheme looks like an extremely intricate structure although only a small fraction of the connections are shown explicitly. However, the black box approach lets one describe the functioning of this system with only a few parameters.

The decision-processing level 2 incorporates most of the intraspinal structures but need not be confined to them. These include, for example, monosynaptic and oligosynaptic reflex loops from the peripheral receptors to α-motoneurons and γ-motoneurons, arcs of recurrent and reciprocal inhibition, and less defined polysynaptic chains with all the complexity of interaction of different loops. (More information on the lower two levels can be found in "Muscle and Its Reflexes" later in this chapter.) It seems absolutely impossible to trace all the connections at this level and to assess their relative functional significance during execution of different motor tasks. However, the framework provided by Bernstein does not require all the information about internal structures of a part of a control system.

The upper level still remains a black box with a homunculus inside, since the available data do not let one make even a tentative draft of the internal structure of the decision-making mechanism. The important roles of certain brain structures, such as cerebellum, basal ganglia, red nucleus, and premotor and motor areas of the cortex, are widely acknowledged (Asanuma, 1973; Evarts, 1973, 1974; Kornhuber, 1974; Conrad, 1978; Brooks, 1979, 1981; Eccles, 1981; Gibson et al., 1985; Dormont et al., 1989) along with the importance of certain neural tracts, including pyramidal and extrapyramidal ones (Ashby et al., 1972; Asanuma et al., 1981; Chapman & Wiesendanger, 1982). However, the process of decision-making in motor control is still waiting for someone to introduce a framework for its analysis. Figure 1.2 does not show feedback signals to level 1 from level 3, implying that the homunculus is able to generate descending commands independently of the peripheral information. It can certainly use this information together with other sources (e.g., visual, auditory, etc.) for planning and/or correcting the descending signals.

The black box approach to motor control studies was for a long time used, on a relatively wide scale, only by Bernstein's disciples in Russia for studying processes of control of posture and movement (see a collective monograph dedicated to Bernstein published in Russian in 1967 and later translated into English: Gelfand et al., 1971b). One of the models, based on the Bernsteinian approach, has proved to be especially influential on the subsequent history of motor control studies. This model, known as the equilibrium-point hypothesis or the λ-model, was introduced by Feldman in the 1960s (Asatryan & Feldman, 1965; Feldman, 1966a,b), and belongs to the class of "mass-spring" models of muscle behavior.

We shall restrict ourselves to analysis of the intermediate level of signal processing in the introduced scheme of voluntary motor control; we will pretend that, on the one hand, we give up the attempt to understand how the upper level generates correct descending commands, and, on the other hand, we know how the output signals are processed by the lower level leading to joint angle and torque changes.

Spring Properties of an Isolated Muscle

Mechanical properties of skeletal muscles have been studied for more than a century. Muscles were studied as isolated physical structures characterized by certain relations between changes in externally applied force and length, as excitable innervated structures able to substantially change their mechanical properties in response to action potentials in the muscle nerve, and as parts of a complex system incorporating feedback loops from receptors sensitive to changes in muscle length, force, rate of length changes, pressure, and so on. Each new step adds to the complexity of the problem, since more and more elements are incorporated into the system, and the system acquires new qualitative features that were impossible to predict at an earlier stage.

Even the isolated skeletal muscle has a rather complicated structure that includes both active and passive elements. Some of the mechanical characteristics of the muscle reflect the characteristics of a sarcomere. For example, Hill's classic equation (Hill, 1938; see chapter 2) describing the relation between force and speed of muscle shortening during tetanic stimulation can be used for a sarcomere (Destcherevsky, 1977).

There are many models of isolated muscle of different levels of complexity that usually include, at a minimum, contractile, dashpot, and elastic elements (e.g., Hill, 1953; McMahon, 1984) providing for different degrees of accuracy in describing the isolated muscle behavior. Figure 1.3 presents an example of such models of isolated muscle. However, we will be interested in certain qualitative features of force-length relationships characterizing the muscle as a black box; we will not try to model the internal structure that causes the observed types of behavior.

A slowly stretched passive muscle (without innervation) starts to noticeably resist the stretching when its length exceeds a certain value called "resting length" of the muscle l_0 (Buchtal, 1942; Ralston et al., 1947; Partridge & Glaser, 1960). It behaves like a nonlinear spring with changing stiffness up to rather large length values (as compared to the range of muscle length changes in the body).

To study the mechanical properties of muscle, one can simulate the effects of innervation, which can be done with electrical stimulation of the muscle nerve. In an intact organism, the muscles usually perform smooth tetanic contractions that can be evoked by a high-frequency (about 25-40 Hz) electrical stimulation (Granit, 1970; Partridge & Benton, 1981). The method of splitting the muscle nerve and stimulating the branches separately, introduced by Rack and Westbury (1969), lets one produce

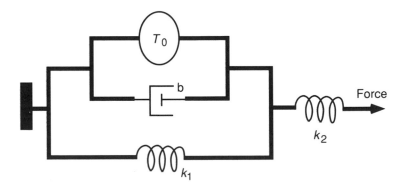

Figure 1.3   An example of an isolated muscle model. An active contractile component (T_o) provides for an increase in muscle force, which is damped by a dashpot element (b). Two elastic elements (k_1 and k_2) are connected in parallel and in series with the contractile element. The resultant dependence of muscle length on force demonstrates properties of a nonlinear damped spring.

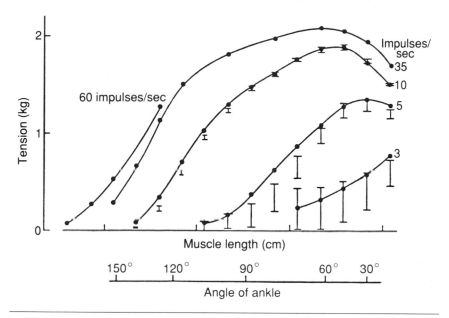

Figure 1.4 Force-length characteristics recorded in a cat soleus muscle on the background of muscle nerve stimulation at different levels. Note a shift of the characteristics to the left with an increase in the stimulation rate.
Note. From "The Effects of Length and Stimulus Rate on Tension in the Isometric Cat So-leus Muscle" by P.M.H. Rack and D.R. Westbury, 1969, *Journal of Physiology*, **204**, p. 455. Copyright 1969 by The Physiological Society. Reprinted by permission.

sustained tetanic contractions at lower frequencies of stimulation, which resemble those recorded in intact muscle nerves.

Force-length characteristic curves during slow stretches and isometric contractions were recorded at different levels of sustained muscle contraction induced by nerve stimulation (Granit, 1955, 1970; Rack & Westbury, 1969; Houk & Rymer, 1981). An example of a family of these curves is presented in Figure 1.4. Note that the electrical stimulation tends to shift the curves to the left. If one describes the muscle as a spring (Weber, 1846; Bernstein, 1926), the effects of electrical stimulation can be considered as shifts in the resting length of the spring.

The next step would be to investigate the force-length characteristic curves of a muscle with intact central connections. Before doing this, however, it seems reasonable to outline some of the characteristic features of intact muscles and their reflex connections.

Muscle and Its Reflexes

Intact muscle is a sophisticated force generator whose design suggests one of the two following conclusions:

• All the nonlinearities of muscle behavior and multiple feedback reflex connections with different time delays are a "boring atavism" complicating the process of motor control. However, the central controller needs to cope with these "inconveniences" and does so by trying to predict their effects and, if possible, eliminating them.

• The unique design of muscle and its reflex connections represents an essential part of the mechanism of motor control. Central programs are built on these features. All the inconveniences, in fact, simplify motor control in a changing environment, make it more reliable, and facilitate acquisition of new motor programs.

The second view was originally introduced by Sherrington (1906) who considered motor control as being based on regulation of parameters of muscle reflexes. Later, this view was developed by Merton (1953) in a form of "servo-hypothesis," and eventually led to the equilibrium-point hypothesis (λ-model; Feldman, 1966a,b, 1979, 1986), which provides, from the author's very subjective view, a productive framework for analysis of different issues in motor control.

Let us first describe muscle reflex connections and try to assess their possible functional role. Skeletal muscles, which enable us to make virtually all voluntary movements, and the surrounding tissues house a variety of receptors (i.e., peripheral endings of neural cells that can generate action potentials in response to some specific stimuli). Each type of nerve ending is particularly sensitive to a certain type of stimuli, for example stretch, tension, pressure, temperature, certain chemicals, and so on, but it can usually be forced to fire with other stimuli as well (e.g., by electrical stimulation). However, each group of receptors, under normal conditions, is very specialized.

Peripheral endings send their axons to spinal ganglia (where the neuron's body is located) and also to the spinal cord through its dorsal roots (Figure 1.5). In the spinal cord, signals generated by the peripheral endings can be transmitted to other cells within the spinal cord and/or to the brain. After traveling through different neural pathways and structures, the signal can return to a muscle located in the vicinity of the signal's origin or to other muscles, including those of other limbs, or it might have no visible effects on muscle activity at all. Such a signal never goes straight to a muscle, though; it undergoes at least minimal neural processing, as in the so-called monosynaptic reflexes.

We can identify the following major receptor groups that are likely to play an important role in processes of motor control:

• Muscle spindles are oriented parallel to the muscle fibers and contain two types of length-sensitive endings. The endings of the first type (group Ia afferents) are sensitive to both muscle length and rate of change, while the second group of endings (group II afferents) are sensitive only to the

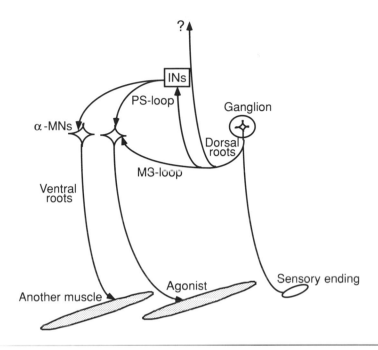

Figure 1.5   This scheme shows some of the connections between a receptor and a muscle housing this receptor (agonist). The body of the receptor neuron is located in a spinal ganglion. Its T-shaped axon sends one of the branches to the sensory ending in the muscle and the other branch through the dorsal roots into the spinal cord. This branch can go straight to α-motoneurons (α-MNs) of the agonist muscle making a monosynaptic connection (MS loop). These connections have been shown only for Ia afferents of muscle spindles. Signals from the receptor can be processed by interneurons (INs) and lead to oligosynaptic or polysynaptic reflex effects on either the agonist or another muscle (PS-loop). They can also go to "higher" neural levels (?) and induce either conscious changes in the motor command or give rise to triggered or pre-programmed reactions, or they can lead to no visible motor effects at all.

length itself. Sensitivity of both types of spindle endings is controlled by a specialized spinal system of small motoneurons (γ-system).

• Golgi tendon organs are oriented in series to the muscle fibers and are sensitive to muscle force changes.

• Articular receptors are located in joint capsules and are sensitive to both joint angle and capsule tension, which in turn depends on torque generated by load and the muscles around the joint.

• Skin and subcutaneous receptors are sensitive to pressure.

• Nociceptive receptors are sensitive to a variety of potentially damaging stimuli including, for example, high temperature or very high pressure (as in the case of a pin prick).

We are not going to discuss separately the possible contribution of each and every kind of peripheral receptor to muscle activation, although this is an interesting problem per se. The "complex system" approach to motor control requires information on general characteristics of muscle behavior defined by the functioning of all these groups of receptors together. Figure 1.5 schematically shows major functional connections between some generalized muscle receptor and the muscle itself. Let us discuss these loops and their possible functions.

Monosynaptic Reflexes

These reflexes are the only ones whose afferent source and reflex pathway are relatively well defined. They originate from primary spindle endings and make only one intraspinal excitatory connection (synapse) with α-motoneurons of the muscle housing the spindle or, sometimes, its agonist (i.e., a muscle that causes joint movements in the same direction). This definition of the agonist, as well as analogous definition of the antagonist as a muscle that causes joint movement in the opposite direction, are far from perfect (cf. "A Special Role for the Biarticular Muscles" in chapter 7 and Zajac & Gordon, 1989). However, we are going to use these loosely defined but commonly used terms since in most cases they do not lead to misunderstanding.

Two methods are commonly used for evoking monosynaptic reflexes: electrical stimulation of the muscle nerve and rapid muscle stretch (e.g., tendon tap). In the latter case, the term *phasic stretch reflex* or just *stretch reflex* is used. General properties of the monosynaptic reflexes are extensively described in a number of works (Hoffmann, 1922; Paillard, 1955; Eccles, 1964; Granit, 1970; Delwaide, 1973; Burke et al., 1983, 1984), so let us fix our attention only on those that seem important for assessing possible contribution of monosynaptic reflexes to motor control:

• Monosynaptic reflexes are characterized by the shortest latencies (i.e., time delays between a stimulus and the first detectable muscle reaction). In human limb muscles, the latencies of the monosynaptic reflexes are of the order of 25 to 40 ms. These time delays are due to conduction time in axons of the receptor cells and α-motoneurons and to a synaptic transmission delay which is relatively short (about 0.5 ms).

• Monosynaptic reflexes are typically phasic; that is, they emerge in response only to rapid *changes* in muscle length, and lead to burst-like activity of muscle fibers and a twitch contraction.

• Monosynaptic reflexes are very poorly controlled by the subject's will, although they have been shown to be controllable by monkeys in experiments with extensive operant conditioning (Wolpaw, 1983; Wolpaw et al., 1983). Human subjects can modulate amplitude of the monosynaptic reflexes indirectly by, for example, activating certain muscle groups, but this is not fair.

• Monosynaptic reflexes demonstrate pronounced modulation just before and during voluntary movement.

• Characteristics of monosynaptic reflexes are changed in certain states of motor disorders (e.g., spasticity).

Oligosynaptic Reflexes

The group of oligosynaptic reflexes includes, by definition, those with a small number of synapses, but more than one. The precise number of synapses is undefined but it is usually assumed to be two or three. These reflexes can lead to both excitatory and inhibitory effects on muscle activity.

For example, one of the best known oligosynaptic connections is from muscle spindle receptors to α-motoneurons of the antagonist muscle, mediated by one interneuron. The latency of inhibitory muscle reactions mediated by these connections is very close to the latency of monosynaptic reflexes evoked by the same stimulus in a muscle housing the receptors. The difference is due to the additional synapse, which adds 0.5 ms to the total transmission time.

Many of the properties of oligosynaptic reflexes are similar to those of the monosynaptic reflexes described earlier (Eccles, 1964; Granit, 1970; Burke et al., 1984; Day et al., 1984b), with the exception of slightly longer conduction time. However, the presence of at least one additional synapse changes the properties of the reflex arc, in particular its reaction to high-frequency stimulation (Eccles, 1964; Granit, 1970). Usually, when more synapses are involved in transmission of a stimulus to α-motoneurons, more pronounced suppression of reflex effects is observed with an increase in stimulation rate.

Since in everyday movements we do not see many twitch-like contractions (and would probably like to avoid them!), these first two groups of reflexes are likely to represent rather artificial phenomena that might be helpful during research and clinical studies but are rarely observed in "normal" life. This certainly does not mean that the *mechanisms* underlying these reflexes are not functioning or are insignificant for everyday motor control. The point is that in everyday life we make smooth movements; most of the peripheral stimuli that induce muscle stretches are subthreshold and do not evoke synchronized monosynaptic reflexes resembling those studied in the laboratory.

Polysynaptic Reflexes

All other muscle reflexes are considered to be polysynaptic (i.e., they involve many synapses in the reflex arc). These reflexes are generally

characterized by noticeably longer latencies and more complex, less reproducible behavior than monosynaptic reflexes and oligosynaptic reflexes.
However, since they involve more interneurons in their transmission loop,
they can potentially provide more information about other levels of the
neural hierarchy that is beyond the reach of monosynaptic reflexes and
oligosynaptic reflexes.

Let us tentatively classify all the reflexes into two groups: *phasic* and
tonic. Phasic reflexes emerge in response to a *change* in the level of a
stimulus specific to a receptor. They usually represent a burst (or a short-
term depression) of muscle activity leading to a twitch or a series of
twitches. Note that monosynaptic reflexes and oligosynaptic reflexes
are phasic.

Tonic reflexes emerge in response to the level of a stimulus itself. They
lead to sustained muscle contractions and/or relatively smooth movements.
These reflexes are always polysynaptic. One should take into account that
a *succession* of stimuli inducing monosynaptic reflexes, that is, a succession
of changes in stimulus level, can lead to a smooth tonic muscle contraction
due to superposition of successive twitch contractions. Let us, however,
clearly dissociate tonic reflexes from tonic muscle contractions.

For example, activity of length-sensitive spindle receptors can lead to
both types of reflexes. If a muscle is quickly stretched, monosynaptic
reflexes and oligosynaptic reflexes can be observed in response to *the
process of stretch*. If the muscle stays at a new, stretched state, these phasic
reflexes quickly disappear. However, if the muscle was active before the
stretch, it is possible to record tonic changes in muscle activity at any
time after the stretch is completed, and the muscle is at a new steady
state. This mechanism is frequently termed *tonic stretch reflex* in contrast
to the phasic stretch reflex, which is a monosynaptic reflex.

The phasic—tonic dichotomy is rather conventional and based on quantitative rather than qualitative differences in the mechanisms. For example,
monosynaptic reflexes are mediated by monosynaptic connections of Ia
afferents with α-motoneurons. Spindle Ia afferents are sensitive to both
muscle length and rate of its change. These monosynaptic connections
with α-motoneurons function during slow stretches or maintenance of a
constant muscle length as well, but do not evoke monosynaptic reflexes.
Therefore, monosynaptic connections of Ia afferents can and are likely to
play a role in tonic reactions. When the rate of muscle stretch reaches
some threshold, monosynaptic reflexes emerge; when it is lower, only
tonic reflexes are observed.

We will return to the notion of tonic stretch reflex later, since the λ-
model considers it the "main" subsystem used by the central nervous
system to control voluntary movements.

Preprogrammed Reactions

There is one more group of semiautomatic reactions to muscle length
changes (or, sometimes, to other stimuli) that may be tentatively called

reflexes. In fact, there are many terms used in the literature for these reactions, including *long-latency reflexes*, *preprogrammed reactions*, *functional stretch-reflex*, M_{2-3}, and *triggered reactions* (Phillips, 1969; Newsom Davis & Sears, 1970; Tatton et al., 1978; Chan et al., 1979a,b; Gottlieb & Agarwal, 1980a; Dietz et al., 1981a).

The most common procedure for eliciting preprogrammed reactions is as follows: A subject maintains a constant position in a joint against a load. He is given an instruction "to return to the starting position as fast as possible in cases of position perturbations." Unexpected rapid load changes give rise to a series of electromyographic (EMG) events. The first one corresponds, according to its latency, to monosynaptic transmission and probably represents the phasic stretch reflex. After that, two peaks (sometimes poorly differentiated) appear with an intermediate latency (M_{2-3}, or preprogrammed reactions), and after that a voluntary reaction.

Phillips (1969) hypothesized that preprogrammed reactions represented a transcortical reflex, and this idea gained support during the succeeding years (Allum, 1975; Marsden et al., 1976, 1977, 1978; Iles, 1977; Chan et al., 1979a,b; Cheney & Fetz, 1984; Capaday et al., 1991; Day et al., 1991). However, these reactions were observed in decerebrated and even spinalized animals (Ghez & Shinoda, 1978; Tracey et al., 1980; Miller & Brooks, 1981). It has also been demonstrated that the difference between the latencies of monosynaptic reflexes and preprogrammed reactions in arms and legs is nearly the same, contradicting the hypothesized essential role of a transcortical loop (Darton et al., 1985).

These reactions can be observed in response to perturbations in a variety of conditions, including such common motor tasks as locomotion or maintenance of the vertical posture. They deserve a special section, and we will return to them later in this chapter (see "Preprogramming in Motor Control") after introducing a language that is going to be used for description and analysis of the problems of motor control.

We have discussed the major classes of muscle reflexes and reflex-type reactions in order to define the terms and be ready for the following analysis. If the motor control is going to be based on regulation of reflex parameters, polysynaptic tonic reflexes look like reasonable candidates since they induce smooth sustaining contractions. Monosynaptic reflexes look like a good tool for studying how the system functions when it is tested with rather unnatural stimuli.

Spring Properties of an Intact Muscle

Recording force-length characteristic curves in an intact muscle with preserved central connections appears to be a complicated methodological problem. First, it is not easy to reliably record changes in muscle length and force in an intact organism. Second, any experimental intervention might disrupt or distort signal processing at some stage, thus corrupting the recorded characteristics. Third, muscle with its central reflex connections receives descending signals that are likely to be able to substantially change muscle reactions to experimentally imposed length changes.

Animal Experiments

In acute animal experiments, many investigators have tried to reach a reasonable trade-off between reliable recording of muscle force and length, reliable control over the descending signals, and minimal disruptions of the signal-transmitting pathways (Matthews, 1959; Rack & Westbury, 1969; Feldman & Orlovsky, 1972). The schemes of these experiments were rather similar. An experimental lesion was introduced, separating the upper neural structures from the lower ones (including the muscle) at a certain level of the neural axis. Stimulating electrodes were placed at the distal stump at the level of the lesion. The distal muscle tendon was separated from its place of attachment and used for monitoring changes in muscle length and force.

Despite considerable differences in the preparations and experimental procedures, the results of these studies were strikingly similar. At a fixed level of "descending command" (controlled by the stimulator) a slow increase in muscle length led, up to a certain value, to a relatively small increase in muscle force. At a certain length, the muscle started to actively resist further lengthening due to autogenic recruitment of the α-motoneurons (Figure 1.6). The measured stiffness (the force/length ratio) increased dramatically. Changes in the experimentally controlled descending signals led to nearly parallel transfers of the whole curve, as in Figure 1.6.

To describe the results, the authors of these experiments invoked the notion of the tonic stretch reflex, which was introduced about 40 years earlier by Liddell and Sherrington (1924). Experimentally recorded force-length characteristic curves were termed tonic stretch reflex characteristics or, later, invariant characteristics (Feldman, 1966a,b, 1974). Muscle length at which the recruitment of autogenic α-motoneurons starts was termed the threshold of the tonic stretch reflex (λ). This parameter became especially important when it was found that any changes in experimentally controlled descending signals led only to nearly parallel transfers of the tonic stretch reflex characteristics, or, to be more precise, to the lack of their intersections (Figure 1.6; an exception has been demonstrated by Nichols & Steeves, 1986). This means that λ appears to be the only parameter of the intact muscle force-length characteristic curve that is necessary to describe any effects of the descending signals. This finding lets one try to dramatically simplify description of the descending control of an intact muscle, considering it a monoparametric spring (cf. Weber, 1846; Bernstein, 1926).

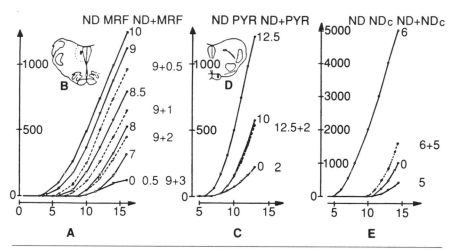

Figure 1.6 Effects of electrical stimulation of different supraspinal formations on the force-length characteristics of cat plantaris (A, C) and gastrocnemius (E) muscles. Stimulation was applied to ipsilateral (ND) and contralateral (ND$_c$) Deiters' nuclei, pyramidal tract (Pyr), and mesencephalon reticular formation (MRF). Note that in all the cases, the stimulation was able only to transfer the characteristics in a nearly parallel fashion.
Note. From "The Influence of Different Descending Systems on the Tonic Stretch Reflex in the Cat" by A.G. Feldman and G.N. Orlovsky, 1972, *Experimental Neurology*, **37**, pp. 481-494. Copyright 1972 by Academic Press. Used by permission.

Note that such a description was possible only when based on the Bernstein approach. Imagine, for example, that all the studies were directed exclusively toward refining our knowledge about central connections of different afferent loops originating from different muscle, tendon, joint, and skin receptors, their interactions, their influences on different interneurons, α- and γ-motoneurons, and possible centrally induced changes in the gains of all the variety of existing feedback loops. This approach would probably yield a body of important information, but it would never lead to the conclusion about monoparametric descending control of an intact muscle. In fact, all the apparent variety and complexity of central circuitry and afferent connections has led to a prejudice that the description of the system as a whole should also be complicated. This might be why the equilibrium-point hypothesis has had to struggle for about 20 years to be considered a serious approach to motor control (Berkinblit et al., 1986a; Feldman, 1986).

Functions of the Tonic Stretch Reflex

What is now termed tonic stretch reflex was originally observed as a certain behavior of a muscle with intact central connections in acute animal experiments (Liddell & Sherrington, 1924; Denny-Brown, 1929). On the other hand, the tonic stretch reflex can be defined more generally as a hypothetical mechanism providing for tonic recruitment (or derecruitment) of autogenic α-motoneurons in response to a slow increase (or decrease) in muscle length. It is likely to represent a combined action of different peripheral receptors via spinal reflex loops (Granit, 1970; Matthews, 1972; Stein, 1974), which is under control from supraspinal structures (Anden et al., 1966; Bergmans & Grillner, 1968; Evarts & Tanji, 1974; Chapman et al., 1983; Fromm, 1983), although direct involvement of supraspinal structures in the tonic stretch reflex loop has been hypothesized (Phillips, 1969; Evarts, 1973; Marsden et al., 1976, 1977). Thus, the tonic stretch reflex appears to be a typical Bernsteinian notion describing a certain input-output relationship for a black box that includes a muscle with its generally unknown central connections. Such an approach has gained support from the observations by Cordo and Rymer (1982), suggesting that muscle spring characteristics are primarily defined by the process of α-motoneuron recruitment.

From the advanced point of view, the tonic stretch reflex can be considered a mechanism providing for the observed shape of the force-length characteristic curves (Matthews, 1959; Feldman, 1974, 1979; Hoffer & Andreassen, 1981). In doing so, it affects the apparent stiffness of the system "muscle-spinal circuitry." Some have argued that this quantity is actually controlled in this process (Nichols, 1974; Crago et al., 1976; Houk, 1976, 1979; Stein & Oguztoreli, 1976). However, modulation of stiffness is an automatic function of the tonic stretch reflex and should be clearly distinguished from independent control of muscle stiffness, that is, supplying a signal from level 1 to level 2 in the scheme in Figure 1.1 that would code stiffness (cf. Feldman, 1979; Houk & Rymer, 1981).

On the other hand, the tonic stretch reflex has sometimes been considered a compensatory mechanism for the variability of the mechanical muscle properties (Rosenthal et al., 1970; Stein, 1974; Nichols & Houk, 1976). Measured stiffness of an intact muscle is in fact a sum of at least two factors, including the mechanical properties of the isolated muscle and the contribution of the neural feedback loops. Although the variability of both components is rather high, their combined action leads to smooth force-angle characteristic curves with a range of relatively constant combined stiffness (Houk, 1976, 1979; Nichols & Houk, 1976). It has also been suggested that muscle reflexes (including tonic stretch reflex) lead to a compensation of yielding and linearization of the inherently nonlinear mechanical properties of the areflexive muscle (Carter et al., 1990).

Human Experiments

The methodological problems encountered in acute animal experiments become even more serious in studies of intact human muscle characteristic curves. One cannot, as a rule, separate one muscle and study its force and length changes in a human subject. This forces switching to another unit of study, the joint, and to another pair of variables, torque and angle.

Every joint in the human body is controlled by at least a pair of muscles, which are usually addressed as agonist and antagonist with respect to a certain movement direction. Most of the experimentally studied joints (e.g., ankle, hip, wrist, and shoulder) are controlled by several agonists and several antagonists whose functional relations may change in different tasks (e.g., Desmedt & Godaux, 1981; van Zuylen et al., 1988; Buchanan et al., 1989; Zajac & Gordon, 1989). It is also impossible, in a human subject, to reliably control the descending supraspinal signals. These factors add to the complexity in interpretation of the data.

Feldman (Asatryan & Feldman, 1965; Feldman, 1966a,b) introduced the following paradigm to cope with these problems. In his experiments, the subjects were asked to occupy a joint position against a load and "not to intervene voluntarily" when the load changed unexpectedly. It is obvious that in these experiments the experimenter relies on the subject's ability to follow a rather vague instruction.

It was not a priori obvious if shifting to the joint level and relying on ideal performance by the subjects would lead to any reproducible findings comparable to the observations in acute animal experiments. However, the first series of experiments (Asatryan & Feldman, 1965; Feldman, 1966a,b) demonstrated families of joint torque-angle characteristic curves strikingly similar to those reported earlier in animal studies (Matthews, 1959). Changes in descending signals were modeled in Feldman's experiments by asking the subjects to occupy different starting positions against the same load. This procedure resulted in nearly parallel transfer of the recorded joint characteristic curves without intersections.

Figure 1.7 shows the results of Feldman's experiments on the elbow joint. The dashed line shows the characteristic curve of a passive muscle in the absence of α-motoneuron recruitment. The solid lines correspond to interpolation of the datapoints obtained with unloading of flexors (upper part) or extensors (lower part) starting from different positions. Similar nonlinear characteristics were later reported by Neilson and McCaughey (1981) who also stressed an increase in the gain of the underlying hypothetical reflex mechanism with an increase in initial muscle force and its decrease with an increase of amplitude of the test stretches. Similarities of the joint torque-angle curves recorded in human subjects and tonic stretch reflex characteristic curves recorded in animals led Feldman to use the same term (tonic stretch reflex) for describing his data (Feldman, 1966a,b, 1979). However, the specific internal mechanisms of these two length-sensitive reflexes are unknown and may differ significantly. Nevertheless, in order to avoid unnecessary proliferation of terms, we will use tonic stretch reflex to describe motor control processes in human subjects without specifying its internal mechanisms, or anatomical structures, or

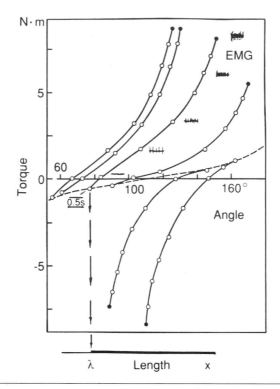

Figure 1.7   Elbow joint compliant characteristics for the flexors (upper curves) and extensors (lower curves) recorded in unloading experiments using the "do not intervene voluntarily" paradigm. Different characteristics correspond to different combinations of initial load and position. The dashed line shows the passive joint characteristic when both muscle groups were relaxed. Tonic EMGs of a flexor muscle are shown for one of the curves. The control parameter λ corresponds to muscle length when an active characteristic starts to deviate from the passive one.
Note. From "Once More on the Equilibrium-Point Hypothesis (λ-Model) for Motor Control" by A.G. Feldman, 1986, *Journal of Motor Behavior,* **18**, pp. 17-54. Copyright 1979 by A.G. Feldman. Reprinted by permission.

its relations to the tonic stretch reflex observed in animals. As such, the "human tonic stretch reflex" is introduced as another abstraction without a clear neuronal counterpart.

Experiments on human subjects using the "do not intervene voluntarily" paradigm were later reproduced in several studies of human subjects (Davis & Kelso, 1982; Vincken et al., 1983; Gottlieb & Agarwal, 1988). All of them demonstrated similar families of joint characteristic curves without intersections, thus supporting the idea of monoparametric description of descending muscle control in human subjects. Recently, Hore et al. (1990) have reported that their subjects were able to voluntarily modulate wrist stiffness even when co-contraction of the antagonist muscles was impossible due to application of local anesthetics to the nerve to the wrist extensors. These observations are, however, not absolutely convincing because there are several wrist flexor muscles, and the subjects could modulate joint stiffness by using different levels of relative involvement of the wrist flexors.

Feldman used only unloadings in his experiments but hypothesized that the same curves might be obtained with the loadings. However, effects of hysteresis had been described in acute animal experiments (Partridge, 1965, Joyce et al., 1969), and later Gottlieb and Agarwal (1986, 1988) described S-shaped torque-angle joint characteristic curves with a point of inflection at the starting position in human experiments, similar but not identical to those of Feldman, which implies possibility of a considerable hysteresis. These findings were reproduced and shown to be independent of certain loading parameters different in the two experimental series (Latash & Gottlieb, 1990a). The S-shape is likely to be different from elastic saturation reported in acute animal experiments (Rack & Westbury, 1969). It implies (although it has never been shown directly) that if a loading is applied to a stationary joint and, after a certain time, an equal unloading is applied, the final joint position can be different from the initial position. Thus, it seems that voluntary human motor performance can be considerably affected by the hysteresis phenomenon (cf. Mussa Ivaldi et al., 1989). However, the hysteresis, by itself, does not contradict the monoparametric description of muscle control, although it should be taken into account by the central control system.

Merton's Servo-Hypothesis of Motor Control

Liddell and Sherrington (1924) were the first to express the opinion that the main function of the tonic stretch reflex is compensation of the influence of load upon muscle length. Later, this view was expanded by Merton (1953; see also Eldred et al., 1953) and advanced as a model of motor control. Merton's servo-hypothesis is based on a notion of the stretch reflex as a length-regulating mechanism (Figure 1.8). According to this view, changes in externally applied load induce changes in muscle length; the muscle length receptors change their level of firing, and the stretch-reflex loops bring about changes in α-motoneuron pool activity. This generates active muscle contractile changes that compensate for the externally induced length changes.

Originally, Merton's model was based on monosynaptic action of Ia muscle afferents on homonymous α-motoneuron pools. However, later, the mechanism underlying the servo-action was reconsidered (Granit,

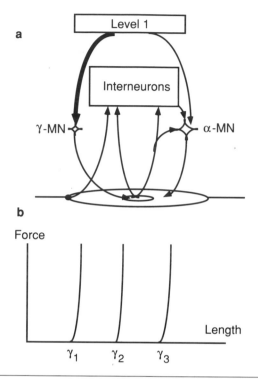

Figure 1.8   An illustration of Merton's γ-model. (a) The "main" descending input to the γ-motoneurons (γ-MN) is shown by a thick line; later, a possibility of action on the α-motoneurons (α-MN) was added (thin line from level 1). (b) The force-length characteristics represent nearly vertical lines implying a very high gain in the corresponding reflex loop. Different central commands (γ₁, γ₂, γ₃) encode different positions (values of muscle length).

1970; Figure 1.8a) and based on the tonic component of the stretch-reflex. We are going to discuss the latter version of the servo-model.

In terms of the tonic stretch reflex characteristics, Merton's model can be illustrated by virtually vertical lines on the force-length plane (Figure 1.8b). Thus, the servo-hypothesis requires very high stiffness of the system, or very high gain in the tonic stretch reflex loop. However, experimental measurement of the tonic stretch reflex gain in decerebrate animals (where the tonic stretch reflex is very pronounced) did not show values high enough to lead to length regulation (Matthews, 1959). Later studies (Vallbo, 1974; Hoffer & Andreassen, 1981) also failed to demonstrate sufficiently high gain in the tonic stretch reflex loop. These findings have shown that the tonic stretch reflex mechanism cannot be considered a position controller.

Merton's model attached special importance to the γ-motoneuron system, which can modulate both threshold and sensitivity of the muscle spindle receptors to muscle length changes, thus shifting and rotating the hypothetical nearly vertical characteristic curves. According to the model, the central motor command initiates a movement by acting on the γ-motoneurons (Figure 1.8a), thus shifting the threshold of functioning of the "load compensating" mechanism and programming a new "position." However, the hypothesis that movement is initiated exclusively by a descending command to the γ-motoneurons has not been confirmed experimentally (Vallbo, 1981).

The elegance of the position-controlling servo-model, however, is very attractive. It was also the first attempt at a system analysis of the input-output characteristics of this part of the motor control system. The first guess was wrong, but it suggested a methodological approach that led later to emergence of other spring-like models, which were also based on particular interpretations of the experimentally recorded tonic stretch reflex characteristics (Feldman, 1966a,b, 1979, 1986; Bizzi et al., 1978a; Bizzi, 1980).

Attempts to resuscitate Merton's model by introducing a possibility of movement initiation by simultaneous action on γ- and α-motoneuron pools (Granit, 1970; Matthews, 1972; Stein, 1974; cf. Akazawa et al., 1982, 1983) did not solve its main problems, since they were also based on a general idea that the function of the tonic stretch reflex is to compensate for the influence of load on muscle length (thus controlling joint position). An alternative approach, which will be described later in this chapter, considers the tonic stretch reflex a mechanism providing for a certain dependence of muscle length upon load. Such a system may be used by a central motor controller even if its gain is not very high.

The α-Model

The α-model introduced by Bizzi and his colleagues in the 1970s (Bizzi et al., 1976, 1982; Bizzi & Polit, 1979; Bizzi, 1980; Abend et al., 1982) is another attempt to simplify the situation by trying to attach a particular importance to a certain anatomical part of the hypothetical tonic stretch reflex mechanism. Merton's servo-model put stress on the γ-motoneuron system as the "most important part" of the tonic stretch reflex loop. Bizzi's α-model is based on the idea that α-motoneuron pool activity is preferentially regulated by the higher motor control systems.

According to the α-model, the central motor command is able to fix, for a muscle, a level of activation of its α-motoneuron pool (Figure 1.9a). Such a system demonstrates spring properties because changes in load lead to shifts in muscle length along a "fixed" characteristic curve (cf. force-length characteristics recorded by Rack & Westbury, 1969). Note that for one muscle, all the characteristic curves originate from the same point (l_0 in Figure 1.9b). For two muscles controlling a joint, two sets of

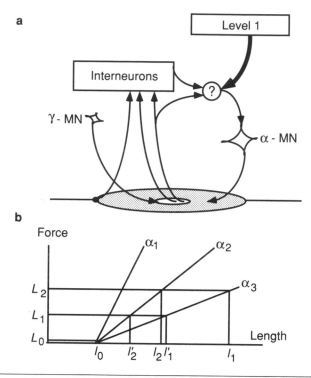

Figure 1.9   An illustration of Bizzi's α-model. (a) The descending command acts at some hypothetical level leading to a constant input to the α-motoneurons (α-MN) and, therefore, to their constant output. (b) The force-length characteristics have different slopes corresponding to different levels of α activation. Note that a change in motor command (α_3 to α_2) leads to a bigger displacement ($l_2 - l_1$) for a heavier load (L_2) than for a lighter one (L_1; $l'_2 - l'_1$). A very light load (L_0) is impossible to move.

curves are defined, each corresponding to the level of activation of one of the two motoneuron pools.

In a sense, the α-model describes a mechanism of control of a deafferented muscle, since it does not incorporate the possibility of an independent influence of feedback signals from the proprioceptors on the α-motoneuron activity. Merton's hypothesis was based on plausible ideas that only quantitatively happened not to correspond to the experimental observations. The α-model seems to contradict some of our knowledge about the muscle and its central connections and thus could not be correct even theoretically. Bernstein in 1935 stressed the theoretical impossibility to independently control the level of α-motoneuron activation, taking into account multiple feedback loops with different time delays. Since that time, many studies have demonstrated changes in the level of autogenic α-motoneuron pool activity or the electromyogram (EMG) following changes in the externally applied load in both animal experiments and human studies using the "do not intervene voluntarily" paradigm (Hoffer & Andreassen, 1981; Vincken et al., 1983; Latash & Gottlieb, 1990a).

Recent articles by Feldman (1986) and Berkinblit et al. (1986a) discuss at length all the theoretical inconsistencies of the α-model and its discrepancies with experimental observations. Among them are the following:

- The fact that all the muscle characteristic curves for different motor commands originate from the same point (Figure 1.9b) implies that the muscle cannot be activated at lengths to the left of l_0 and is always active to the right of this point.

- Centrally induced changes in motor command for a fixed load would lead to smaller load displacements for smaller loads (cf. length changes for loads 1 and 2 in Figure 1.9b).

- The model does not describe how zero load (L_0) can be moved.

It might look strange that very different models emerged on the basis of the same body of experimental data, namely observations of the tonic stretch reflex characteristic curves. However, the experimentally observed curves cannot usually be reliably traced at very low loads, when the reflex components are comparable to the passive components (see Figure 1.6, 1.7; Pompeiano, 1960; Feldman, 1966a,b; Houk et al., 1970). So, it is hard to say if the experimental curves originate from the same or different points.

The α-model was based mostly on observations in deafferented animals (Bizzi et al., 1976; Bizzi, 1980) when the reflex pathways were disrupted, and the control system had no choice but to act directly on the α-motoneuron pools. Certainly, force-length characteristic curves observed in such experiments correspond to different levels of α-motoneuron pool activation, but they are different from the characteristic curves of muscles with intact central connections and cannot be called tonic stretch reflex characteristics since the tonic stretch reflex was eliminated by the experimental procedure.

Despite obvious inconsistencies, the α-model has proven to be very resilient, probably because it provides an attractive framework for investigators studying EMG patterns of movements. Since EMGs represent a direct consequence of signals from the α-motoneuron pools after certain processing (Brown et al., 1982; Gottlieb et al., 1989a), they may be considered, from the viewpoint of the α-model, a direct reflection of central control signals. Therefore, EMG recordings get a new important meaning. Many studies consider the EMGs from this viewpoint and try to model central motor commands so that their timing and amplitudes correspond to those parameters of the recorded EMGs (Hallett et al., 1975; Lestienne, 1979; Wallace, 1981; Bizzi et al., 1982; Enoka, 1983). This "pattern-imposing" approach (see chapter 4) is in fact based on the α-model, although this is not explicitly acknowledged in most of the more recent studies

NORTHERN MICHIGAN UNIVERSITY.
MARQUETTE, MICHIGAN 49855

(Sanes & Jennings, 1984; Ghez & Gordon, 1987; Schmidt et al., 1988; Sherwood et al., 1988; Gottlieb et al., 1989a,b).

The γ-model and the α-model seem to share the same drawback: an attempt to draw too close a parallel between function and anatomical structure (cf. Bernstein's warning against equating coordination and localization, Bernstein, 1935). An alternative approach is to analyze input-output relations of the system without considering the anatomy and/or neurophysiology. After the general properties of the system are defined, the experimental data and anatomical findings can be incorporated into the model. This approach was used by Feldman and led to the λ-model.

The λ-Model

The λ-model emerged as a formal language for describing a body of experimental data on observations of single-muscle force-length characteristic curves in animals (Matthews, 1959) and single-joint torque-angle characteristic curves in human subjects (Asatryan & Feldman, 1965; Feldman, 1966a,b). The most important finding was the lack of intersections of the curves recorded with "fixed" descending command at different initial limb positions (see "Spring Properties of an Intact Muscle" in this chapter). These curves were termed *invariant characteristics*. Although "fixing" the descending command could not be controlled ideally (except in animal studies with lesions at a suitable level of the neural axis) the results themselves and their reproducibility corroborated the used approach.

The lack of intersections of the invariant characteristics let Feldman introduce a monoparametric description of the process of control of one muscle. The chosen variable represented the threshold (λ) of the tonic stretch reflex, that is, the muscle length at which autogenic recruitment of α-motoneurons starts during a slow muscle stretch. Actual characteristic curves recorded in the experiments represented a sum of active and passive components, so λ could be defined as a point of deviation of an observed actual curve from the passive curve for the same muscle (see Figure 1.7).

Figure 1.10 illustrates the λ-model. Central motor command defines a value of the threshold of the tonic stretch reflex (λ_1, λ_2, or λ_3, Figure 1.10a) thus choosing a force-length characteristic curve for the muscle. Actual muscle length and force will also depend on the peripheral load. Peripheral load can be described by a curve on the same plane. Three examples of load characteristics are shown in Figure 1.10b including isometric (1), isotonic (2), and elastic (3) external loads. The point of intersection of the invariant characteristic and load characteristic curve is an equilibrium point (EP) that defines actual values of muscle length and force under

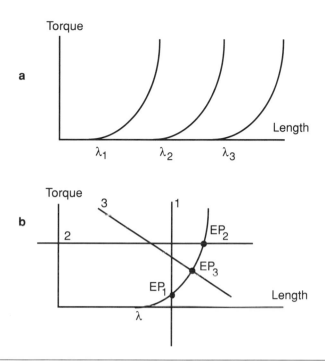

Figure 1.10   Feldman's λ-model does not specify any particularly important structure at the decision-processing level and can be illustrated by Figure 1.2. It assumes that changing a central command to a muscle leads to a parallel shift of the force-length characteristic (a). The threshold of the characteristic (λ) is the only parameter supplied from level 1 to level 2. Actual values of muscle length and force correspond to an equilibrium point (EP) defined by both central and load characteristics (b). Isometric (1), isotonic (2), and intermediate (3) load characteristics are shown leading to different EPs for the same central motor command (λ).

static conditions. The notion of the equilibrium point is central to the λ-model (also called the *equilibrium-point hypothesis*) because it stresses spring properties of the system.

If one of the two curves shifts, their intersection, and therefore the equilibrium point, shifts as well, leading to changes in the values of muscle length and/or force. Figure 1.11 illustrates equilibrium point shifts in cases of shifts of load and/or central characteristic curves. If, initially, central command to a muscle can be described as λ_1 and external load is L_1, muscle length (l_1) will correspond to an equilibrium point (EP$_0$). A change in the load to L_2 without a change in the central command will lead to an equilibrium point EP$_1$ while a change in central command to λ_2 without a change in load will lead to a new equilibrium point EP$_2$, both corresponding to a new muscle length l_2. Note that speed at which

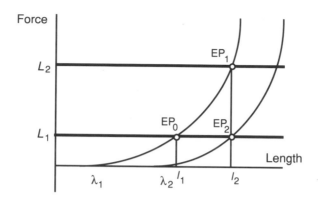

Figure 1.11   Changes in muscle length and/or force can be induced by a change in either central motor command, load characteristic, or both. For example, if the initial state is characterized by EP_0 corresponding to a central command λ_1 and load L_1, a change in the muscle length from l_1 to l_2 can result from a change in the load (L_2, EP_1) or from a change in the central command (λ_2, EP_2).

muscle length and torque change will depend on mechanical properties of the peripheral apparatus and the rate of the invariant characteristic shift (Feldman, 1979, 1986). Even during slow movements, the actual values of muscle length and force will lag behind the point of curve intersection. This lag is dependent upon the rates of invariant characteristic shifts and will be analyzed later (see chapter 2).

If one considers a pair of antagonistic muscles controlling a joint, their central control can be described with a pair of variables corresponding to values of λ for the two muscles. Let us address them as λ_{fl} (flexor, the agonist) and λ_{ext} (extensor, the antagonist). Torque-angle characteristics of a pair of muscles are illustrated in Figure 1.12. The upper graph shows a couple of muscle characteristics when either one of the muscles is active (the flexor at muscle lengths to the right of λ_{fl}, and the extensor at muscle lengths to the left of λ_{ext}) or both muscles are inactive (at muscle lengths between λ_{fl} and λ_{ext}). In the lower graph, both muscles can be active simultaneously (at muscle lengths between λ_{fl} and λ_{ext}). Note that activation of the antagonist leads to development of torque directed against the agonist muscle torque and/or movement in a joint in the opposite direction. That is why the curve for the antagonist is turned upside down so that it corresponds to negative values of torque. Behavior of the joint will be defined by the algebraic sum of the two invariant characteristics and the load characteristic. In the length range where both muscles are activated, the resultant joint characteristic can be moved along the length axis by unidirectional shifts of λ_{fl} and λ_{ext}, or its slope can be changed by contradirectional shifts of λ_{fl} and λ_{ext} due to the nonlinear form of the invariant

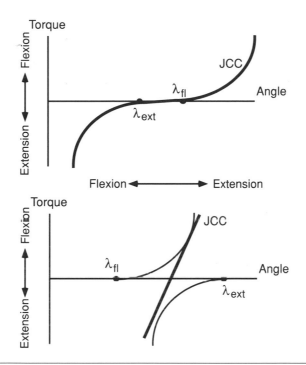

Figure 1.12   A joint is controlled by a pair of muscles that will be addressed as the flexor and the extensor. Joint compliant characteristic (JCC, thick lines) is an algebraic sum of force-length characteristics (invariant characteristics) for the two muscles. A muscle is active when its length exceeds the threshold λ. That is, the flexor is active to the right of λ_{fl}, and the extensor is active to the left of λ_{ext}. Two basic cases are demonstrated: when the zones of activation for the two muscles do not overlap (above) and when both muscles can be active simultaneously (below).

characteristics (Feldman, 1979). Figure 1.13 illustrates the results of unidirectional (upper graph) and contradirectional (lower graph) shifts of λ_{fl} and λ_{ext}.

Another pair of variables can be used for describing central regulation of a joint (Feldman, 1980a,b; Feldman & Latash, 1982a). These variables represent half a sum and half a difference between λ_{fl} and λ_{ext} and were originally introduced in order to draw close parallels with physiological notions of reciprocal activation and coactivation (Granit, 1970):

$$r = \frac{1}{2} (\lambda_{fl} + \lambda_{ext}),$$

$$c = \frac{1}{2} (\lambda_{fl} - \lambda_{ext}).$$

Simultaneous shift of both λ_{fl} and λ_{ext} in one direction obviously increases the level of activation for one of the muscles and decreases it for the

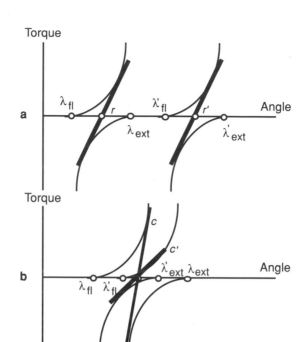

Figure 1.13   Voluntary motor commands to a joint can be described with λ shifts for the muscles (λ_{fl} and λ_{ext}) or with shifts of another pair of variables, a reciprocal command (r) and a coactivation command (c). The r command (a) corresponds to unidirectional shifts in λ_{fl} and λ_{ext} and leads to a shift of the joint compliant characteristic along the angle axis without changing its slope (compare r and r'). The c command (b) corresponds to contradirectional shifts in λ_{fl} and λ_{ext} thus effectively changing the slope of the joint compliant characteristic (compare c and c') without a comparable effect on its position.

other muscle (Figure 1.13a), which can be associated with the reciprocal excitation r of one of the muscles. Shifts of both λs in opposite directions increase or decrease activation of both muscles simultaneously (Figure 1.13b), which can be associated with changes in a coactivation command c. The pairs (λ_{fl}, λ_{ext}) and (r, c) are equivalent and either one can be used for analysis of joint control. Using the (r, c) pair for description of single-joint motor control has been indirectly corroborated by observations of two groups of motor cortex cells, one causing joint movement and another causing co-contraction of antagonist muscles, thus changing joint stiffness (Humphrey, 1982).

The λ-model not only introduces a consistent and simple way of describing single-joint control but also provides an explanation for a body of data demonstrating equifinality of movements in cases of transient external perturbations (Bizzi et al., 1976, 1978a,b, 1982; Polit & Bizzi, 1978; Cooke,

1980; Kelso & Holt, 1980; Schmidt, 1980a,b; Brown & Cooke, 1981; Simmons & Richardson, 1984; Hasan & Enoka, 1985; Latash & Gottlieb, 1990a). Since the final state of a joint and the final torque around it depend only on the final position of the central joint and load characteristics (Figure 1.11), any transient changes in load cannot alter the final equilibrium point. Let me be fair and note that the other spring models, including the servo-model and the α-model, can also explain the phenomenon of equifinality (Bizzi et al., 1978a, 1982; Bizzi, 1980).

As far as the mechanisms are concerned, the λ-model assumes that the tonic stretch reflex, providing for the characteristics, incorporates all the reflex loops that can be influenced at the levels of γ-motoneurons, α-motoneurons, and interneurons (Feldman, 1974, 1979). It does not single out one anatomical structure as "the most important" as was done in the γ-model and the α-model. Moreover, the feedback loops from receptors other than muscle receptors (e.g., skin and subcutaneous receptors), which can influence the hypothetical tonic stretch reflex (Lundberg, 1979; Kniffki et al., 1981; Paintal & Walsh, 1981) can also play an important role in defining the shape of the invariant characteristics. This may explain, in particular, the reported violations of equifinality in the experiments with skin anesthesia (Day & Marsden, 1981, 1982); if any of the factors influencing the tonic stretch reflex are altered, the equilibrium of the system may change. Another reported violation of equifinality in the experiments with unexpected introduction of viscous loads (Sanes, 1986) was probably due to a change in central motor commands by the subjects when they experienced the perturbation.

Many experiments have demonstrated a variety of reflex connections whose role can differ depending on the motor task (long-latency reflexes or preprogrammed reactions; see "Preprogrammed Reactions," "Preprogramming in Motor Control," and Hammond, 1954; Marsden et al., 1976, 1977; Gottlieb & Agarwal, 1980a,b; Lee & Tatton, 1982), position of the whole body and its segments (Magnus, 1924; Latash & Gurfinkel, 1976; Nashner, 1976, 1979; Gurfinkel & Latash, 1979), presence of a pathology (Kudina, 1980; Gottlieb et al., 1982; Myklebust et al., 1982; Latash, 1986), and other factors. Relations of these reflex responses to the hypothetical tonic stretch reflex mechanism are not always obvious, and one should not conclude that any muscle reaction to length changes is necessarily mediated by the tonic stretch reflex mechanism. The tonic stretch reflex, constituting the core of the λ-model, is only a part of the motor control system, a part with a well-defined autonomic function that is, however, controlled by descending commands; its effects can combine with the effects of other unrelated reflexes sharing the same final common path (α-motoneurons).

Originally, the λ-model was based on data from "static" experiments when the experimenter tried to "freeze" the descending signals. Probably because of this, the model has frequently been considered an artificial language applicable only to analysis of static positions (Day & Marsden, 1982; Enoka, 1983; Marsden et al., 1983; Atkeson, 1989). However, even the original version of the λ-model contained a notion of centrally controlled shifts of the joint invariant characteristics as the basis of voluntary motor control. Later, to analyze movements at different speeds, the model was expanded, taking into account dynamic sensitivity of muscle spindle receptors (Adamovitch & Feldman, 1984; Adamovitch et al., 1984; Abdusamatov & Feldman, 1986; Feldman, 1986; Abdusamatov et al., 1987). Incorporation of the dynamic component into the λ-model (Feldman, 1986; Feldman et al., 1990a) has also led to assuming a better defined neural mechanism associated with the variable λ. In particular, λ has been associated with a subthreshold depolarization of the α-motoneuron membrane. Although this recent addition looks controversial, it provides a clear neural substrate for many experimentally observed phenomena. These problems will be discussed in chapter 3.

Hypothetical Supraspinal Mechanisms

There have been attempts at drawing direct relations between the findings of experiments with recording single cell activity in various supraspinal structures with the notions of equilibrium-point control and, in particular, the λ-model. Patterns of neuronal activity in the supplementary motor area, motor cortex, putamen, and red nucleus have suggested relations to characteristics of voluntary movement at a rather abstract kinematic level being compatible with the notions of equilibrium-point control or end-point trajectory control for multijoint movements (see also chapter 7) (Tanji & Kurata, 1982; Georgopoulos et al., 1982, 1985, 1989; Kalaska et al., 1983; Gibson et al., 1985; Kennedy et al., 1986; Alexander, 1987; Houk & Gibson, 1987; Alexander & Crutcher, 1990a,b; Werner et al., 1991). In particular, it has been shown that patterns of activity of cortical cells coded trajectory independently of required forces and patterns of muscle activation (Alexander & Crutcher, 1990a; Crutcher & Alexander, 1990), which is expected from the equilibrium-point control (see chapter 5).

Some of the studies have suggested very direct relations to the λ-model. In particular, Fromm (1983) recorded activity of motor cortex cells and found that it was related to joint force if joint angle was fixed, and to joint angle if the force was kept constant. This exactly corresponds to what might be expected from hypothetical cells controlling λ. Humphrey (1982) observed two groups of motor cortex cells, one causing joint movement and another causing co-contraction of antagonist muscles, thus changing joint stiffness, which corresponds well to what is expected from hypothetical cells controlling r and c. More recently, Georgopoulos et al. (1989) have reported changes in activity of cortical cells, leading to a rotation of a "neuronal population vector" coding direction of a planned movement.

Houk and Gibson (1987) have suggested that activity of red nucleus neurons can participate in motor control by coding position of the force-length muscle characteristics and velocity of their shifts. Earlier, Gibson et al. (1985) reported that activity of separate red nucleus neurons could be related to beginning, speed, and amplitude of voluntary movements. These are the basic variables assumed to be used in the equilibrium-point control. Later, Houk (1989; see also Sinkjaer et al., 1990) suggested a more general model of control of endpoint position based on interactions between the cerebellar and rubral nuclei. This model has been based on positive feedback loops between the nucleus interpositus and magnocellular red nucleus and nucleus reticularis tegmenti pontis. These loops were assumed to be under an inhibitory control from the Purkinje cells. Releasing of the feedback loops by switching off a set of Purkinje cells has been assumed to lead to emergence of adjustable pattern generators leading to a movement. Sets of Purkinje cells have been associated with "motor programs."

Generally, the results of many of the studies of the patterns of neuronal activity in supraspinal structures create a feeling of being too good to be directly interpretable in the framework of the λ-model. They lead one to believe that cells of various brain regions encode λs or corresponding hypothetical variables used for control of multijoint movements (see chapter 7). Although this is not impossible it seems rather unlikely.

The Notion of Shifting Invariant Characteristics

Descending motor control of a joint, in the framework of the λ-model, is realized by translations of force-length characteristic curves (invariant characteristics) of agonist and antagonist muscles controlling the joint. This results in changes in form and position of the joint invariant characteristic. Let us first consider a simpler case of descending control of only one muscle (Figure 1.14).

Initial muscle state is characterized by the position of its invariant characteristic (λ_1) and load characteristic (L_1), which together define the equilibrium state of the muscle (EP_1). If a central controller wants to change muscle force and/or length, it shifts the muscle curve to a new position (λ_2) so that the new equilibrium point may correspond to the same value of muscle force (EP_1 in isotonic conditions, L_1), to the same value of muscle length (EP_2 in isometric conditions, L_2), or to changes in both length and force (EP_3 in conditions of an elastic load L_3). The load characteristic is externally imposed and defines accessible combinations of muscle force and length. For example, in isotonic conditions, no change in descending command is able to increase static muscle force.

The shift of the invariant characteristic can be performed at different rates (Feldman, 1979, 1986), which will define, with participation of other

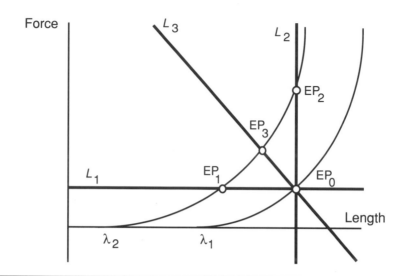

Figure 1.14 According to the equilibrium-point hypothesis, active movements result from a voluntary motor command changing the threshold (λ) of the muscle force-length characteristic and thus shifting it. A shift in λ from λ_1 to λ_2 can lead to different peripheral effects depending on the load characteristic. In isotonic conditions (L_1), muscle length changes (compare EP_1 and EP_0), in isometric conditions (L_2), the force changes (compare EP_2 and EP_0), and in an intermediate case (L_3), both force and length change (compare EP_3 and EP_0).

factors, the speed of the resultant movement. Instantaneous "jumps" of the invariant characteristics represent a degenerated case of motor control and are certainly not the way the λ-model works, although it has been criticized for this (Enoka, 1983; Agarwal & Gottlieb, 1986).

In real life, any muscle has at least one antagonist acting at the same joint. Availability of two muscles, and therefore, two independently controllable variables, adds one more degree of freedom for the central controller. For one muscle, there is only one invariant characteristic going through a point on a force-length plane. For a joint, there is a possibility of independently changing slope of the joint compliant characteristic by contradirectional shifts of λ_{fl} and λ_{ext} (which correspond to changes in the coactivation variable c).

Figure 1.15 illustrates possible combinations of central commands to a pair of muscles for moving from an initial point EP_1 to a final point EP_2. In the first case (a), the level of coactivation is the same after the movement, while in the second case (b) it is noticeably increased, which is reflected in the slope of the final joint invariant characteristic. The third case (c) shows that a shift in λ of only one muscle may be considered as a combination of r and c commands. The process of shifting the invariant characteristics can be described with time functions [$\lambda_{fl}(t)$; $\lambda_{ext}(t)$] or equivalently

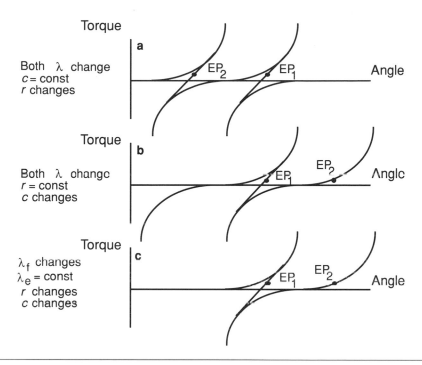

Figure 1.15   An active movement in a joint can result from different combinations of shifts of the control variables λ for the flexor and extensor muscles (λ_{fl} and λ_{ext}) or r and c. The three parts of the figure illustrate shifts in the equilibrium point (EP) for different combinations of central commands when the external load is constant (isotonic conditions). (a) A movement is induced by a parallel shift in λ_{fl} and λ_{ext}, i.e., a "pure reciprocal" command (c = constant, r changes). (b) A movement is induced by a contradirectional shift in λ_{fl} and λ_{ext}, i.e., a "pure coactivation" command (r = constant, c changes). (c) A shift in λ_{fl} without a shift in λ_{ext} may be considered as a combination of reciprocal and coactivation commands.

[$r(t)$; $c(t)$]. At a first approximation, we are going to discuss shifts of the controlled variables at constant rates and use the letter ω for the rate of the shifts.

Adding to the model several agonists and several antagonists complicates the description, especially if one takes into account more than one degree of freedom for movements in many of the joints. The analysis would probably require separate consideration of each degree of freedom, introducing coefficients of individual muscle involvement for this particular degree of freedom, and performing algebraic summation of the invariant characteristics for all the muscles, taking these coefficients into account. However, in the framework of the λ-model, the number of independently controlled variables will always be equal to the number of participating muscles, thus significantly simplifying both the process of control and its description.

Analysis of relatively fast movements (where one cannot ignore the dynamic sensitivity of muscle spindles) requires introducing a new variable in the λ-model, namely velocity of muscle length changes. Feldman and his colleagues (Adamovich et al., 1984; Feldman, 1986; Abdusamatov et al., 1987) introduced, for this purpose, a notion of a velocity-length phase plane (V-L plane, Figure 1.16). Each point on the V-L plane corresponds to a combination of certain values of muscle length and velocity of its changes. The V-L plane can be divided by a straight line into zones of muscle activation (autogenic α-motoneuron recruitment) and silence. This line crosses the abscissa axis at a point corresponding to the current value of the controlled variable, that is, threshold of the tonic stretch reflex (λ). Points 1, 2, and 3 in Figure 1.16 correspond to active agonist α-motoneurons, although at point 1 muscle length is smaller than λ. Points 4, 5, and 6 correspond to silent agonist α-motoneurons, although at point 6 muscle length is bigger than λ.

Analysis of fast movements on the V-L plane has provided a necessary link between kinematic description in the λ-model framework and interpretation of the EMG patterns accompanying voluntary movements. In particular, the possibility of emergence of the well-known biphasic and triphasic patterns (for reviews see Mustard & Lee, 1987; Gottlieb et al., 1989a) observed during fast single-joint movements has been demonstrated (Abdusamatov & Feldman, 1986; Abdusamatov et al., 1987; Feldman et al., 1990a; Latash & Gottlieb, 1991a). However, the model has still been criticized for its inability to account for certain experimentally observed phenomena, including changes in the EMG patterns with changes in task parameters (Enoka, 1983; Marsden et al., 1983) and variability of the single-joint movements (Darling & Cooke, 1987). Therefore, some of the purposes of this book are to expand the area of applicability of the λ-model as a particular example of the dynamic approach to motor control, to demonstrate that this model provides reasonable interpretation of the findings during both fast and slow movements (including the EMG patterns), and to use the same approach for analysis of other aspects of single-joint movements, including motor learning and variability.

Let me suggest the following hypothetical answer to the question of what parameters are controlled by the nervous system, which has been proposed by Feldman and which I agree with: The independently controlled variable for an intact muscle is the threshold of the tonic stretch reflex. Descending control of single-joint movements can be described as time shifts of this variable for participating muscles.

Let us now return to a group of motor reactions that is on the border between muscle reflexes and voluntary movements, the preprogrammed reactions or M_{2-3}. They lead us to consider a more general problem of preprogramming in voluntary movements.

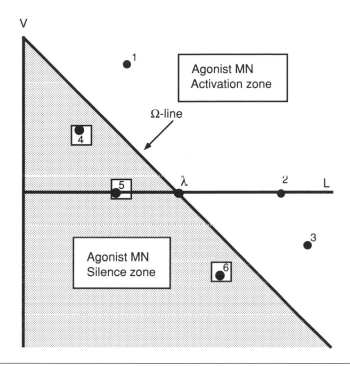

Figure 1.16   Each point on the muscle phase plane (V-L plane) corresponds to a combination of values of muscle length (L) and speed of its changes (V). The V-L plane is assumed to be divided by a straight line (Ω-line) into zones of autogenic α-motoneuron activation (points 1, 2, and 3) and silence (points 4, 5, and 6). The point of intersection of the Ω-line with the length axis corresponds to a current value of λ.

Note. From "An Equilibrium-Point Model for Fast Single-Joint Movements: I. Emergence of Strategy-Dependent EMG Patterns" by M.L. Latash and G.L. Gottlieb, 1991, *Journal of Motor Behavior*, **23**, pp. 163-177. Reprinted with permission of the Helen Dwight Reid Educational Foundation. Published by Heldref Publications, 4000 Albemarle St., N.W., Washington, D.C. 20016. Copyright © 1991.

Preprogramming in Motor Control

Hammond (1954) was the first to describe such reactions to sudden human limb perturbations at latencies exceeding those of the monosynaptic stretch-reflex but lower than voluntary reaction times. These reactions depended heavily on the instruction to the subject. Similar reactions have been observed since then in a variety of experimental paradigms in different human and animal muscles (e.g., Phillips, 1969; Newsom Davis & Sears, 1970; Melvill Jones & Watt, 1971a,b; Tatton et al., 1978; Colebatch et al., 1979; Dietz et al., 1981a). Let us address them as M_{2-3} or preprogrammed reactions.

M_{2-3} Is Not a Stretch Reflex

Several experimental findings do not let one consider M_{2-3} as a kind of stretch reflex. Depending on the instruction to the subject, M_{2-3} could be observed in a muscle that is shortened by perturbation (e.g., Marsden et al., 1979; Nashner et al., 1979; Nashner & Cordo, 1981). The amplitude of the M_{2-3} responses did not correlate with the amplitude of the applied perturbation if the latter could not be predicted by the subject (Marsden et al., 1981; Rothwell et al., 1982b, 1986). Compensation of the perturbation due to M_{2-3} could vary in different trials from 0% to 100% or even to overcompensation (Houk, 1978).

The fact of independence of these responses of the stimulus magnitude suggests that the stimulus represents a nongraded signal to the response generation and the response magnitude is defined prior to the stimulus based on some other factors. In this connection, the M_{2-3} can certainly be called triggered or preprogrammed reactions (cf. Chan & Kearney, 1982; Bonnet, 1983; McKinley et al., 1983).

A Hypothesis: *The M_{2-3} responses represent preprogrammed time functions* $\lambda(t)$ *for the participating muscles, that is, they reflect an advance planned fast shift of the joint compliant characteristic.* By generating these functions in advance, the preprogramming reduces the reaction time to a perturbation at the cost of suboptimal magnitude of the corrective response.

In a paradigm including "compensation as fast as possible" of an unexpected perturbation, a subject can synthesize the $\lambda(t)$ functions for a compensating reaction in advance since the experiments are carried out in standardized conditions.

Another Hypothesis: *The analysis of the M_{2-3} responses to perturbations applied during other motor tasks suggests that execution of any motor task is associated with preprogramming of fast compensatory reactions (functions $\lambda(t)$) to conceivable external perturbations.* This suggestion is mainly based on common sense because execution of the familiar movements in everyday life is inevitably connected with certain unpredictable but frequently encountered changes in the external conditions of movement execution. These changes include, in particular, rapid changes in the external load or obstacles on the movement trajectory. Therefore, one may assume that any motor task is always associated with preprogrammed compensations of unexpected loadings or unloadings. Magnitude of the preprogrammed reaction can be generated according to the previous experience of the subject. This preprogramming can lead, in particular, to M_{2-3} responses to unexpected load changes during smooth movements (Marsden et al., 1972, 1975, 1976, 1977, 1978, 1979). The latency of these responses and their double-peak pattern correspond well enough to similar features of the preprogrammed reactions observed in the position-hold paradigm.

The hypothesized role of the cerebellum in processing sensory information during control of voluntary movements (see Houk & Gibson, 1987) makes some of the findings of the experiments on inferior olive activity during different voluntary movements interpretable in the present framework. In particular, it has been shown that responsiveness of inferior olive cells to sensory stimuli can be changed when the limb experiences an unexpected perturbation during a voluntary movement (Gellman et al., 1985; Lou & Bloedel, 1986) or when a perturbation is introduced into the step cycle during locomotion (Armstrong et al., 1982; Lou & Bloedel, 1986). The climbing fibers seem to signal the occurrence of a peripheral somatosensory event without detailing the intensity and duration of the stimulus (Rushmer et al., 1976; Armstrong et al., 1982; Gellman et al., 1985). So, there are information inputs providing a nongraded triggering signal that can be used for releasing a preprogrammed motor reaction. Theoretically, preprogramming can occur at different levels of the motor hierarchy. Therefore, the hypotheses on transcortical loops, spinal loops, and cerebellar loops participating in the preprogramming are not mutually exclusive.

Experimental Findings. Let us now discuss features of the preprogrammed responses described in different papers:

• It has already been noted that these reactions depend significantly on the instruction to the subject. Namely, the responses take place when the instruction is "to compensate as fast as possible . . ." and they are absent in the case of "do not pay attention. . . ." This result is obvious from the advanced view since the subject is free to preprogram or not to preprogram.

• Emergence of the preprogrammed reactions in a perturbation-shortened muscle is also quite understandable since the subject can preprogram any combination of $\lambda(t)$ functions for any muscle or muscle group (depending on the given instruction) independently of the influence of a future perturbation on the muscle length.

• Because amplitude of the preprogrammed movements should be defined prior to the perturbation, random changes of the perturbation amplitude cannot lead to any correlation of the preprogrammed response amplitude with that of the perturbation. Reproduction of the same perturbation amplitude in a series of trials should lead to an improvement of the compensation due to the preprogrammed responses, which was observed experimentally (Nashner, 1976; Nashner & Grimm, 1978). Since preprogramming of compensatory reactions is based on the previous experience, experiments with random changes of the perturbation amplitude could rather lead to a correlation of preprogrammed response amplitude with perturbation amplitude in the preceding trial (or with average value for several preceding trials); however, such analysis has not been performed.

- Desmedt and Godaux (1978) have shown that the preprogrammed responses can be induced by perturbations of a smooth movement and are virtually absent during perturbations of a ballistic (very fast, programmed in advance) movement. In case of a smooth movement, deflection from the planned trajectory caused by a perturbation plays the role of a signal for a preprogrammed fast corrective movement with a preset amplitude. A smooth change of the $r(t)$ (see chapter 2) is replaced by a rapid jump to a new value at a close to maximal rate ω_{max} giving rise to the preprogrammed responses. In case of a ballistic movement, the rate of change in r may already be close to ω_{max}, and any preprogrammed changes in r cannot significantly influence the ongoing muscle activation pattern. Note also that control of ballistic movements by setting final values of the control variables at the highest available rate leads to equifinality independently of transient perturbations (e.g., Cooke, 1980; Brown & Cooke, 1981). Therefore, the preprogrammed corrections become redundant, and perhaps execution of such movements is not accompanied by preprogramming at all.

- High-frequency muscle vibration leads to a pronounced suppression of the muscle monosynaptic reflexes (e.g., De Gail et al., 1966; Delwaide, 1973; Desmedt & Godaux, 1978) and can influence certain other reflexes (e.g., Ertekin & Akcali, 1978; Gregoric et al., 1979). However, the preprogrammed responses remain unchanged on the background of the vibration (Lee & Hendrie, 1977; Hendrie & Lee, 1978; Agarwal & Gottlieb, 1980). Vibration-induced changes in muscle afferent activity are not directly related to preprogramming and influence only the amplitude of a signal for the preprogrammed response playback. On the other hand, the amplitude of the preprogrammed response does not depend on the magnitude of a signal giving rise to the response, which explains the lack of vibration influence upon the M_{2-3} reactions.

Afferent Source Cannot Be Defined

Consideration of the preprogrammed reactions as triggered responses to a signal provided by the perturbation suggests that the search for the "afferent source" of these responses is likely to provide unreliable results. In fact, if the perturbation provides only a triggering signal, the source of this signal is not really significant. It is only essential to provide necessary information about occurrence of the perturbation. In this context, signals for the preprogrammed reaction can be provided by virtually any group of peripheral receptors bearing information on changes in load, position, pressure on the skin, and, in certain experimental situations, also by visual, auditory, and vestibular receptors (see later in this chapter).

Recently, a concept of multisensority has attracted attention as a parallel processing of afferent information from different sources providing some resulting "value" important for processes of both motor control (e.g., Lundberg, 1979) and kinesthetic perception (Feldman & Latash, 1982a,b). If one accepts this concept for the preprogrammed reactions, elimination of one (or several, but not all) of the afferent sources should not be expected to lead to the elimination of the responses, although it can influence certain features including, for example, latency, since velocity of signal transmission and, possibly, time delays due to the central processing may be different for different afferent systems.

Not surprisingly, experiments with selective blocking of the transmission along certain afferent systems (e.g., Marsden et al., 1979; Traub et al., 1980b; Bawa & McKenzie, 1981) have not provided conclusive information on the role of these afferents in the generation of the preprogrammed responses (cf. Loo & McCloskey, 1985). Note that dependence of the preprogrammed responses on different sensory information sources have been demonstrated in studies of influence of visual and vestibular perception on these responses (Berthoz et al., 1979; Craik et al., 1982; McKinley & Smith, 1983).

Monkey experiments (Wylie & Tyner, 1981) with lifting of an unknown load have also demonstrated presence of short-latency (as compared with the voluntary reactions) responses leading to "smoothening" of limb movements in cases of unexpected load changes. Deafferentation of the limb led to disappearance of these components; the reactions became more sharp and delayed. In these experiments, the preprogrammed responses disappeared after *total* deafferentation of the limb, and the residual compensation was based only on visual information.

Long-term training of monkeys could lead to the disappearance of the M_{2-3} responses with a concomitant increase in a short-latency, presumably monosynaptic response M_1 (Christakos et al., 1983). One can tentatively suggest that the M_1 changes can also be preprogrammed (or conditioned; cf. the studies by Wolpaw, 1983; Wolpaw et al., 1983) using influence on the mechanism of presynaptic inhibition or, directly, on the α-motoneuron pool.

Two Unusual Examples of Preprogrammed Responses

In the previous section, it was proposed that any motor program (with the possible exception of ballistic movements) is associated with preprogramming of certain motor reactions for rapid compensations of possible movement perturbations. Let us analyze two examples of "nonstandard" preprogrammed responses whose relation to the motor task context is especially pronounced.

Thumb Flexion Experiments. A particular example of the prepro-grammed response inversion has been described by Marsden et al. (1979). Their subjects performed flexion of a thumb against a constant load pro-vided by a torque-motor. The load could be increased or decreased unex-pectedly during movement. An increase of the load led to an increase in flexor activity with a characteristic for the M_{2-3} response latency; a decrease of the load led to a flexor activity subsidence (see also Mardsen et al., 1972, 1976, 1978). These reactions can be explained, as suggested earlier, by preprogramming of a corrective ballistic movement and its playback in response to a signal provided by the load changes. However, if the unloading was provided not by a decrease in the motor torque but by lifting the hand by the wrist so that the thumb was moved away from the lever, flexor activity was increased with the same characteristic latency, bringing about a movement that tried to restore the thumb position on the lever.

Trying to reveal the afferent source of the observed motor inversions, the authors performed a series of experiments with anesthesia of the distal hand and arm skin, which did not influence the observed pattern. Limitation of proximal limb movements brought about elimination of the "strange" responses. It was concluded that the essential role in the ob-served pattern was played by afferent signals from the proximal limb joints and muscles. However, signals from the proximal limb segments did not by themselves give rise to the inversion, as demonstrated in the experiments when a subject held the entire device in his or her hand (so that a similar lift by the wrist did not lead to moving the thumb from the lever).

These results are quite understandable from the position of preprogram-ming in response to changes in the conditions of motor task execution. Moving the thumb from the lever made it impossible for the subject to carry out the given instruction "to move the lever with the thumb." In this situation, the preprogrammed reaction should primarily try to restore the thumb position, thus giving rise to the flexion observed in the experi-ments. Information on the perturbation was provided by several sources, including, probably, afferents from the proximal limb segments. Informa-tion from these afferents was presumably sufficient to provide a signal for the preprogrammed response, which was why anesthesia of the distal skin did not eliminate the effect.

When a subject held the entire device in his or her hand, the same information from the proximal limb did not bring about significant changes in the conditions of movement execution (i.e., the information did not mean that a preprogrammed response was required).

Grab Reflex Experiments. An elegant example of the preprogrammed reactions was presented by Traub et al. (1980b) in experiments with the grab reflex. A subject was given an instruction to position his or her thumb and index finger just near the glass "as if going to grab it." Although no command occurred, the instruction clearly implied preprogramming of a grabbing movement by the subject. This movement was actually observed, with a characteristic for the M_{2-3} response latency, when the subject's arm was unexpectedly lifted so that the glass remained below the hand.

Different experiments with anesthesia led the authors to conclude that the afferent source of the observed reaction was in the proximal segments of the limb. In fact, information from these segments has proven to be sufficient as a signal for the preprogrammed reaction, but it has not proven to be necessary. It is probable that the same response could have been observed in a naive subject in response to an unexpected loud sound.

Automatic Corrections of the Vertical Posture

Unexpected perturbations of the vertical posture bring about compensatory reactions with latencies intermediate between the phasic stretch reflex and voluntary reactions. These reactions were observed during maintenance of the vertical posture (Nashner, 1976, 1979; Nashner & Woollacott, 1979; Nashner et al., 1979; Marsden et al., 1981; Allum, 1983) and during walking (Nashner et al., 1979; Nashner, 1980; Dietz et al., 1984).

Note that maintenance of the vertical posture is probably the most common motor task. It is reasonable to presume that the mechanism of vertical posture control is well "defended" against unexpected changes in external conditions. This means that different preprogrammed corrections of the vertical posture are ready to be initiated in response to certain peripheral signals without any special instructions. We have up to now considered mainly preprogrammed ballistic motor programs. However, any other set of functions $\lambda(t)$ can be preprogrammed corresponding to a foreseen perturbation of a given motor task. In particular, a rather complicated problem of maintaining the vertical posture requires relatively complex corrective reactions involving activation of different muscle groups (see also the following section).

Experimental Findings. Marsden et al. (1981) investigated the responses of different postural muscles to perturbations applied to different parts of the body. Reactions with an intermediate latency (such as M_{2-3}) were evoked by the perturbations applied to different joints independently of whether activity of the recorded muscle affected position in the perturbed joint. Experiments with anesthesia demonstrated that the afferent source of these reactions was at the place of application of the perturbation rather than in the reacting muscle. The authors have drawn the dubious conclusion that the reactions observed in response to a perturbation applied to a joint controlled by contractions of the recorded muscle can be considered "reflex," while the reactions to perturbation of other joints have a different nature.

However, the similarity of the reactions in both situations suggests their common nature. According to the advanced view, these reactions represented preprogrammed motor commands realized when the peripheral signals informed about a perturbation *independently of the afferent source.*

In the same study, the authors recorded the preprogrammed responses in arm muscles in response to a perturbation of the vertical posture when the subjects stood grasping some object for additional support. When the object had low inertia, the responses could invert and emerge in the antagonistic muscle groups, and the overall pattern of the movement resembled those observed in a man holding a cup of tea in his hand in response to a postural perturbation. The authors advanced a complex explanation of the reaction with possible participation of cerebellar and cortical structures.

This inversion can, however, also be explained using the concept of preprogrammed responses, since the task "to maintain an equilibrium of an object in the hand during postural perturbations" is a common one. If the subject knows in advance that the object he grasps cannot be helpful in correcting posture perturbations, the corrections can be provided by preprogrammed activation of other muscle groups. If the subject "relies" on the object, arm muscle activity will be preprogrammed. This activity, in cases of low-inertia objects, will not be able to compensate for the perturbation leading to more pronounced postural disturbances as compared with the "normal" corrections. Probably everyone can recollect similar situations from everyday life, when a support used for postural stabilization appears to be "low-inertial" (unreliable). For example, if you stand in a bus and grab a metal rail, a certain combination of learned activity changes in postural and arm muscles helps you to maintain vertical posture during bus accelerations and decelerations. If the rail is broken (frequently the case on Russian buses), the learned pattern of activity changes may lead to aggravation of the effects of the perturbation rather than to their compensation.

Let us consider one more experimental finding suggesting preprogramming of postural reactions of an intermediate latency: These reactions could be observed in muscles either lengthened or shortened by perturbation (Nashner, 1979; Nashner & Woollacott, 1979). Nashner and Cordo (1981) compared latencies of voluntary posture corrections and the preprogrammed reactions. Voluntary reaction times were reproducibly larger although long-lasting training could reduce them up to a point when their values coincided with those for the preprogrammed reactions. Perhaps when the subjects repeated the same voluntary correction in response to a signal the correction was eventually transferred into the preprogrammed class.

Reflex Reversals

Reflex reversals, that is, reflex activation of different muscle groups due to changes of certain factors (e.g., position of a limb, position of the body, etc.) while stimulus location and intensity are maintained constant have been known since the classic experiments by Pfluger on decapitated frogs. Since that time, similar phenomena have been observed in snails (Jordan, 1905), insects (Bassler, 1976), cats (Magnus, 1909a,b; Lisin et al., 1973; Forssberg et al., 1975; Duysens & Pearson, 1976), and humans (Lisin et al., 1973; Latash & Gurfinkel, 1976; Gurfinkel & Latash, 1978; Feldman & Latash, 1982c; Duysens et al., 1990; Lacquaniti et al., 1991).

In the first studies, reflex reversals usually led to activation of a muscle from an antagonistic pair that was relatively more stretched (Von Üexkull, 1904, 1909; Magnus, 1909a,b). This finding let Von Üexkull formulate a rule bearing his name, according to which reflex activation is primarily observed in a relatively more stretched muscle. However, since that time, many exceptions to this rule have been observed (e.g., Sherrington, 1908; Graham Brown, 1911).

Relation of Motor Reversals to Locomotion.   Basing on the investigations of the ipsilateral flexor and crossed extensor reflexes and their dependence upon limb position, Sherrington (1910) was the first to propose that the mechanism of the reflex reversals can be closely related to generation of locomotor movements. Since that time, reflex reversals have been observed in many experimental conditions both in static postures and during the locomotor movements in response to electrical stimulation of skin nerves and areas. Reversals of ipsilateral (Lisin et al., 1973; Duysens & Pearson, 1976; Forssberg et al., 1976, 1977), contralateral (Duysens & Loeb, 1980; Duysens et al., 1980), and propriospinal (Halbertsma et al., 1976; Miller & van der Meche, 1976; Miller et al., 1977) influences have been reported in cats and dogs.

In experiments on healthy humans, reversals of the tonic vibration reflex were observed in static positions differing in joint angles and support pressure on the foot (Latash & Gurfinkel, 1976; Gurfinkel & Latash, 1978, 1980). The analysis of these results has also led to proposal of their relation to the mechanism of locomotor movement generation.

A Hypothesis.   Locomotion is one of the most commonly used movements in everyday animal and human activity. *Therefore, it is conceivable that the mechanism of locomotor movement generation is always in a preprogrammed state.* Let us presume that the locomotor generator is virtually always ready to generate a preprogrammed motor command according to the existent body configuration and other factors. Reflex loops of a number of relatively complex polysynaptic reflexes including the flexor, the crossed extensor, and the tonic vibration reflexes may involve the structures of the preprogrammed locomotor generator giving rise to activation of certain muscle groups depending on body position. Note that reversals of more simple monosynaptic reflexes have not been observed in similar experimental conditions.

Motor Reversals Are Preprogrammed Responses.   So, muscle responses during reversals are considered to be preprogrammed rather than reflex, and the stimuli play the role of a signal to play back the preprogrammed activation pattern. This viewpoint is corroborated, in particular, by the fact that the observed responses are not confined to muscle groups adjacent to the stimulation site as has been demonstrated for reversals during skin stimulation and during local vibration of shin or thigh muscles. Thus, the activation pattern appears to be essentially autonomous of the stimulus-evoked afferent inflow and, in a sense, reflects not the stimulus but the state of the preprogrammed mechanism.

The experiments with tonic vibration reflex reversals in humans demonstrated that the reversal of reflex activity in thigh muscles could be observed in standing subjects depending on position of the hip joint (Latash & Gurfinkel, 1976). Similar stimulation and similar changes of the joint position did not bring about the reversals if the subject was in a supine position; in this situation, only the autogenic tonic vibration reflex was observed. This finding can be explained if one considers that locomotor movements in a supine position are absolutely unusual, and the locomotor generating mechanism is not likely to be preprogrammed in this position.

Corrective Stumbling Reaction.   A particular pattern of reflex responses has been observed during cat locomotion associated by the authors with overcoming an unexpected obstacle—the corrective stumbling reaction (Forssberg et al., 1975, 1976, 1977; Andersson et al., 1978; Forssberg, 1979; Grillner, 1979). This pattern could be observed during weak mechanical stimulation of skin areas of the paw or of the leg (even with an air puff) or during short electrical stimulation of skin nerves or dermatomes.

Any of these stimuli, when applied during the swing phase, gave rise to a flexor reaction with the hindlimb transfer over a hypothetical obstacle. The same stimulation applied during the stance phase could give rise to an extensor reaction. "Sensibility" of this reaction and its relative independence of the stimulus suggest that it is in fact a preprogrammed response of a mechanism responsible in everyday life for the compensatory reactions during real stumbling. Note that the existence of such a mechanism was predicted by Bernstein in 1947.

Afferent Source Cannot Be Defined.   Scrupulous search for an afferent source of the reversals of the crossed extensor reflex (Rossignol & Gauthier, 1980) has demonstrated that this problem probably cannot be solved in principle. Since playback of a preprogrammed command is initiated by a signal indicating a change in the conditions of movement execution, it is essential only to obtain sufficient information independent of its source. The experiments with anesthesia of afferent information sources of certain modality with persistence of a reversal pattern (cf. Magnus,

1909a,b, 1910) do not give any conclusive information concerning the role of the excluded afferent signals; they say only that the intact information sources are sufficient.

Rossignol and Gauthier (1980) performed successive denervation of a limb and observed a dependence of the crossed response to skin nerve stimulation upon position of the nonstimulated limb. Usually, if the limb contralateral to the stimulus was initially flexed, the crossed extensor response was observed, and if the contralateral limb was extended in all three main joints, the crossed flexor response took place. Independent of the order of denervation, the reversal was preserved if some minimal amount of afferent signal sources was intact. In experiments on two animals, the "summed" denervation could be total, and the reversals were still present in both cases

In the same study, the authors performed successive sectionings of the spinal dorsal roots. Independent of the order of root section, the reversal of crossed responses took place if some minimal number of roots were intact. All these results suggest that one can hardly single out one information source that would be crucial for the reversals, although Baxendale and Ferrell (1981) argue that the reversals of this type are crucially dependent on information from the hip joint receptors.

It has been hypothesized that execution of any motor task is associated with preprogramming of certain motor patterns depending on the task and on probable perturbations that can occur during the task execution. The preprogrammed patterns provide for the fastest start of compensatory responses to the perturbation. These motor reactions lead to rather crude approximate corrections that can further be adjusted with a voluntary reaction. Playback of preprogrammed $\lambda(t)$ patterns begins with a signal indicating peripheral changes that are likely to hinder accomplishment of the motor task. These reactions cannot as a rule be unambiguously associated with activity of certain afferent systems because their generation requires getting sufficient information independent of its source.

The preprogrammed reactions can occur in cases of familiar movements and familiar ("everyday") perturbations like unexpected loading or unloading of a muscle, phasic perturbation of vertical posture, unexpected obstacle on a leg trajectory during locomotion, and so on. However, training in standardized conditions can lead to preprogramming of other reactions or of reactions to unusual perturbations, as has been demonstrated by Nashner and Cordo (1981) in experiments with measuring latencies of automatic posture corrections and voluntary posture corrections.

Note that preprogrammed reactions can be of very different structure, starting from the simplest program for a ballistic movement giving rise to the well known pattern of M_{2-3}, up to coordinated activity of different muscle groups (as in the "grab reflex," Traub et al., 1980, or in the corrective stumbling reaction).

Here is an example of a very complex but obviously preprogrammed reaction.

Imagine an officer giving commands to soldiers: "lie down, stand up, lie down, stand up, lie down, stand up, stand up!" It is obvious that some of the participants of this experiment would lie down in response to the last command. After numerous repeats of the presented sequence, the motor reactions become preprogrammed, and the voice of the officer plays the role of a triggering signal.

Marsden et al. (1977) have demonstrated that preprogrammed responses such as M_{2-3} are suppressed under the influence of alcohol. Every person with some experience of alcohol intoxication can recollect that the most prominent features of motor disturbances include suppression of automatic corrections—for example, drunk people have trouble keeping upright while riding in an accelerating or decelerating vehicle. They also stumble and fall down frequently. It is possible that alcohol suppresses the preprogramming mechanisms at different levels of the motor hierarchy.

In this chapter, the λ-model has been presented as an analytical black box approach. The λ-model is based on central control of a variable (λ) that does not have even a hypothetical relation to activity of certain neuronal populations. This does not mean, however, that λ is not measurable. It just requires a proper instrument. For example, we would have a hard time finding a neuron or a population of neurons that code movement amplitude or time, although these two variables can easily be measured with a ruler and a watch. In the next chapter, we are going to discuss a method of measuring λ that could make it another observable within human motor behavior.

Chapter 2

Analysis of Joint Compliance

The equilibrium-point hypothesis considers control of a joint as central modulation of spring properties of a system "joint plus its reflexes." Compliance (or stiffness) is a major characteristic of a spring, and therefore, measuring or reconstructing joint compliance becomes essential for experimental investigation of predictions and consequences of the equilibrium-point hypothesis. Unfortunately, compliance of an intact joint with its central connections is not readily observable. Fortunately, compliance can be reconstructed, although with certain assumptions and simplifications. Time changes of the reconstructed joint compliance during voluntary movements bear a very important meaning to patterns of central control variables. These patterns reconstructed during various single-joint movements make λ an observable of human motor behavior.

Equations of Mass-Spring Systems

The notion of joint compliance has become central to a group of models of motor control, including those described in chapter 1. Compliance, by definition, is a characteristic of a spring, or a spring-like mechanical system, describing magnitude of displacement in response to a unit of force. As such, it is inversely related to stiffness, which describes magnitude of force necessary to cause a unitary displacement:

$$k = \frac{\Delta F}{\Delta x},$$

$$c = \frac{\Delta x}{\Delta F}, \qquad (2.1)$$

$$k = \frac{1}{c},$$

where ΔF is a change in an externally applied force, Δx is displacement, k is a coefficient of stiffness (or just "stiffness") and c is a coefficient of compliance (or just "compliance").

Besides a spring component, all the mass-spring models of muscle behavior also include an inertial component (mass). If there is a change in external force, this inertial component behaves according to Newton's laws of mechanics:

$$F = m\frac{d^2x}{dt^2},$$ (2.2)

where x is length of the spring, m is a coefficient of proportionality termed *mass*, and t is time.

It is necessary to introduce at least one more element to damp oscillations that would inevitably occur in response to a change in an external force applied to a mass-spring system. This element is usually modeled with the help of the notion of viscosity. Viscosity is an ability of a mechanical system to generate force directed against a velocity vector:

$$F = b\frac{dx}{dt},$$ (2.3)

where b is a coefficient of viscosity.

Figure 2.1 schematically shows a simple mechanical system consisting of a spring with a known stiffness k, a damping element with a known coefficient of viscosity b, and an inertial component with a known mass m (cf. Figure 1.3). In general, behavior of such a system can be described by Equations 2.1, 2.2, and 2.3 combined together. Let us take into account that, generally speaking, all three coefficients may represent time functions. Then, we get:

$$F(t) = m(t)\frac{d^2x(t)}{dt^2} + b(t)\frac{dx(t)}{dt} + k(t)[x(t) - x_0(t)],$$ (2.4)

where $x_0(t)$ is a "zero" length of the spring that theoretically can also change with time.

However, the system illustrated in Figure 2.1 implies linear movements induced by a force vector. In a living body, most of the movements are rotations around the joints. Let us consider a pin joint with only one degree of freedom and the same three elements (Figure 2.2). Equation 2.4 will undergo minor transformations:

$$T(t) = M(t)\frac{d^2\alpha(t)}{dt^2} + b(t)\frac{d\alpha(t)}{dt} + k(t)[\alpha(t) - \alpha_0(t)],$$ (2.5)

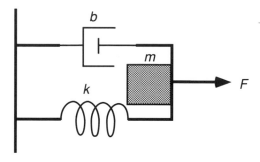

Figure 2.1 Behavior of a simple mass-spring-dashpot scheme can be character-ized by three parameters: m is inertia, k is stiffness, and b is viscosity. Corre-spondingly, total muscle force (F) can be considered as a combination of three components. On the other hand, F can be described in the framework of the equilibrium-point hypothesis as a function of λ and muscle length.

where T is torque around the joint, α is joint angle, α_0 is a "zero" joint value, and M is moment of inertia. This equation is analyzed with different degrees of simplification in most of the mass-spring models of single-joint motor behavior. It is familiar to any physicist. Experimental analysis of the systems described by Equation 2.5 looks relatively straightforward: An experimenter tries to control time changes of one of the two measurable variables, torque $T(t)$ or angle $\alpha(t)$, measures time changes of the other variable, and, with a sufficient number of measurements, is able to calcu-late time changes of the coefficients $M(t)$, $b(t)$, and $k(t)$ with some degree of accuracy. There are numerous ways of performing such experiments, including application of singular small torque perturbations at certain times and measuring length changes (Ma & Zahalak, 1985; Gottlieb et al., 1986), using sinusoidal perturbations over a range of frequencies (Joyce et al., 1974; Agarwal & Gottlieb, 1977; Rack et al., 1978; Cannon & Zahalak, 1982), or using randomized perturbations with subsequent correlation of torque and angle changes (Hunter & Kearney, 1982; Lacquaniti et al., 1982a; Weiss et al., 1988; Bennett et al., 1989).

Attempts at direct application of Equation 2.5 for modeling and analysis of voluntary joint movements lead to considerable complications if one tries to incorporate at least minimal information about muscles and their central connections, severely limiting potential applicability of Equation 2.5. The first of these problems is that Equation 2.5 models a damped loaded spring with an instantaneous reaction to external load or central command changes. Thus, this approach ignores reflex time delays, which are likely to play an important role in movements (especially fast ones) of real single-joint motor systems (see chapter 1). Another problem is due to at least two different mechanisms giving rise to viscous behavior of intact muscles (Feldman et al., 1990a). First, there is a well-known relation

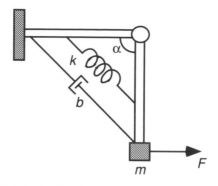

Figure 2.2   A mass-spring model of a single-joint system with damping. Spring (*k*) and damping (*b*) elements are shown for only one of the antagonist muscle groups.

between muscle force and the speed of its length changes (Hill's equation, Hill, 1938, 1953):

$$T - T_0 = b_0 \frac{dx}{dt},$$

where T_0 and b_0 are constants.

Besides that, dynamic sensitivity of muscle spindle afferents makes the tonic stretch reflex (which, according to the λ-model, forms the basis of single-joint motor control) sensitive not only to muscle length but also to the speed of its changes. Formally, this also implies presence of a viscous component in muscle reaction to load (or length) changes. Let us consider these problems in detail.

At Least Three Springs!

Before stating what is wrong with the so benign-looking Equation 2.5, let us reconsider the information about muscle and its reflexes that has been briefly summarized in chapter 1. What will happen if the length of an intact activated muscle is rapidly changed by some external factor? One can single out at least three components in the muscle reaction:

- A muscle will oppose the length change by an increase in its force due to purely mechanical factors, as if it were a rubber band.
- There may be a distinct short-latency monosynaptic reaction leading to a phasic muscle contraction also opposing the external force but short in duration (loop 1 in Figure 2.3).

- There will be a tonic increase in muscle force due to a hypothetical length-sensitive mechanism commonly associated with the notion of tonic stretch reflex (loop 2 in Figure 2.3).

The lower part of Figure 2.3 illustrates all three components with different values of all three coefficients m, b, and k. Now, imagine that a rapid step-like perturbation is applied at a time t_0 leading to a virtually instantaneous increase in external torque (load). The muscle will start to lengthen. During the first several tens of milliseconds, this process will depend only on the first factor (i.e., muscle reaction as a rubber band). After a time delay corresponding to the latency of monosynaptic reaction for the muscle (t_1), there will be a burst of activity leading to a short phasic contraction also resisting the perturbation. Then, the time will come for

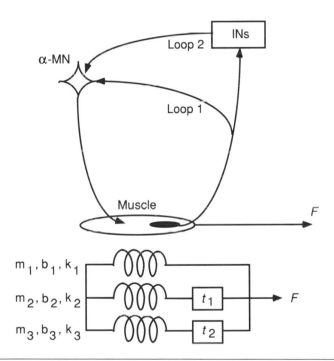

Figure 2.3   There are at least three mechanisms with different time delays that affect muscle reaction to a change in an external load. First, the muscle reacts "as a rubber band." Let us consider this reaction instantaneous. This reaction leads to a change in muscle length, which is sensed by the spindle endings (shown as a black spot). The changes in spindle activity lead to reflex changes in muscle force via monosynaptic connections (loop 1) with a characteristic time delay t_1 and via the hypothetical polysynaptic length-sensitive reflex (loop 2) with a characteristic delay t_2. Total muscle reaction can be modeled with the help of three springs (with inertial and dashpot components, not shown) and different time delays (lower figure).

the longer latency tonic reflexes, which will lead to development of tonic muscle contraction also resisting the perturbation. Let us assume that the latency of the tonic stretch reflex is t_2.

So, if one applies a perturbation (a change in external load) at a time $t = 0$ and measures joint angle changes in response to the perturbation, the kinematics will be defined by different factors at different times from the moment of perturbation. A measurement at a time t $(0 < t < t_1)$ would only supply information about the muscle's "rubber-band" reaction and is equivalent to testing a deafferented muscle. A measurement at a time t' $(t_1 < t' < t_2)$ would also supply information on the monosynaptic reactions (which is, by the way, not easy to extract!), that is, on the muscle with intact monosynaptic reflex connections but without the hypothetical longer latency tonic reflex connections. And eventually, at a time t'' $(t'' > t_2)$, measurement would supply information relevant to all three components.

Now, what will happen if an experimenter applies a perturbation at t_0 and measures muscle length immediately after the perturbation? The answer is obvious: Muscle length changes will reflect properties of only the first of the three mechanisms providing for spring muscle properties. This conclusion is independent of a mathematical procedure that may be used for further processing of the data, no matter how sophisticated it is, because if a measurement does not convey information about a certain factor, no further processing is able to extract this information.

The system does not look that simple any more, and it seems absolutely necessary to modify Equation 2.5 in order to take into account the three mechanisms providing for muscle reaction to external perturbations. The major problem is not in the presence of the three springs illustrated in Figure 2.3, but in their different characteristic time delays. Nevertheless, let us try to analyze all three components and write corresponding equations. We will, at first, consider all three springs as being independent and having different coefficients of viscosity, stiffness, and inertia.

It is also important to realize what is the cause and what is the consequence in muscle reactions. If there is an external change in torque around a joint, it will be a cause for muscle length (or joint angle) changes due to the first, purely mechanical mechanism. Let us ignore time delays in muscle reaction to perturbation defined by this component:

$$T_0(t) - M_0(t)\,\frac{d^2\alpha(t)}{dt^2} - b_0(t)\,\frac{d\alpha(t)}{dt} = k_0(t)[\alpha(t) - \alpha_{00}(t)], \qquad (2.6)$$

where α_{00} corresponds to zero angle for the first, "peripheral" spring. We will use α_{01} and α_{02} for zero angles for the two other springs. Note that the viscous component $b_0(t) \dfrac{d\alpha(t)}{dt}$ in Equation 2.6 can be considered an equivalent to Hill's equation.

Changes in muscle length will cause monosynaptic reactions, and torque changes become a consequence! Taking into account three components of muscle torque, external, inertial, and viscous, and the delay t_1, we get

$$T_1(t + t_1) - M_1(t + t_1) \frac{d^2\alpha(t + t_1)}{dt^2} - b_1(t + t_1) \frac{d\alpha(t + t_1)}{dt} = k_1(t)[\alpha(t) - \alpha_{01}(t)].$$

(2.7)

Changes in muscle length will also provide for the tonic stretch reflex leading to a tonic torque change. The following equation describes behavior of this third spring:

$$T_2(t + t_2) - M_2(t + t_2) \frac{d^2\alpha(t + t_2)}{dt^2} - b_2(t + t_2) \frac{d\alpha(t + t_2)}{dt} = k_2(t)[\alpha(t) - \alpha_{02}(t)].$$

(2.8)

Now we are ready to substitute Equation 2.5 with another one taking into account all three components of muscle reaction to perturbation. It seems fair to suppose that $M_0(t) = M_1(t) = M_2(t) = M(t)$. Then,

$$
\begin{aligned}
&T_0(t) - M(t) \frac{d^2\alpha(t)}{dt^2} - b_0(t) \frac{d\alpha(t)}{dt} + T_1(t + t_1) \\
&- M(t + t_1) \frac{d^2\alpha(t + t_1)}{dt^2} - b_1(t + t_1) \frac{d\alpha(t + t_1)}{dt} + T_2(t + t_2) \\
&- M(t + t_2) \frac{d^2\alpha(t + t_2)}{dt^2} - b_2(t + t_2) \frac{d\alpha(t + t_2)}{dt} \\
&= k_0(t)[\alpha(t) - \alpha_{00}(t)] + k_1(t)[\alpha(t) - \alpha_{01}(t)] + k_2(t)[\alpha(t) - \alpha_{02}(t)],
\end{aligned}
$$

(2.9)

where external, inertial, and viscous torque components of the three springs are superimposed so that total external torque is $T(t) = T_0(t) + T_1(t) + T_2(t)$.

This equation illustrates how complicated such a system is and how hard it is to analyze its behavior without introducing simplifications. Obviously, one cannot perform such an analysis by directly corresponding values of $\alpha(t)$ to $T(t)$.

Until now, we have not considered which variables are specified by the central control system during control of voluntary movements. According to the equilibrium-point hypothesis, motor control is based on central regulation of parameters of the human tonic stretch reflex. Equation 2.8 represents action of this loop and, therefore, the components in Equation 2.9 with an argument $(t + t_2)$ in no circumstances can be ignored since such a simplification would mean ignoring the basic mechanism of control. From this view, Equation 2.5 seems to be absolutely inadequate for studying the processes of motor control, although it might be useful for analyzing issues dealing with mechanical properties of the peripheral apparatus under different states of muscle activation.

Introducing Central Control

Regulation of parameters of the tonic stretch reflex arc can lead to two basic effects upon the joint compliant characteristic: translation parallel to the angle axis and change in the slope. In chapter 1, two variables were introduced corresponding to these two effects, r and c. Changes in the reciprocal variable r lead to the joint compliant characteristic translations along the angle axis while changes in the coactivation variable c induce changes in the joint compliant characteristic slope. This can be formally expressed as

$$f\,[c(t)]T(t) = \alpha(t) - r(t) \tag{2.10}$$

where f is a monotonically decreasing function.

According to the equilibrium-point hypothesis, central control can be described with time functions $c(t)$ and $r(t)$. This will give rise to translations and/or rotations of the joint compliant characteristic. The movement will result from a disparity between current joint torque and angle values and those corresponding to a new position of the joint compliant characteristic. Since there are no a priori restrictions on $r(t)$ and $c(t)$ functions, they need not necessarily lead directly to a new desired position of the joint compliant characteristic but may demonstrate a nonmonotonic behavior that might be useful for defining not only a new final equilibrium state but also the dynamics of getting to it (cf. Hogan, 1984; Hasan, 1986a; Latash & Gottlieb, 1991a,c; Lan & Crago, personal communications).

Getting back to Equation 2.9, note that central control by specifying functions $r(t)$ and $c(t)$ is analogous to defining time changes of $\alpha_{02}(t)$ and $k_2(t)$. It seems to be the right time to confess that Equation 2.9 is also wrong. Once again, the problem is in the neurophysiology of the system providing for the hypothetical mechanism of control and, in particular, in its time delays. Figure 2.4 illustrates the hypothetical tonic stretch reflex arc. Note that the latency of the tonic stretch reflex is a sum of three components: transmission in the afferent path, central delay, and transmission in the efferent path. It is virtually impossible to assess the central delay time. For the present purposes, let us assume that this time is zero and add half of the total time to the afferent transmission time and the other half to the efferent transmission time. Then, $t_{aff} = t_{eff} = t_{lat}/2$ where t_{lat} is the total latency of the tonic stretch reflex.

The central control signals are supposedly supplied to a central processor (INs in Figure 2.4). If at time t_1, instantaneous position of the joint

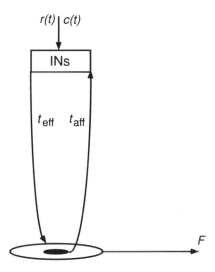

Figure 2.4 Let us assume that the latency in the hypothetical reflex loop consists of two equal components, afferent and efferent conduction time, $t_{aff} = t_{eff}$. The central control signals $r(t)$ and $c(t)$ are supplied to the central processor (INs). They interact with afferent information reflecting activity of the peripheral receptors generated t_{aff} ago. As a result, an efferent volley is generated that will affect the muscle state after t_{eff}.

compliant characteristic is described by central variables c_1 and r_1, the generated efferent signal will reflect the position of the joint compliant characteristic and an afferent volley that was in fact generated $t_{lat}/2$ ago. When the efferent signal arrives at the muscles, it conveys outdated information reflecting the position of the joint compliant characteristic that was centrally generated $t_{lat}/2$ ago and afferent information that was generated by the peripheral receptors t_{lat} ago.

Equation 2.9 takes into account a hypothetical delay in the tonic stretch reflex arc t_2 (analogous to t_{lat}). However, it fails to reflect the delay between generation of an afferent volley and its interaction with the central command, which would be when the afferent signal arrives at the central processing unit (interneurons [INs] in Figure 2.4). This factor can easily be introduced by changing the argument of $\alpha_{02}(t)$ and $k_2(t)$ functions:

$$T_0(t) - M(t)\frac{d^2\alpha(t)}{dt^2} - b_0(t)\frac{d\alpha(t)}{dt} + T_1(t + t_1)$$

$$- M(t + t_1)\frac{d^2\alpha(t + t_1)}{dt^2} - b_1(t + t_1)\frac{d\alpha(t + t_1)}{dt} + T_2(t + t_2)$$

$$- M(t + t_2)\frac{d^2\alpha(t + t_2)}{dt^2} - b_2(t + t_2)\frac{d\alpha(t + t_2)}{dt}$$

$$= k_0(t)[\alpha(t) - \alpha_{00}(t)] + k_1(t)[\alpha(t) - \alpha_{01}(t)] + k_2(t + \frac{t_2}{2})[\alpha(t) - \alpha_{02}(t + \frac{t_2}{2})].$$

(2.11)

Now, let us be satisfied with this equation and move forward to describing experimental methods of its analysis. These methods will certainly be based on a number of assumptions and simplifications, but we will try to make sure that these simplifications are explicitly stated, their possible effects on the quality of the results are assessed, and, most importantly, they do not eliminate the basic features of the hypothetical motor control process.

Reconstruction of Static Joint Compliant Characteristics

Experimental analysis of joint compliance and possible influence of central control signals started with so-called "static" experiments (Asatryan & Feldman, 1965; Feldman, 1966a,b). In these experiments, the subjects were instructed to maintain the position of a joint against an external load. The subject received an instruction "not to intervene voluntarily," and it was assumed that the subject did not change the voluntary motor command when changes in the external load were introduced. These changes led to joint movements. Joint angle and torque were measured after the limb reached a new equilibrium state and were plotted as points on a torque-angle plane. Interpolation of the points was considered a joint compliant characteristic.

Similar experiments were later performed by other researchers (Davis & Kelso, 1982; Vincken et al., 1983; Gottlieb & Agarwal, 1988; Latash & Gottlieb, 1990a). In all these experiments, the experimenters assumed and relied upon the subject's ability to "fix" the voluntary motor command. This assumption has always been a weak point in interpreting the joint compliant characteristic recordings. Attempts to prove that the subjects were in fact fixing a central control signal (Gielen & Houk, 1984) have not been strong enough to satisfy the skeptics (Gottlieb & Agarwal, 1988).

Let us return to Equation 2.11 and try to simplify it so that it would describe the static "do not intervene voluntarily" paradigm. First, since all the measurements were performed at equilibrium states, all the components of the monosynaptic reaction can be ignored, since it leads to transient, relatively short-lasting changes in joint torque and angle. Let us once again remember that ignoring the effects of monosynaptic *reflexes* does not mean ignoring the effects of monosynaptic *connections* that are likely to take part in the tonic length-sensitive changes in muscle force. Second, at the equilibrium $d\alpha/dt = 0$ and $d^2\alpha/dt^2 = 0$. Third, at the final equilibrium state, joint torque and angle are constants, and one may assume that, since the central command is presumably constant, all the other coefficients are constant with time as well. So, the differences in time arguments do not matter in this case. Note, however, that the fact that k_0 and k_2 are independent of time does not mean that they are constants since they may depend on joint angle (a nonlinear spring). Then, we get

$$T = k_0[\alpha - \alpha_{00}] + k_2[\alpha - \alpha_{02}]. \tag{2.12}$$

A passive (e.g., deafferented) muscle would demonstrate a behavior defined only by the peripheral spring:

$$T = k_0[\alpha - \alpha_{00}]. \tag{2.13}$$

The last two equations are simple enough to allow experimental reconstruction of corresponding dependencies and assessment of the values of the coefficients α_{00}, α_{02}, k_0, and k_2. Feldman was the first to do this in human experiments, and corresponding curves were presented in chapter 1. Similar curves were also reconstructed in animal experiments (Matthews, 1959; Feldman & Orlovsky, 1972). They formed the basis of the λ-model and allowed Feldman to hypothesize that control of a muscle is monoparametric. However, the "do not intervene voluntarily" paradigm in its original form allows reconstruction of joint compliant characteristics *only* when the voluntary motor command is constant, because it is based on measurements of joint torque and angle at a new *equilibrium state*. The λ-model has been criticized for being unable to account for the movement dynamics (Day & Marsden, 1982; Enoka, 1983; Atkeson, 1989) because it seemed impossible to reconstruct joint compliant characteristics *during* voluntary movements, and the notion of a shifting joint compliant characteristic turned into an illusory abstraction.

Recently, a method has been introduced in our laboratory allowing the reconstruction of joint compliant characteristics during voluntary movements that is related to the "do not intervene voluntarily" paradigm (Latash & Gottlieb, 1990b, 1991c). Before discussing the method, its potential, and relation to the formidable Equation 2.11, let us briefly summarize the methodological problems with reconstructing joint compliant characteristics in movement dynamics and the ways of solving them.

How to Reconstruct Joint Compliant Characteristics During Voluntary Movements

In static experiments, it is assumed that the subject does not change the voluntary motor command, and therefore does not shift the position of the joint compliant characteristic when the external load changes. These external changes lead to changes in the joint position. Joint angle and torque are measured at new equilibrium states and plotted as points on a torque-angle plane. Interpolation of the points defines a static joint compliant characteristic.

Attempts to experimentally reconstruct joint compliant characteristics during voluntary changes in the hypothetical central variables meet a number of methodological problems.

1. According to the λ-model, single-joint central motor control can be described as two time functions, $r(t)$ and $c(t)$, the first being related to position of the joint compliant characteristic and the second to its slope. Let us, for the sake of simplicity, temporarily accept a simple linear shift of r as a hypothetical control signal for single-joint movements:

$$r(t) = r_0 + \omega t, \qquad (2.14)$$

where r_0 is the initial value of r at $t = 0$, and ω is a constant.

It takes time (Δt) for the peripheral motor apparatus to come to a new steady state after a change in the external load. If central motor commands (e.g., r) change, as is assumed to happen during voluntary movements, the new state at which torque and angle are measured would correspond to a new value of r. That is, if a perturbation is applied when $r(t) = r_1$, the measurements will be made Δt later at a different value of r (cf. with Equation 2.14):

$$r_2 = r_1 + \omega \Delta t. \qquad (2.15)$$

Thus, initial and final values of torque and angle pertain to different static joint compliant characteristics (corresponding to r_1 and r_2) and cannot be directly compared.

2. The experimentally reconstructed joint compliant characteristic is a sum of at least two components, Equation 2.12. The first one is due to compliant properties of the joint in the absence of any changes in recruitment of α-motoneurons. The second one is defined by autogenic reflex recruitment (or derecruitment) of α-motoneurons due to a change in muscle length. The variables λ_{ag}, λ_{ant} (the subscripts refer to agonist and antagonist muscles), r, and c, introduced for describing motor control processes in the framework of the λ-model, relate only to the second component (Feldman, 1980a, 1986). Therefore, if a movement starts from a relaxed state, that is, in the initial absence of α-motoneuron activity, values of all these variables are undefined. This corresponds to an undefined constant r_0 in Equation 2.14, thus creating a problem if one wants to define time changes in r introduced by a voluntary motor command.

3. Fast perturbations, applied during a movement, often give rise to phasic reactions (monosynaptic, oligosynaptic, and/or preprogrammed) (Allum, 1975; Marsden et al., 1976; Iles, 1977; Gottlieb & Agarwal, 1980a,b; see "Preprogrammed Reactions" and "Preprogramming in Motor Control," chapter 1). These reactions lead to transient angle and torque changes so that a stable steady state is achieved only after at least 200 to 300 ms. However, after this time the central variable r will have changed (see point 1) creating difficulties in interpretation of the measurements. Slow perturbations, even if they evoke no phasic reflex responses, would end after a considerable change in r.

4. Fast movements must overcome considerable inertial and possibly viscous torques from the limb itself (Gottlieb et al., 1989a,b). An external device (e.g., strain gauge) does not measure these components and supplies information on only a part of the total torque developed by the muscles.

In order to cope with these problems, a procedure was designed that included recording of angle and torque changes during movements performed on the background of a slowly changing torque. While understanding that the reader's patience has its limits, we would like to go slightly deeper into the method of reconstructing joint compliant characteristics because it illustrates both the problems with experimental analysis of the equilibrium-point hypothesis and its predictive power. It will also serve as an example of how the data are actually "cooked" in the lab.

A series of experiments with relatively slow movements can be termed *quasi-static* since muscle torque components dependent on the first and second derivatives of angle were ignored during most of the analysis. On the other hand, these experiments are different from purely static experiments in that there was a change in the central motor command, and this change was experimentally reconstructed and analyzed. During the experiments with faster movements, the inertial component was also taken into account. It was assessed by multiplying measured acceleration by an assessed moment of inertia of the system "limb plus manipulandum." Viscous torque component has not been incorporated into the analysis because of the lack of reliable assessments of the coefficient of viscosity and its changes during voluntary movements. However, later we will try to assess possible errors introduced by this simplification.

The subjects were instructed and trained to perform a standardized and consistent time profile of a voluntary motor command by practicing a simple voluntary elbow movement. After training, the subjects were given an instruction, very similar to the one used earlier in the "do not intervene voluntarily" paradigm, requesting reproduction of the same time profile of the learned voluntary command while ignoring changes in the externally applied load. Both unidirectional elbow flexions and oscillatory elbow movements at different frequencies were analyzed.

The subjects reproduced the learned voluntary motor program against an extending torque bias generated by a torque motor. Therefore, all the movements were assumed to start from the same initial value of r (r_0), determined by the nonzero initial muscle torque (cf. with point 2). There is usually a slight coactivation of antagonist muscles while holding a joint position against an external torque bringing about some initial value of c (c_0). Generally speaking, r_0 and c_0 are defined only as a pair (cf. "The λ-Model" and "The Notion of Shifting Invariant Characteristics" in chapter 1) and one parameter cannot be uniquely related to external torque. Although we did not know the value of c_0, we assumed that it was the same throughout the course of one experimental series but did not test this assumption (as was done in all previous static experiments with reconstruction of joint compliant characteristics). Reproducibility of the experimentally reconstructed c-related coefficient k_2 (see later in this chapter) is consistent with and assumed to indirectly corroborate this assumption.

During a movement, the torque could unexpectedly change in either direction (loading and unloading) or it could stay constant. Changes in external torque occurred over 300 to 1,000 ms. They were fast enough to induce considerable torque change over the course of a movement, but slow enough to be significantly longer than (and let us ignore) an unknown delay in the hypothetical reflex loop forming the basis of the λ-model. Earlier experiments with perturbation of a stationary limb (Latash & Gottlieb, 1990a) showed that monosynaptic reflexes and/or triggered reactions are not evoked if perturbations are applied gradually over 300 ms or more.

We assumed, as in all such experiments, that the subjects tried their best and were able to follow the instructions. Consistent movement trajectories, in the absence of external torque changes, were taken as evidence of a similar consistency in the motor command. After sufficient practice, subjects could generate consistently repeatable kinematic profiles, and we assume, therefore, a standard voluntary motor command. Figure 2.5 shows a hypothetical, monotonic function $r(t)$ as such a profile; however, the method is insensitive to the form of r changes. Three movement commands are illustrated, differing only in their reaction time from the signal to start moving. Torque changes always started at the same moment after the signal tone while central variable changes started with different time delays due to different reaction times of the subject.

The individual records from a set of movements were realigned so that time zero (t_0) corresponded not to the tone but to the beginning of acceleration changes due to the voluntary motor command (Figure 2.5b). This point can be detected easily on the individual records. We used first deflection of the acceleration trace rather than deflection of the agonist EMG because, for slow movement, the first deflection of the acceleration is easier to detect. Deflection of the acceleration trace starts after the start of central variable changes. This delay is due to transmission of neural signals, the dynamics of muscle contraction, and inertial properties of the limb. We assume that this delay is constant if the initial conditions (joint angle and torque before the movement) are the same in successive trials and the subjects are reproducing the same time profile of the central motor command. Then t_0 in all the trials corresponds to the same value of r (r_0).

If one measures torque and angle values at different times (t_i) after t_0, these values, in different trials, would correspond to the same values of r_i (and c_i) or to the same position (and slope) of a hypothetical shifting joint compliant characteristic. If all the assumptions are correct, and the

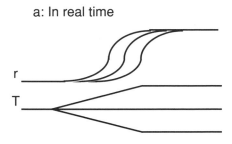

a: In real time

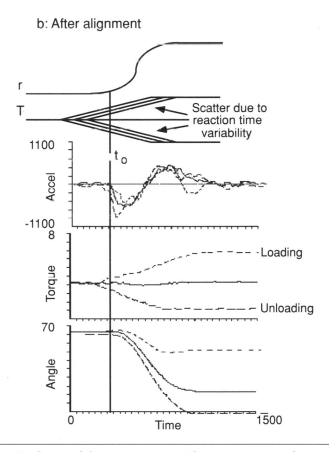

b: After alignment

Figure 2.5 A scheme of the experiments with reconstruction of joint compliant characteristic and virtual trajectories during voluntary movements.
Note. From "Reconstruction of Joint Compliant Characteristics During Fast and Slow Movements" by M.L. Latash and G.L. Gottlieb, 1991, *Neuroscience*, **43**, pp. 697-712. Copyright 1991 by Pergamon Press. Reprinted by permission.

subject was successfully reproducing the same time profile of motor command in successive trials, plotting the values of torque and angle at a particular t_i in the torque-angle plane would give a set of points pertaining to the same joint compliant characteristic but measured at different levels of external torque. At each value of t_i, these points should be described by Equation 2.12. This exactly corresponds to sets of data-points used for plotting joint compliant characteristics in classic static experiments.

A joint compliant characteristic for a time after t_0 can be obtained by linear regression analysis of a set of datapoints from one set of trials. Each regression line (joint compliant characteristic) can be characterized by two parameters: slope and intercept. The slope will correspond to joint stiffness (related to c) while the intercept will correspond to position of the joint compliant characteristic (r). Linear regression analysis for the sets of datapoints measured at different times after t_0 will give changes of both r and c over the time course of the movement, which is the ultimate goal of these experiments.

Separate analysis of the unperturbed trials yields an average trajectory of the learned movement. This trajectory can be compared with changes in r (expressed in angle units) derived from the regression analysis. Then, one can compare changes in both the central control variable (a "virtual" trajectory, see below) and the "actual" trajectory plotted on the same scale.

To check the method, we also applied it to the analysis of joint compliant characteristics obtained in static experiments with the traditional "do not intervene voluntarily" instruction. In the static series, it was not necessary to realign the trials, since there were presumably no changes in the central motor command and movement was due solely to the changing external torque. Therefore, values of torque and angle in individual trials were measured at different times after the beginning of the perturbations.

Getting back to Equation 2.11, the quasi-static experiments are similar to the static ones in ignoring the dynamic components of the right side of this equation, and they differ from the static experiments in introducing a possibility of centrally induced changes in the parameters of the tonic stretch reflex "spring," k_2 and α_{02}. In the present context, we will also ignore possible contribution of the passive spring and reduce Equation 2.11 to

$$T(t + t_2) = k_2(t + \frac{t_2}{2})[\alpha(t) - \alpha_{02}(t + \frac{t_2}{2})],$$

or, using a different pair of symbols:

$$f\ [c(t + \frac{t_2}{2})]T(t + t_2) = \alpha(t) - r(t + \frac{t_2}{2}). \tag{2.16}$$

The experimental procedure is based on a slightly different equation:

$$f\,[c(t)]T(t) = \alpha(t) - r(t), \tag{2.17}$$

since the regression analysis was applied to the joint angle and torque values measured at the same moments of time [cf. with Equation (2.10)]. However, since $T(t)$ changes linearly and slowly, the introduced error

$$\frac{T(t + t_2) - T(t)}{T(t)}$$

is likely to be small, especially if one takes into account that the characteristic values of t_2, although not precisely defined, are likely to be of the order of 50 ms (Feldman, 1979; Houk, 1979; Sinkjaer et al., 1988). Note, however, that if we try to use the same approach for step-like external torque changes, the errors introduced by ignoring t_2 might be very large due to considerable differences in $T(t)$ and $T(t + t_2)$. We have also ignored a delay $t_2/2$ between the generation of an afferent volley corresponding to a position in the joint and central motor command, which provides for a joint compliant characteristic interacting with the volley and defining the motor output. As a result of these two simplifications, we are going to reconstruct a joint compliant characteristic whose form will be somewhat distorted and whose position will be shifted along the angle axis by $t_2/2$. However, this time is also small as compared with the movement time.

Shifting Joint Compliant Characteristics During Slow Movements

Figure 2.6 represents an example of the datapoint sets and regression lines. All the regression lines are shown at a common scale in the lower part of Figure 2.6. The regression equations may be represented as

$$T = k_1 - k_2\alpha, \tag{2.18}$$

where T is the torque, α is the angle, k_1 is the intercept, and k_2 is the slope.

Because there is a separate equation for each value of time (t_i), we can represent the coefficients k_1 and k_2 as time functions. Plotting these functions (Figure 2.7a) demonstrates monotonic time changes in both k_1 and k_2.

The Notion of Virtual Trajectory

The idea that there is some central representation of voluntary movement is rather old. Different terms have been used to address this notion, including *movement engram* (Bernstein, 1936) and *virtual trajectory* (Hogan, 1984). These terms emphasize different aspects of motor control, from

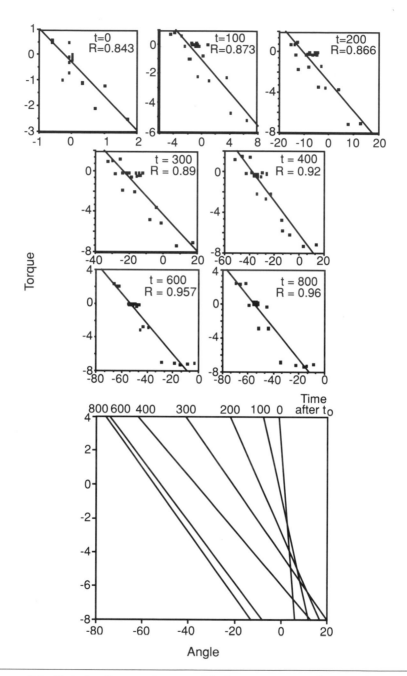

Figure 2.6 Data for the experiments with slow movements for one of the subjects. All the regression lines are superimposed in the lower graph.
Note. From "Reconstruction of Joint Compliant Characteristics During Fast and Slow Movements" by M.L. Latash and G.L. Gottlieb, 1991, *Neuroscience*, **43**, pp. 697-712. Copyright 1991 by Pergamon Press. Reprinted by permission.

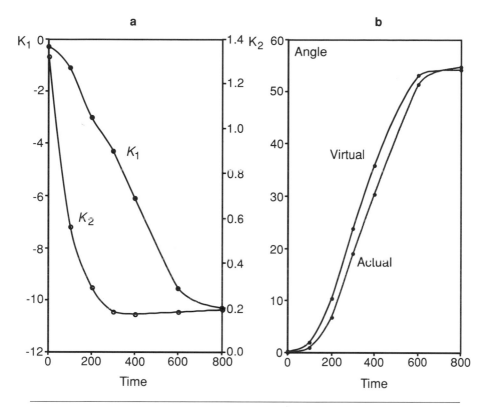

Figure 2.7   Time changes in the coefficients of the regression equations (cf. Figure 2.6). K_1 and K_2 (a) and virtual and actual trajectories (b) during a slow elbow flexion.

Note. From "Reconstruction of Joint Compliant Characteristics During Fast and Slow Movements" by M.L. Latash and G.L. Gottlieb, 1991, *Neuroscience*, **43**, pp. 697-712. Copyright 1991 by Pergamon Press. Reprinted by permission.

planning a whole succession of complex motor acts to some kind of central anticipatory "tracking" of an ongoing simple movement. In particular, one may suppose that any voluntary movement may be associated with time shifts of some central variables whose values define, after certain processing, movement kinematics. The term *virtual trajectory* seems appropriate for time patterns of these central variables.

Hasan (1986b) illustrated this notion with a "cat and mouse" situation: The peripheral apparatus is a cat trying to catch a mouse whose movements are defined centrally (virtual trajectory). Hasan has quite legitimately posed a question: Is there a mouse? It has never been observed experimentally and seems to be an abstraction rather than an observable rodent (although an alternative is discussed in the epilogue of this book).

There is empirical evidence suggesting that complex motor acts can be based on a central program realized at different speeds (Terzuolo & Viviani,

1979; Carter & Shapiro, 1984; Schmidt, 1988). However, until recently, there were no empirical data demonstrating that single motor acts can also be regarded as controlled with standardized motor programs that can be realized at different speeds.

In the framework of the λ-model, central control of single-joint movements can be described as time functions $r(t)$ and $c(t)$ where r and c are reciprocal and coactivation commands respectively (see Equation 2.15). For isotonic movements, any combination (r, c) defines a position and a state of co-contraction of the antagonist muscles controlling the joint that would be attained after the limb is given a chance to come to a new equilibrium. It seems reasonable to associate an imaginary succession of these equilibrium states corresponding to $r(t)$ and $c(t)$ functions with a virtual trajectory of the limb. Note that a change in peripheral conditions of movement execution (e.g., external load, limb inertia, damping) may lead to a change in the limb trajectory but not in the virtual trajectory. According to this definition, virtual trajectory represents a feature of the central motor program but not of an actually observed movement.

So virtual trajectory can be defined as *a path that would be followed by a massless limb without damping and changes in the external load.*

Virtual and Actual Trajectories of Quasi-Static Movements

In the experiments, the joint compliant characteristics have been described by two parameters, k_1 and k_2, related to intercept and slope, Equation 2.18. This equation is equivalent to Equation 2.17 if

$$k_1 = -\frac{r}{f[c]}$$

and

$$k_2 = -\frac{1}{f[c]}.$$

The first important feature of k_1 behavior is its nearly linear change over time. This observation is consistent with assumptions introduced by Feldman and his colleagues (Abdusamatov & Feldman, 1986; Abdusamatov et al., 1987; Feldman et al., 1990a) for modeling EMG patterns of single-joint movements, who postulated a linear shift of the centrally regulated variable r.

Note, however, that in Equation 2.18 k_1 is measured in units of torque, while during an isotonic movement, it is more natural to measure the reciprocal command in units of angle (trajectory). From Equation 2.18 one gets

$$\alpha = \frac{k_1}{k_2} - \frac{1}{k_2} T. \qquad (2.19)$$

Therefore, to calculate a hypothetical central command r in angle units, one should divide k_1 by k_2. The variation of r with time can be considered a measure of a virtual trajectory. Figure 2.7b shows the results of this operation plotted on the same graph with the "average unperturbed trajectory" (see previous). Both trajectories start at approximately the same point. The virtual trajectory starts to change somewhat faster so that the actual trajectory lags behind it. By the end of the movement, the actual trajectory catches up.

Stiffness Changes

Let us now consider the behavior of the second coefficient k_2. It has the units of stiffness and depends only on the coactivation component c. Its apparent changes during the movements (Figure 2.7a) suggest a significant initial drop in the stiffness, after which it is constant. This seems to contradict demonstrations of muscle coactivation during isotonic movements (Lestienne, 1979; Mustard & Lee, 1987; Gottlieb et al., 1989a). Note, however, that although joint movements in our experiments were deliberately restricted to low speeds, inertial forces were not zero and were likely to play their most significant role during the initial phase of movement. For example, at the moment of peak acceleration (100-200 ms after t_0; cf. Figure 2.5), inertial torque was comparable with the externally applied torque (about 3 Nm in Figure 2.5). The presence of a significant inertial component of torque implies that the measured component is smaller than the total torque and therefore the actual stiffness is less than estimated by the regression curves.

This effect was likely to be most pronounced during the initial phase of the movements, when absolute values of angle displacements were small and the inertial components of torque were relatively high. To check this assumption, we reprocessed the data with and without taking into account the inertial torque component (Figure 2.8). Including the inertial torque into analysis has led to a smaller initial drop in k_2 and a slightly inflected virtual trajectory (cf. thin solid and dashed lines in Figure 2.8). The new virtual trajectory is very similar to the one modeled by Hogan (1984) for movements of similar duration. There is still an initial drop in k_2 that may be due to the crudeness of our estimation of the inertial torque, ignoring viscous torque component, and/or other factors.

Static Experiments

The proposed explanation of the changes in k_2 over time should also apply to the static experiments. In the static experiments, the same measurements were made without voluntary changes in the motor command

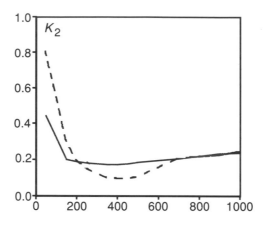

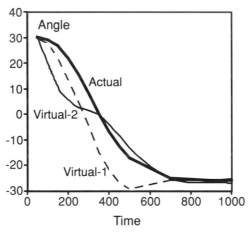

Figure 2.8 The K_2 coefficient and virtual trajectories were reconstructed with (solid thin lines) and without (dashed lines) taking into account the inertial component of muscle torque. Taking into account the inertial torque component has led to a smaller initial drop in K_2 and a slightly inflected virtual trajectory. Actual trajectory is shown by the bold line. The time scale is in ms, angle scale is in degrees, and K_2 scale is in Nm/deg.

Note. From "Reconstruction of Joint Compliant Characteristics During Fast and Slow Movements" by M.L. Latash and G.L. Gottlieb, 1991, *Neuroscience*, **43**, pp. 697-712. Copyright 1991 by Pergamon Press. Reprinted by permission.

(Figure 2.9). Note that the changes in k_1 were much smaller in the static than in the quasi-static experiments, but the time course of k_2 was similar in both. Small variations in k_1 provide an indirect support for the most vulnerable assumption of the "do not intervene voluntarily" experiments, that is, for the ability of the subjects to actually preserve their central command in conditions of external perturbations.

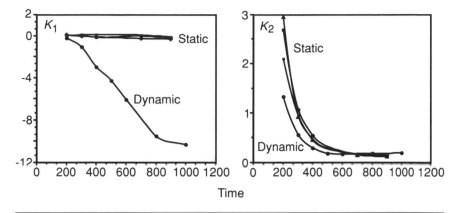

Time

Figure 2.9 As compared with the experiments with voluntary movements, changes in K_1 during the static experiments are very small while K_2 shows a similar initial drop. Three sets of data were analyzed for the static experiments (black symbols) differing in the initial joint position. K_1 scale is in Nm, K_2 scale is in Nm/deg.
Note. From "Reconstruction of Joint Compliant Characteristics During Fast and Slow Movements" by M.L. Latash and G.L. Gottlieb, 1991, *Neuroscience*, **43**, pp. 697-712. Copyright 1991 by Pergamon Press. Reprinted by permission.

Shifting Joint Compliant Characteristics During Fast Movements

The general idea of these experiments is similar to the one described in the previous section and used for reconstruction of joint compliant characteristics during the quasi-static experiments. The basic difference is in trying to incorporate the effects of the considerable inertial torques that take place during fast single-joint movements (cf. Corcos et al., 1989; Gottlieb et al., 1990a). Getting back to the basic equation (2.11), we are going to simplify its right side with the assumptions that:

- the moment of inertia of the limb, M, is constant and independent of time;
- we are going to ignore the viscous components;
- the external torque is changing relatively slowly, and therefore we may compare values of torque and angle measured at the same moments of time, $T(t)$ and $\alpha(t)$, thus ignoring the unknown time delay in the tonic stretch reflex arc.

Although the first assumption is self-explanatory, the second may need some explanation (Figure 2.10). According to existing assessments (Viviani et al., 1976; Cannon & Zahalak, 1982; Gottlieb et al., 1986; MacKay et al., 1986), the coefficient of viscosity for a quasi-static human elbow joint is approximately 0.01 Nm s/° . The moment of inertia of the system "arm plus manipulandum" in our setup was about 0.003 Nm s²/°. External torque was about 3 Nm. Let us now take into account that typical values of peak acceleration for very fast elbow flexions are about 10,000 °/s², and typical values of peak velocity are about 500 °/s. Therefore, if there are no dramatic increases in the coefficient of viscosity during a fast movement, peak viscous torque is likely to be of the order of 5 Nm, and peak inertial torque is likely to be about 30 Nm. During the initial phase of a fast movement (point A in Figure 2.10), the external component is likely to dominate due to low values of acceleration and velocity. In the vicinity of points B and D, the inertial component dominates. The viscous component is likely to play an important role in the vicinity of point C, when the speed is maximal and acceleration is zero, and maybe later in the movement (after point D, but before the movement termination). So, we are aware of possible distortions of the results at certain phases of the movement. However, these distortions are likely to be relatively minor during the A-B interval, in the vicinity of point D, and at the end of the movement.

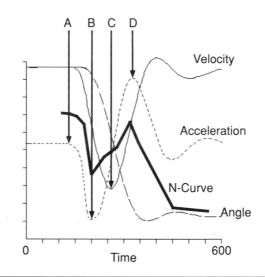

Figure 2.10   A scheme of typical changes in kinematic variables and virtual trajectory (N-curve) during fast single-joint movements. Viscous component of muscle torque is likely to play a major role in the vicinity of point C, while its role is likely to be small as compared with the inertial components in the vicinities of points B and D. Therefore, ignoring the viscous torque component has not presumably qualitatively influenced the N-shape of the virtual trajectory, although its path from point B to point D could have been affected.

Note. From "Reconstruction of Joint Compliant Characteristics During Fast and Slow Movements" by M.L. Latash and G.L. Gottlieb, 1991, *Neuroscience*, **43**, pp. 697-712. Copyright 1991 by Pergamon Press. Reprinted by permission.

Justification for the third assumption was discussed previously. Equation 2.12 is now reduced to:

$$T(t) - M(t) \frac{d^2\alpha(t)}{dt^2} = k_2(t + \frac{t_2}{2})[\alpha(t) - \alpha_{02}(t + \frac{t_2}{2})].$$ (2.20)

For now we are also going to ignore the $t_2/2$ time delay, as has been done for the quasi-static experiments. However, later we will be forced to reconsider this delay and take into account that the reconstructed virtual trajectories should be shifted with respect to the actual trajectories.

Examples of the sets of datapoints and regression lines are shown in Figure 2.11. According to the assumptions, the regression lines represent joint compliant characteristics reconstructed for a certain time t_i. As in the previous section, coefficient k_1 related to the joint compliant characteristic intercept is equivalent to r, and coefficient k_2 related to the joint compliant characteristic slope is equivalent to c. Reconstructing joint compliant characteristics for different t_i yields time changes of k_1 and k_2. Figure 2.11 shows time changes of k_2, the virtual, and the actual trajectory. There is a considerable increase in k_2 (stiffness) in approximately the middle of the movement. The virtual trajectory has a nonmonotonic N-shape. At first, the virtual trajectory is ahead of the actual one; it then reverses direction, and the actual trajectory takes the lead. Eventually, at the end of the movement, both trajectories meet.

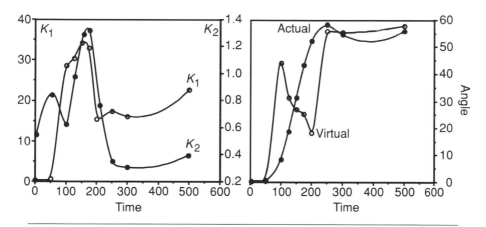

Figure 2.11 Time changes in the coefficients for the regression equations, virtual and actual trajectories. Note an N-shaped form of the virtual trajectory and a phasic increase in K_2 in the middle of the movement.

Note. From "Reconstruction of Joint Compliant Characteristics During Fast and Slow Movements" by M.L. Latash and G.L. Gottlieb, 1991, *Neuroscience*, **43**, pp. 697-712. Copyright 1991 by Pergamon Press. Reprinted by permission.

N-Shaped Virtual Trajectories

In modeling motor control systems, there is always a trade-off between a desire to make the most parsimonious model and the model's predictive power. An increase in the number of variables or in complexity of their behavior would obviously allow better description of the actual system, but it also makes the model less attractive.

From this standpoint, the equilibrium-point hypothesis (alias λ-model), which is based on central control of only one variable for each muscle (threshold of the tonic stretch reflex, λ), is one of the most attractive models of single-joint motor control. Single-joint movements are controlled by at least two muscles. According to the "law of maximal parsimony" (also known as Occam's razor), Feldman (1980a,b; Feldman & Latash, 1982a) suggested that single-joint control can be described by a pair of variables (r, c) that have direct relations to common physiological notions of reciprocal activation and coactivation (cf. "The λ-Model" in chapter 1). These variables can also be directly related to position (r) and slope (c) of the joint compliant characteristic.

In an attempt to describe the emergence of the electromyographic patterns during single-joint movements, Feldman and his colleagues (Abdusamatov & Feldman, 1986; Abdusamatov et al., 1987; Feldman et al., 1990a) applied the law of maximal parsimony and suggested the following scheme (Figure 2.12a): A single-joint movement is controlled by a ramp change in r to a new level with a simultaneous increase in coactivation (c). An alternative model in the framework of the equilibrium-point hypothesis has been introduced recently (Latash & Gottlieb, 1991a; Figure 2.12b; see also chapter 5). The major difference from Feldman's approach is the possibility of independently controlling activation of the antagonist muscle with a delayed shift in λ_{ant}. However, the resulting patterns of r and c changes differ dramatically from those assumed by Feldman. In particular, r demonstrates a nonmonotonic N-shaped behavior, and c has a peak somewhere in the middle of the movement.

The experimental reconstruction of virtual trajectories has corroborated the second of the two patterns demonstrating N-shaped virtual trajectories for smooth, fast single-joint movements. They have also demonstrated a transient threefold increase in the joint compliant characteristic slope (joint stiffness) somewhere in the middle of the movement (cf. with c changes in Figure 2.11). We conclude that our "guess" of the central motor control process underlying fast single-joint movements, advanced in a model of EMG emergence (Latash & Gottlieb, 1991a; see also chapter 5) was closer to the truth. The virtual trajectory shown in Figure 2.11 is similar to those proposed by Hogan (1984) and Hasan (1986a) based on certain optimization criteria (cf. chapter 8).

It appears that the motor control system does not always obey the law of maximal parsimony. There might be secondary considerations, like

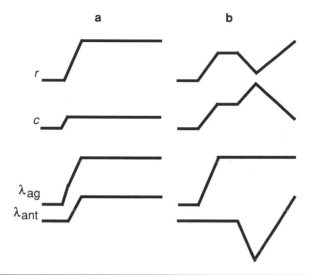

a   b

r

c

λ_{ag}
λ_{ant}

Figure 2.12 Two hypothetical patterns of central commands for a fast single-joint movement. (a) Feldman's version implies monotonic changes in all the variables. (b) The other version is more complicated, having a possibility of an independent control on the process of braking, thus leading to a nonmonotonic N-shaped virtual trajectory (changes in r) and a transient increase in joint stiffness (changes in c). Compare with Figure 2.11.
Note. From "Reconstruction of Joint Compliant Characteristics During Fast and Slow Movements" by M.L. Latash and G.L. Gottlieb, 1991, *Neuroscience, 43*, pp. 697-712. Copyright 1991 by Pergamon Press. Reprinted by permission.

damping terminal oscillations, that may make the control patterns more complex.

Virtual Trajectories of Oscillatory Movements

Feldman (1980a) has suggested that there exist two types of control patterns for single-joint oscillatory movements. The first one is used at relatively low frequencies (up to 1 Hz) and consists of discrete shifts of λ of the flexors and extensors leading to a smooth combination of discrete flexion and extension movements. At higher frequencies, stiffness of the joint is held relatively constant at a certain level depending on the movement frequency. This is accomplished by controlling the slope of the joint compliant characteristic with the help of the c variable. The r variable shifts back and forth between certain limits corresponding to the desired amplitude of the oscillatory movement (Figure 2.13a). On the other hand, if one considers an oscillatory movement as a smooth combination of

targeted flexion and extension movements, the experiments with reconstruction of N-shaped virtual trajectories suggest a different control pattern (Figure 2.13b). It implies phasic changes in joint stiffness at a doubled
frequency leading to a peak during each flexion and each extension movement, and the possibility of N-shaped nonmonotonic changes in r during
both flexions and extensions.

We have recently reconstructed virtual trajectories and joint stiffness
changes during elbow oscillatory movements in a horizontal plane over
a fixed distance at different frequencies and against different inertial loads
(Latash, 1992). The subjects were instructed to move between two targets
shown on a monitor screen at a frequency specified by a metronome
against an extending torque bias. During a three-minute session, the external torque could slowly and unexpectedly increase or decrease. The instruction ("do not intervene voluntarily") and method of analysis were
similar to those used for reconstruction of joint compliant characteristics
during discrete movements.

Figure 2.14 shows examples of virtual trajectories and averaged unperturbed actual trajectories for movements at different frequencies against

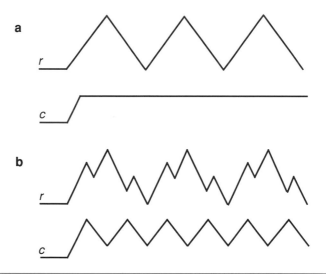

Figure 2.13   Two control patterns for single-joint oscillatory movements at a
high frequency. (a) A steady increase in joint stiffness (command c) is accompanied by alternating shifts of the joint compliant characteristic (command r) between the desired end positions. (b) Each flexion and extension movement is
accompanied by an N-shaped virtual trajectory (r) and a transient increase in
joint stiffness (c).

Note. From "Virtual Trajectories, Joint Stiffness, and Changes in Natural Frequency During Single-Joint Oscillatory Movements" by M.L. Latash, 1992, *Neuroscience,* **49,** pp. 209-
220. Copyright 1992 by Pergamon Press. Reprinted by permission.

a small inertial load. For the low movement frequencies, the virtual trajectory is approximately in phase with the actual trajectory. An increase in the frequency leads to an increase in the difference between the peaks of virtual and actual trajectories. At a certain critical frequency (about 1.5 Hz), the virtual trajectory is nearly flat. Further increase in movement frequency leads to a phase lag of π between the virtual and actual trajectories. The difference between the peaks of the virtual and actual trajectories continues to monotonically increase. Figure 2.15 illustrates these findings. Note also that, at higher movement frequencies (i.e., at higher velocities), there are visible inflections of the virtual trajectories resembling the N shaped curve (Figure 2.14). Joint stiffness demonstrates transient increases at a doubled frequency near the times of peak joint velocity, more pronounced at higher movement frequencies (Figure 2.14). A threefold increase in the inertial load led to qualitatively similar patterns of changes in both joint stiffness and virtual trajectory. However, the critical frequency dropped (about 0.75-1 Hz) and peak joint stiffness increased for the same

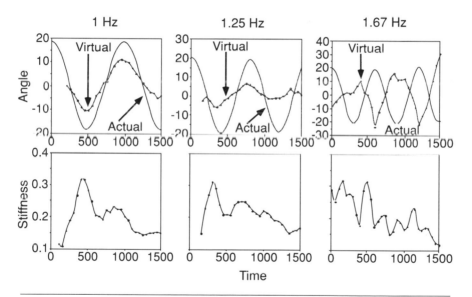

Figure 2.14   Examples of actual and virtual trajectories (above, in degrees) and joint stiffness changes (below, in Nm/deg) reconstructed during oscillatory elbow movements as different frequencies. Note an in-phase relation between the trajectories during 1 Hz oscillations, an out-of-phase relation for the 1.67 Hz movements, and smaller amplitude of changes in the virtual trajectory for the 1.25 Hz movements. Changes in joint stiffness are smaller toward the end of the analysis epoch because of some yielding by the subject.

Note. From "Virtual Trajectories, Joint Stiffness, and Changes in Natural Frequency During Single-Joint Oscillatory Movements" by M.L. Latash, 1992, Neuroscience, **49**, pp. 209-220. Copyright 1992 by Pergamon Press. Reprinted by permission.

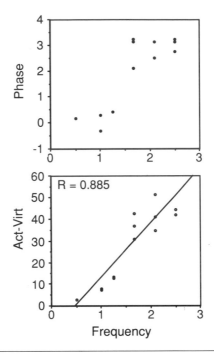

Figure 2.15   Phase shift between the virtual and actual trajectories (in rad, above) and the peak difference in their amplitude (in degrees, below) during oscillatory elbow movements at different frequencies (in Hz).
Note. From "Virtual Trajectories, Joint Stiffness, and Changes in Natural Frequency During Single-Joint Oscillatory Movements" by M.L. Latash, 1992, *Neuroscience*, **49**, pp. 209-220. Copyright 1992 by Pergamon Press. Reprinted by permission.

movement frequencies. Assessment of the natural frequency of the system "limb + manipulandum" was performed using value of joint stiffness averaged across a movement cycle. Figure 2.16 shows the assessed values of the natural frequency and actual movement frequency for small and high inertial loads. Note that the critical frequencies correspond closely to the natural frequency of the system.

These observations make physical sense. At the natural frequency, the system is supposed to require minimal central interference, and we have indeed observed minimal changes in the virtual trajectory at frequencies close to the assessed natural frequency of the system. Considerable changes in the average joint stiffness with movement frequency and inertial load are also likely to have a purpose. First, an increase in stiffness increases the natural frequency of the system. This leads to a decrease in the difference between the actual movement frequency and the natural frequency when movement frequency increases. On the other hand, an increase in joint stiffness counteracts changes in the natural frequency

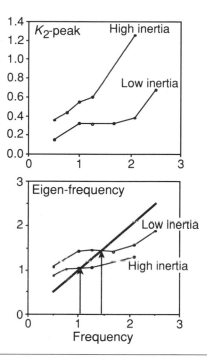

Figure 2.16 Above: The dependence of joint stiffness peak value on frequency of oscillatory elbow movements for low and high inertia. Below: The dependence of estimated Eigen-frequency on movement frequency. Arrows show estimated critical frequencies. They are close to the frequencies at which phase transition was observed between actual and virtual trajectories. The frequency scales are in Hz, the K_2 scale is in Nm/degree.
Note. From "Virtual Trajectories, Joint Stiffness, and Changes in Natural Frequency During Single-Joint Oscillatory Movements" by M.L. Latash, 1992, *Neuroscience*, **49**, pp. 209-220. Copyright 1992 by Pergamon Press. Reprinted by permission.

that would be otherwise introduced by an increase in the system's inertia. The purpose of the system may be to keep the natural frequency as close as possible to actual movement frequency by the accessible means, that is, by changing joint stiffness (cf. the importance of natural frequency for patterns of different types of voluntary movements stressed by Turvey et al., 1988). A similar conclusion has recently been drawn by Bingham et al. (1991), who studied simultaneous rhythmic movements of two limbs against different inertial loads and observed corresponding differences in the stiffness values for the left and right joints.

Observations of an abrupt phase transition between the virtual and actual trajectories (Figure 2.15) are somewhat similar to the results of experiments studying phase relations between two limbs during bimanual tasks (Kelso et al., 1979a,b, 1983; Kelso, 1984; Kay et al., 1987; Kelso & Schoner, 1988; Schmidt et al., 1990). Only two stable states have been found by Kelso et al.: in-phase and out-of-phase. A slow change in the movement frequency led to an abrupt jump from an out-of-phase mode into an in-phase mode, just as in our experiments. I will return to these ideas in chapters 6 and 7 while discussing different aspects of the "dynamic pattern generation" approach.

Virtual trajectory is certainly harder to measure than, for example, joint angle, but it should eventually be accepted as another observable variable of motor behavior. If one considers virtual trajectory a behavioral observable, phase transitions between it and another behavioral observable (i.e. actual movement trajectory), are illustrations of behavioral discontinuities introduced by slow changes in a higher level control parameter, movement frequency. Relative phase shift between these two trajectories may be considered an order parameter (or collective variable, see chapter 6). In this context, the present results are illustrations of relatively rapid changes in the order parameter introduced by slow changes in a control parameter (cf. Turvey, 1990b).

Recordings of surface EMGs of two elbow flexors (biceps and brachioradialis) showed a monotonic increase with movement frequency of both peak and integrated EMGs without any special behavior at the critical frequencies. These monotonic EMG changes illustrate limited value of the EMG recordings in motor control studies. Generally speaking, the EMGs are consequences of both changes in the central commands and tonic stretch reflex action (see chapter 5). The recorded burst-like muscle activity at the critical frequency was most likely due exclusively to the reflex action while the virtual trajectory was nearly flat.

The last several sections were designed to create an impression that the equilibrium-point hypothesis provides an appropriate and productive framework for experimental studies of movement dynamics. By saying this I suggest that the hypothesis leads to testable predictions that are in principle disprovable. It even leads to "a quarrel in the family" when competing control patterns are advocated by different groups based on the equilibrium-point hypothesis (Latash, 1989a; Feldman et al., 1990a; Latash & Gottlieb, 1991a,c). A large part of the next chapter is devoted to discussing this "quarrel," which originated from a relatively recent addition to the original black box version of the λ-model.

Chapter 3

The Equilibrium-Point Hypothesis and Movement Dynamics

To analyze patterns of muscle activation during changes in muscle length, notions of muscle phase plane and activation zones are introduced. Any changes in muscle length lead to trajectories of the muscle's "current state" on the phase plane. The level of muscle activation (revealed by electromyogram) is assumed to be proportional to how deep the muscle current state is into the activation zone. Analysis of the phase-plane trajectories allows qualitative interpretations of certain muscle reflexes and electromyographic patterns during voluntary movements. A neurophysiological interpretation of λ as subthreshold depolarization of α-motoneurons seems somewhat confusing and leads to difficulties in interpretation of certain observations.

Phase Plane and Activation Zones

The λ-model, from its inception, has been the target of severe criticism from both theoretical and experimental points. Original publications by Feldman gave rise to the impression of an oversimplified model attempting to account for very complicated phenomena of voluntary motor behavior. It took 20 years to understand that the λ-model has only created a feeling of simplicity while being in fact intrinsically very deep and complicated.

One of the major points of criticism (Day & Marsden, 1981, 1982; Marsden et al., 1983; Gielen & Houk, 1984; Atkeson, 1989) originated from the fact that the experimental basis of the λ-model included only static experiments, that is, reconstruction of invariant or compliant characteristics under a presumably constant supraspinal drive (Matthews, 1959; Asatryan & Feldman, 1965; Feldman, 1966a,b; Feldman & Orlovsky, 1972). The seeming lack of a dynamic component in the λ-model has also led to the assumption that it is unable to account for the experimentally observed EMG patterns during voluntary movements (Enoka, 1983). The need to account for the movement dynamics suggested a new step in the development of the λ-model that led to the emergence and elaboration of a concept of muscle activation zones (Abdusamatov & Feldman, 1986; Feldman, 1986; Abdusamatov et al., 1987; Feldman et al., 1990a). This step has led to a considerable revision of the concept of the tonic stretch reflex that underlies the process of muscle activation. In particular, the original term *tonic* stretch reflex becomes hardly applicable because of rapid changes in muscle length during fast voluntary movements, which lead to phasic changes in spindle afferent activity and, consequently, in α-motoneuron recruitment patterns.

A hypothetical tonic stretch reflex that, in the framework of the λ-model, gives rise to the muscle invariant characteristics was originally introduced as a functional rather than morphological concept. The overall characteristics of muscle force changes in response to length changes were considered to result from action of these reflex connections. Length changes of an intact muscle are likely to give rise to changes in activity of a variety of peripheral receptors, including muscle spindles, Golgi tendon organs (for example, due to a "purely peripheral" change in muscle force induced by length change), articular receptors (due to a change in joint position, joint capsular tension, or both), and cutaneous and subcutaneous receptors. However, Feldman has recently suggested (Feldman et al., 1990a) that there exists "the main path" of the tonic stretch reflex that is supposedly mediated by receptors that are especially sensitive to muscle length changes, that is, by muscle spindle endings (Granit, 1970). Primary endings of muscle spindles are sensitive to both muscle length and velocity of change (Matthews, 1933; Katz, 1950; Brown et al., 1967). Therefore, their dynamic sensitivity provides a natural mechanism for integrating movement dynamics into the λ-model. Reflex connections from other receptors are supposed to change sensitivity of α-motoneurons to the "main reflex path" by changing their membrane potential. This membrane potential has been directly associated with λ.

An attempt to single out "the most important" peripheral source of the hypothetical tonic stretch reflex has both good and bad sides. On the one hand, the concept of the tonic stretch reflex becomes better morphologically defined, potentially allowing incorporation of a body of neurophysiological data into the λ-model (see later in this chapter) and probably suggesting new neurophysiological approaches to study the reflex. On the other hand, attempts to attach a particular importance to some morphologically defined structures on the basis of far from exhausting information has frequently proven to be misleading (see, for example, the descriptions of the α-model and Merton's servo-model in chapter 1). This step also requires reassessment of the previous data that used to provide support for the λ-model, including those performed under the "do not intervene voluntarily" paradigm (see chapter 2 and Asatryan & Feldman, 1965; Feldman, 1966a,b, 1974, 1986).

The "original" and "modern" versions of the λ-model are illustrated in Figure 3.1 (see also Figure 3.3 and Figure 3.4). The original black box version considers λ a centrally supplied variable while action of *all* the peripheral receptors, including those from antagonist and distant muscles, defines muscle reaction to external load and/or length changes (i.e., provides for the form of the invariant characteristics). In the modern version, only autogenic reflex connections from muscle spindles give rise to the invariant characteristic while the rest of the peripheral receptors, including antagonist muscle spindles, are able to change λ, thus making it both centrally and peripherally influenced.

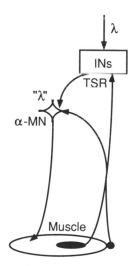

Figure 3.1   According to the original black box version of the λ-model, λ is the only centrally supplied variable for a muscle. It defines position of the muscle invariant characteristic whose shape is defined by the tonic stretch reflex (TSR) mechanism. The central signals interfere with the signals from peripheral receptors (black spots) and eventually change the level of α-motoneuron (α-MN) depolarization ("λ"). "λ" can also be changed by afferent activity and therefore cannot be considered an independently controlled central variable.

At some point it will be desirable to provide a direct link between the functional concepts of the λ-model and anatomical and morphological findings. The question is whether the present knowledge of the reflex, intraspinal, and descending connections is sufficient to form a basis for a good guess. The original black box version of the λ-model still seems more appealing to me, and later in this chapter, I am going to describe in parallel both versions. Note that the fact that λ is introduced as a black box abstraction and does not have a clear neurophysiological counterpart does not mean that λ is not measurable. For example, virtual trajectories of single-joint movements and joint stiffness changes described in chapter 2 are interpolations of reconstructed (quite close to "measured") r and c variables, or, equivalently, λs of the agonist and antagonist muscles. Kinematic and kinetic parameters of movement are certainly measurable, although one cannot single out a neuron that would code distance, velocity, acceleration, or muscle force.

Coexistence of two concepts of λ leads to similar coupling of concepts for the r and c variables. The original definitions of r and c as intercept and slope of the joint compliant characteristic do not allow any direct neurophysiological or morphological analogies. They suggest a relation to the notions of reciprocal activation and coactivation reflected, in particular, in the terms *reciprocal command* and *coactivation command*. However, this relation has been until recently considered general and descriptive rather than implying involvement of certain neurophysiological mechanisms. In the modern version of the λ-model, a direct relation of the r command to the mechanism of reciprocal inhibition, which involves a well-defined group of Ia-interneurons, is suggested.

Dynamic sensitivity of primary spindle endings leads to introducing a concept of muscle activation zones in both versions of the λ-model (cf. "The Notion of Shifting Invariant Characteristics" in chapter 1). Let us consider a phase plane (Figure 3.2) where the axes correspond to length of a muscle and velocity of its changes (V-L plane). If the threshold of α-motoneuron recruitment to a very slow lengthening of the muscle is λ, the muscle will be active at all the points on the abscissa axis to the right of λ and will be silent at all the points to the left of λ. If muscle length is being increased at a certain speed, the dynamic sensitivity of primary spindle endings can induce recruitment of the α-motoneurons at points to the left of λ (e.g., point B in Figure 3.2). Similarly, if the muscle is being shortened, its α-motoneurons can be silent even at points to the right of λ (e.g., points C and D in Figure 3.2). These effects can be represented in a simple form with the help of a straight line (Ω-line) dividing the V-L plane into zones of muscle activation and silence. Let us introduce a notion of muscle's "current state," which corresponds to a particular point on the V-L plane and, therefore, involves both muscle length and velocity of its changes.

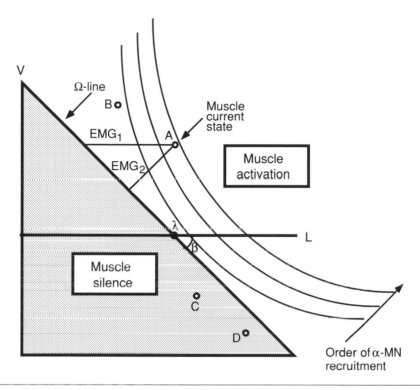

Figure 3.2 At any moment of time, muscle state can be characterized by instantaneous values of its length and speed corresponding to a point (muscle current state) on a phase plane (V-L plane). The V-L plane is divided by a straight line (Ω-line) into the zones of muscle activation and silence. The EMG can be assessed as distance from the muscle current state to the Ω-line measured either parallel to the L axis (EMG_1) or perpendicular to the Ω-line (EMG_2). In the first case, a notion of a dynamic tonic stretch reflex threshold can be introduced as $\lambda^* = \lambda - \mu V$, and the EMG is proportional to the difference between muscle length and λ^*. These two methods become equivalent if the angle between the Ω-line and L axis (β) is constant. Curved thin lines illustrate thresholds of recruitment of individual motoneurons.

Any joint movement leads to changes in muscle length and can be described, for a muscle, as a trajectory of its current state on the V-L plane. If this trajectory enters the zone of α-motoneuron silence, muscle activity disappears; if the trajectory enters the zone of activation, recruitment of autogenic α-motoneurons starts and the muscle demonstrates electrical activity. One may tentatively suggest that the level of muscle activation is proportional to how deep is its current state into the activation zone. This assumption has been formalized in two different ways, allowing modeling of the EMG patterns during fast voluntary movements. Let us start with the version suggested by Feldman et al. (1990a).

One can formally introduce a parameter λ^* describing a new, velocity-dependent threshold of autogenic α-motoneuron recruitment:

$$\lambda^* = \lambda - \mu \frac{dx}{dt}, \qquad (3.1)$$

where μ is a constant having the dimension of time and x is muscle length. Then, level of muscle activation (EMG) can be associated with the difference between instantaneous values of muscle length and λ^*, that is, horizontal distance between the muscle's current state and the Ω-line (EMG$_1$ in Figure 3.2). Feldman and his colleagues have also suggested the existence of "threshold lines" for different α-motoneurons inside the activation zone of the V-L plane (curved lines in Figure 3.2). These lines define thresholds for recruitment of individual motoneurons, and their pattern can be used for describing some of the experimental findings with α-motoneuron recruitment including the size principle of Henneman (Henneman et al., 1965) and its reported violations (Desmedt & Godaux, 1981; Garnett & Stephens, 1981; Haar ter Romeny et al., 1982). For example, if the current state of a muscle enters the activation zone, whether due to a centrally induced shift in λ or because of peripheral changes in muscle length, it will cross the recruitment lines in a fixed order leading to a fixed order of α-motoneuron recruitment. The process of de-recruitment will similarly follow the inverse order. It is conceivable that a change in the motor task for a muscle acting around a joint with several degrees of freedom may lead to a reorganization of the recruitment lines on the V-L plane. For example, if the muscle is a "prime mover," the lines generally follow in the order of increase of the α-motoneuron size. If the muscle is not a prime mover, the order for certain motoneurons may be reversed.

The alternative version (EMG$_2$ in Figure 3.2; Latash & Gottlieb, 1991a; see also chapter 5) suggests the shortest distance from the muscle's current state to the Ω-line as a measure of muscle activation. The two versions of calculating EMG become virtually identical (i.e., they differ by a constant coefficient) if the slope of the Ω-line is constant:

$$\text{EMG}_2 = \text{EMG}_1 \sin\beta, \qquad (3.2)$$

where β is angle between the Ω-line and the abscissa axis. However, if β is allowed to change, which means a change in relative effectiveness of the length- and velocity-dependent components (e.g., due to selective activation of the dynamic γ-motoneurons), predictions of the two versions start to differ. Note, however, that rotations of the Ω-line have not been analyzed in either model.

Is λ a Measure of α-Motoneuron Membrane Depolarization?

Let us discuss, in detail, a relatively recent addition to the λ-model by Feldman (Feldman, 1986; Feldman et al., 1990a), which considers λ a measure of subthreshold depolarization of the α-motoneuron membrane for a pool of motoneurons sending their axons to a muscle. This step makes λ a well-defined neurophysiological variable that can, at least theoretically, be measured with common neurophysiological methods. It also allows a relatively straightforward incorporation into the λ-model of such widely studied mechanisms as reciprocal and Renshaw inhibition (see "Reciprocal and Renshaw Inhibition").

This addition, however, strips λ of its meaning of an independently controlled variable since it becomes sensitive to afferent signals that can affect the level of subthreshold α-motoneuron membrane depolarization. It also forces us to reconsider all the previous results of the experiments measuring joint compliant characteristics using the "do not intervene voluntarily" paradigm (see chapter 2). Indeed, if λ is affected by level of activity of peripheral receptors, it is going to change following a change in the external load even if the subject is wholeheartedly trying his or her best to freeze a voluntary motor command.

Figure 3.3 illustrates a typical set of datapoints used for reconstructing joint compliant characteristics in the experiments when the subject is asked to "fix" a central motor command and not to react to unexpected changes in the external load. Note that different points in Figure 3.3 correspond to different values of joint angle and torque and, therefore, to different levels of activity of peripheral receptors sensitive to joint angle and torque. In fact, virtually all the peripheral receptors change their level of firing in response to a load and position change. These changes in the peripheral receptor firing are likely to bring about changes in the level of depolarization of the α-motoneurons of all the muscles controlling the joint and possibly muscles of other joints. Therefore, all the datapoints in Figure 3.3 are likely to correspond to different levels of α-motoneuron depolarization and cannot be directly used for interpolating an invariant characteristic corresponding to a fixed value of a central control variable.

In order to save the situation, Feldman argues that the experimentally reconstructed invariant characteristics, presumably corresponding to a constant supraspinal drive, do not represent iso-λ-curves, but can, however, be used by the control system if it is able to take into account a predictable contribution of the peripheral receptors and adjust its central command correspondingly. So, the reconstructed invariant characteristic is assumed to correspond to a constant central motor command but not to a constant λ (or, more accurately, to a combination of λs for all the muscles controlling the joint). However, the λ-language is still thought to be used by the central controller.

The next step in adjusting the λ-model was to separate action of autogenic spindle afferents upon the α-motoneurons from all the rest of the reflex connections. The model is still based on regulation of reflex parameters, namely of the threshold of a tonic stretch reflex. This reflex is now considered to incorporate all the components of α-motoneuron reaction

Force

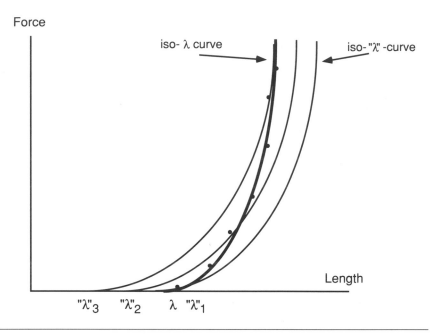

Figure 3.3 If a subject is asked to occupy a position against a load and "not to intervene voluntarily" when the load is changed, a muscle invariant characteristic can be reconstructed (an extrapolation of the black circles). It is assumed to correspond to a constant central motor command (an iso-λ curve). If a parameter ("λ") is identified with subthreshold depolarization of the α-motoneurons, it can be affected by activity of the peripheral receptors, and changes in external load and/or joint position will lead to its change. Then, different points on an experimentally reconstructed characteristic will correspond to different values of "λ" although the subject is not intervening voluntarily. Thus, λ remains a reflection of central motor command while "λ" reflects both central and peripheral factors.

to muscle stretch mediated by the spindle afferents of the agonist muscle. All the rest of the reflex connections, including those from Golgi tendon organs, cutaneous and subcutaneous receptors, and spindle endings of antagonist and distant muscles, exert their influence on the level of α-motoneuron membrane depolarization (i.e., on λ).

This raises a number of rhetorical questions.

First, if λ is no longer a measure of central motor command, what is?

Second, how does one justify an artificial-looking classification of all the afferent reflex connections on α-motoneurons into those changing λ and mediating the "main" reflex?

Third, is it advantageous for the controlling system to have to take into account future possible changes in λ mediated by the peripheral receptors?

It seems that most of the questions and seeming contradictions result from unclear terminology. Let us consider Figure 3.4. In the "older"

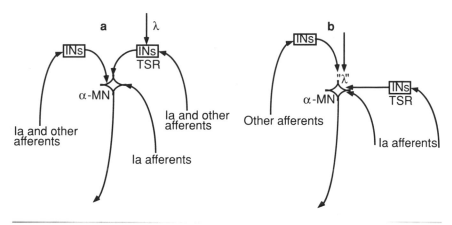

Figure 3.4   According to the original black box version of the λ-model (a), λ is the only centrally supplied variable for a muscle. It defines position of the muscle invariant characteristic, whose shape is defined by the tonic stretch reflex (TSR) mechanism. There are other possibilities to affect α-motoneuron (α-MN) firing including both monosynaptic and polysynaptic reflexes. According to the modern version of the λ-model ("λ" model, b), the shape of the invariant characteristic is defined by both monosynaptic and polysynaptic reflex effects of the agonist muscle spindle afferents. The threshold of the invariant characteristic ("λ") can be shifted by both central commands and reflex influences from all the other afferents, including those from the nonspindle receptors and spindles of other muscles.

version of the λ-model (a), λ is associated with a formal central input to the black box executing the command and incorporating all the segmental structures and peripheral receptors. As such, it certainly affects the resting potential of the α-motoneuron membrane but does not define it unambiguously since other factors can affect the membrane potential as well. Let us use a different symbol for the α-motoneuron membrane potential: "λ." If the control system cares about time changes in "λ" (Figure 3.4b), it is likely to be able to control them by adjusting time patterns of the descending signals so that they compensate for changes in "λ" introduced by the peripheral reflex inputs. This is likely to be a difficult task.

Note that in Figure 3.4, λ is still a centrally supplied variable while "λ" is not, being equally affected by inputs from peripheral receptors. Another ambiguous point in the newer version of the λ-model is the concept of "λ" being affected by all the reflex inputs aside from those from the agonist muscle spindle afferents. A set of studies by Jankowska and her colleagues (Jankowska & Roberts, 1972; Jankowska, 1979; Jankowska et al., 1981a,b, 1983; Jankowska & McCrea, 1983) has demonstrated convergence of signals from proprioceptors of different types, including muscle spindle endings, on the same spinal interneurons. These findings suggest that α-motoneurons are likely to receive mixed information from different proprioceptors, and therefore, it looks artificial to divide it into autogenic spindle, providing for the invariant characteristic form, and all the rest, shifting its threshold "λ."

Let us return to Figure 3.3, illustrating a typical set of data from the "do not intervene voluntarily" experiments. Each datapoint corresponds to the same value of λ but different values of "λ." If the subject follows the instructions and does not change his or her voluntary motor command during the experiment, the set of points provides for an iso-λ but not an iso-"λ" curve. We still assume that "λ" lacks many of the attractive features of λ, including independent central control, and we are going to use the λ-language in its original form.

Let us once again explicitly state our understanding of λ as a formally introduced variable describing muscle behavior in certain experimental conditions and its relation to the known anatomical and functional reflex connections. During normal functioning of the body with the intact neuromuscular system, λ is supplied by some hypothetical central nervous structures and defines position of the invariant characteristic, whose form reflects action of all the reflex connections from all the peripheral receptors, eventually leading to changes in α-motoneuron firing. As such, λ is the only central variable whose time changes affect muscle behavior. This does not mean, however, that during abnormal functioning of an intact system, or functioning of a pathologically changed system, λ remains the only variable affecting the system's behavior. Some of the muscle reactions in abnormal situations can be mediated by structures not involving a hypothetical "tonic stretch reflex loop" as shown in Figure 3.4a and are more likely to be superimposed upon the λ-mediated behavior than to be a part of it. These problems will be discussed later in this chapter.

Muscle Reactions to Length Changes

Changes in muscle length, whether due to a voluntary movement or induced by an external influence, give rise to changes in activity of a variety of peripheral receptors. Traditionally, it is assumed that most of the α-motoneuron reactions to muscle length changes are mediated by the afferents of muscle spindles although effects from other afferent groups have also been documented (Lloyd, 1943, 1946; Hunt, 1952; Granit, 1970; Watt et al., 1976).

If one wants to perform a voluntary movement in a joint, the central command can, according to the λ-model, be expressed with time functions $\lambda(t)$ for the participating muscles. Then, the hypothetical tonic stretch reflex starts playing its role and leads to changes in α-motoneuron activity. The movement itself and associated changes in muscle length are secondary to activation of α-motoneurons. These muscle length changes lead to a change in activity of all the peripheral receptors whose combined action is described as the tonic stretch reflex.

Therefore, there coexist two processes leading to changes in muscle activation and, eventually, affecting the movement: (a) a primary one, representing centrally induced changes in λs; and (b) a secondary one due to movement-induced changes in activity of peripheral receptors. Analysis of EMG patterns and movement kinematics should take into account both processes. Note that in many models of EMG emergence, central signals are supposed to be the only factor defining the EMGs (Hallett et al., 1975; Lestienne, 1979; Wallace, 1981; Bizzi et al., 1982; Enoka, 1983; Gottlieb et al., 1989a,b) thus limiting predictive ability of the models and requiring introduction of rather complicated control patterns. These models and an alternative one based on the original version of the λ-model will be discussed in chapters 4 and 5.

Muscle reactions to externally imposed changes in muscle length were discussed in chapter 1 ("Muscle and its Reflexes"). Remember that depending on the speed of imposed length changes, we can observe reactions of autogenic α-motoneurons at different latencies. A slow change in muscle length leads only to tonic EMG changes presumably mediated by the tonic stretch reflex, which forms the basis of the λ-model. Faster length changes can also give rise to the so-called "preprogrammed" reactions. Very fast changes in muscle length induce phasic monosynaptic responses of autogenic α-motoneurons mediated by primary spindle afferents.

Monosynaptic Reflexes

According to the modern version of the λ-model ("λ"-model), any change in activity of α-motoneurons is interpretable as either a shift in "λ" or action of the "main" tonic stretch reflex loop. The main reflex is assumed to be mediated by autogenic projections of muscle spindle afferents, while all the rest of the reflex projections and descending signals change "λ". Since α-motoneurons are the "common final path," any reflex or voluntary changes in muscle activity are interpretable in the λ-language.

Figure 3.5 illustrates emergence of monosynaptic reflexes in response to a rapid stretching of a muscle in the framework of the "λ"-model. Spindle endings are the most sensitive receptors to muscle length changes, and monosynaptic reflexes may be, at a first approximation, considered results of changes in exclusively spindle afferent activity (Figure 3.5a; see also "Monosynaptic Reflexes," chapter 1). Although muscle length changes lead to reactions of other afferents as well, and therefore may lead to shifts in "λ," let us consider these effects minor. Let us imagine that, before the stretching, muscle length and the speed of its changes correspond to a point ("current state") in the zone of silence (point 1 in Figure 3.5). Rapid stretching of the muscle leads, on the muscle phase

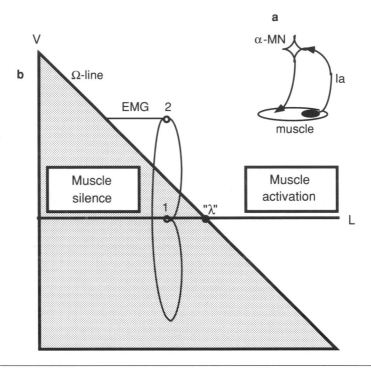

Figure 3.5   (a) A monosynaptic reflex is evoked by a synchronized volley in muscle afferent fibers (primarily group Ia), which directly acts on the α-moto-neuron (α-MN) membrane, inducing an efferent volley, which leads to a muscle contraction. The reflex can be induced by a rapid muscle stretch (tendon tap) or by an equivalent electrical stimulation of the afferent fibers (H-reflex). (b) If, originally, muscle length is smaller than "λ," and therefore its current state is in the zone of silence on the V-L plane, rapid muscle stretch can lead to a transient shift of the muscle current state into the activation zone and emergence of a phasic EMG burst.

plane, to a rapid increase in speed, which may bring the muscle's current state into the activation zone (point 2), and then a rapid decrease in speed, bringing the current state back close to the initial point. The EMG can be associated with horizontal distance from the current state to the Ω-line (Feldman et al., 1990a). Therefore, the reflex will represent a rapid transient increase in muscle activity.

Pathological Reflexes

Polysynaptic pathological reflexes, which are frequently induced by activity of other receptor groups (e.g., skin receptors in the case of Babinski's reflex), may be associated with abnormal shifts in "λ." According to the

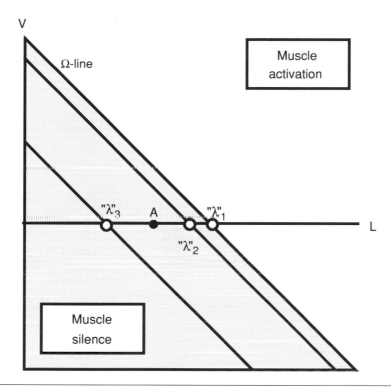

Figure 3.6   Emergence of the pathological reflexes according to the modern version of the λ-model ("λ"-model). Initially, muscle current state corresponds to a point A in the zone of muscle silence (the central command is "λ"₁). Stimulation of cutaneous and other receptors can lead to a change in "λ" and to a corresponding shift of the Ω-line. In healthy subjects, this shift ("λ"₂) is likely to be subthreshold and not lead to muscle activation. In certain pathologies, this shift ("λ"₃) can be bigger and lead to muscle activation (point A is to the left of "λ"₃ and, therefore, in the activation zone).

"λ"-model, changes in activity of skin receptors lead to shifts in "λ" and, therefore, to a shift of the Ω-line. In the absence of pathology, these shifts are supposed to be relatively ineffective and do not lead to activation of a relaxed muscle (Figure 3.6; cf. points "λ"₁ and "λ"₂). In certain pathologies, for example in spasticity (see chapter 9), this mechanism of "λ" shifts may become much more effective and lead to larger shifts of the Ω-line bringing about muscle activation (point "λ"₃, Figure 3.6).

These interpretations lead, however, to a number of inconsistencies with clinical and experimental observations. They originate from the central assumption of the "λ"-model that everything that comes out of the α-motoneurons is mediated through "λ" (Figure 3.4b) and therefore should be under voluntary control. For example, Figure 3.7 implies that it is

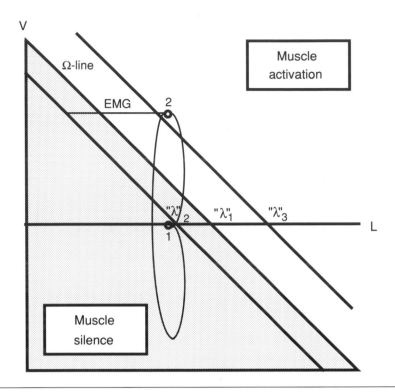

Figure 3.7   Emergence of the monosynaptic reflexes according to the "λ"-model. Initially, muscle current state corresponds to a point 1 in the zone of muscle silence. Tendon tap leads to a short-lasting stretching of the muscle that can bring its current state into the zone of muscle activation (point 2). Peak EMG can be assessed as the distance from point 2 to the Ω-line. A subthreshold change in "λ," according to this scheme, can lead to both an increase (cf. "λ"$_2$ and "λ"$_1$) and a decrease (cf. "λ"$_3$ and "λ"$_1$) of the peak amplitude of the monosynaptic response.

possible to voluntarily modulate amplitude of monosynaptic reflexes by voluntary subthreshold shifts of "λ" leading to corresponding shifts of the Ω-line (cf. amplitudes of peak EMGs for "λ"$_1$ and "λ"$_2$). However, subjects cannot modulate monosynaptic reflexes by pure "thinking" if it is not associated with activation of agonist, antagonist, or some distant muscles. Spastic patients with residual voluntary control of affected muscles cannot voluntarily modulate their pathological reflexes (note, however, that they are frequently able to provoke muscle spasms; see chapter 9), although this should not be a problem since both voluntary commands and pathological reflexes result in shifting "λ."

The original black box version of the λ-model has room for other influences on the α-motoneuron activity not mediated by the λ-mechanism thus avoiding the mentioned inconsistencies of the "λ"-model. Let us return to the diagram in Figure 3.4a. In the black box version, α-motoneurons are the point of summation of several possible sources of excitation. One of them is a hypothetical tonic stretch reflex whose threshold (λ) is mediated voluntarily. Another one is monosynaptic connections of muscle spindle afferents on α-motoneurons. Pathological reflexes may represent still another potential source of α-motoneuron excitation that is usually subthreshold in the intact body but can become effective in certain pathological states. According to this scheme, there are no hard-wired relations between voluntary shifts in λ and monosynaptic or pathological reflexes, although some effects can be seen as results of summation at the α motoneuron level. For example, voluntary activation of a muscle can lead to an increase in monosynaptic reflexes, while voluntary activation of an antagonist can lead to suppression of the reflexes through the mechanism of reciprocal inhibition. During normal everyday movements, not involving athletic competitions, we rarely experience muscle length changes at rates inducing monosynaptic reflexes (for example, Latash & Gottlieb, 1990a have observed monosynaptic reflexes only when the time of an externally imposed elbow loading was not exceeding 20 ms). From this view, monosynaptic reflexes look like quite artificial phenomena, a source of addiction for the experimenters, but probably limited in their relation to voluntary motor control.

Reciprocal and Renshaw Inhibition

These two mechanisms of inhibition of the α-motoneurons have been most widely studied with respect to both normal functioning of the spinal cord and its pathological changes, for example in spasticity (Renshaw, 1941; Granit, 1970; Hultborn, 1972; Lance, 1980b; Pierrot-Deseilligny, 1983; see also chapter 9). They both represent direct postsynaptic inhibition of the α-motoneurons by hyperpolarization of their membranes with one well-defined interneuron in each of their loops.

The Renshaw or recurrent inhibition is mediated by small spinal interneurons named Renshaw cells that receive, in particular, inputs from axon collaterals of α-motoneurons of a pool. Renshaw cells send their axons to α-motoneurons of the same pool and hyperpolarize their membranes, thus decreasing their sensitivity to excitatory inputs. Renshaw cells are also under descending supraspinal control. These supraspinal inputs are sometimes considered an important part of the descending control system for voluntary movements (Hultborn, 1972; Hultborn & Pierrot-Deseilligny, 1979).

Reciprocal inhibition gives rise to a decrease in activity of a pool of α-motoneurons when antagonist muscle spindles are activated (e.g., by stretching the antagonist muscle). Its reflex path contains one interneuron named Ia-interneuron, which receives excitatory projections from Ia spindle afferents and projects to α-motoneurons of an antagonist pool, hyperpolarizing their membranes. These neurons are also under descending control (Lundberg, 1966, 1975; Grillner, 1975) and were shown to be active during natural movements in humans and animals (Feldman & Orlovsky, 1975; Baldissera et al., 1981). Ia-interneurons are also inhibited by Renshaw cells (Hultborn, 1972).

Since both these systems exert their effects by hyperpolarizing the α-motoneuron membrane, they can be readily incorporated into the modern version of the λ-model, which associates "λ" with the α-motoneuron membrane potential (Feldman et al., 1990a). For example, an increase in reciprocal inhibition between two antagonistic muscles and their α-motoneuron pools would lead to an increase in reflex changes in the α-motoneuron activity induced by changes in muscle length and, therefore, increase the slope of the invariant characteristic for both agonist and antagonist muscles (cf. points C and A in Figure 3.8) corresponding to an increase in joint stiffness (cf. Nichols, 1989). On the other hand, an increase in recurrent inhibition of an α-motoneuron pool would lead to suppression of the reflex α-motoneuron activation induced by muscle stretching and, therefore, decrease the slope of the invariant characteristic (or stiffness) for a muscle controlled by this pool (cf. points B and A in Fig. 3.8).

Feldman et al. (1990a) describe the possibility of independent descending control of Renshaw cells and Ia-interneurons, thus giving the central control system two more degrees of freedom for controlling a joint. They hypothesize that a "reciprocal inhibition" command specifies angular ranges in which joint stiffness is enhanced while its action can be attenuated by a "recurrent inhibition" command. These two commands are added to the available central "menu," which has already included r and c commands (see chapters 1 and 2).

Let us not forget, however, that all these conclusions are made in the framework of the "λ"- rather than λ-model. And, as has already been noted, r and c, which are assumed to be simple algebraic functions of "λ" for the agonist and antagonist (Feldman et al., 1990a), are considered direct reflections of membrane potential of agonist and antagonist α-motoneuron pools. As such, they should be affected by activity of Renshaw cells and Ia-interneurons. Therefore, separation of reciprocal inhibition and recurrent inhibition commands from the (r, c) pair looks rather artificial.

Once again, the problems seem to be generated by the terminology. In the original version of the λ-model, r and c commands describe inputs to a joint as to a black box and, therefore, take all the mechanisms affecting activity of corresponding α-motoneuron pools into account including those of reciprocal and recurrent inhibition. If there were a possibility of independent control over the two basic inhibitory mechanisms under discussion, it would pose a serious challenge to the whole concept of the λ-model, making muscle control nonmonoparametric.

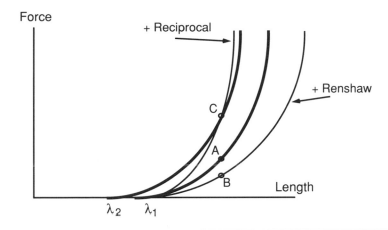

Figure 3.8 Assume that a subject is taking part in a classical "do not intervene voluntarily" experiment. His central command to a muscle is λ_1, and the equilibrium point is A. If the subject can independently increase the effectiveness of the system of Renshaw inhibition, this would lead to a decrease in the level of α-motoneuron excitation and, consequently, to a drop in muscle force (point B). As a result, the system would become "less stiff." An increase in reciprocal inhibition would lead to an increase in muscle stiffness (compare points C and A). If independent control over reciprocal or Renshaw inhibition existed, intersections of the curves could have occurred (point C belongs to a curve corresponding to a command $\lambda_1 + \text{RecIn}$ and to a command λ_2). Thus, central control could not be described as monoparametric.

Let us try to answer a couple of questions: Does central control of Renshaw cells and Ia-interneurons exist? Is it used in addition to the "regular" λ-control during normal voluntary movements?

A body of experimental data suggests that the answer to the first question is yes (see reviews in Lundberg, 1975; Grillner, 1975). In particular, Feldman and Orlovsky (1975) have demonstrated that lengthening of an antagonist muscle can lead to a shift in λ of the agonist leading to lateral translation of the muscle invariant characteristic. However, if independent control over interneurons mediating these two kinds of inhibition were possible, it might also lead to changes in shape of the invariant characteristic as suggested by Figure 3.8, where point C belongs to two invariant characteristics, "λ_1 plus reciprocal inhibition" and λ_2. Therefore, reconstructing invariant characteristics in two different conditions (different voluntary motor commands in humans or stimulation of different supraspinal formations in animals) could lead to intersection of the characteristics at some point. However, these intersections have not been observed (Matthews, 1959; Feldman, 1966a,b, 1974; Feldman & Orlovsky, 1972; Davis & Kelso, 1982; Vincken et al., 1983; Gottlieb & Agarwal, 1988; Latash & Gottlieb, 1990a) with the only exception provided by Nichols and Steeves (1986) in conditions of a specific stimulation of the magnocellular red nucleus in premammillary preparations. We may conclude that there is likely to be a possibility of descending control of Ia-interneurons and Renshaw cells but it is unlikely to be independent, at least during normally occurring voluntary movements. In other words, λ remains a formally introduced central variable that takes into account contribution of all the descending signals.

Recurrent and reciprocal inhibition are certainly important for normal functioning of the system of motor control as demonstrated, in particular, by their apparent changes in certain pathologies (Herman et al., 1973; Lance, 1980b; Ashby & Verrier, 1976; Davidoff, 1978; Merritt, 1981; Pierrot-Deseilligny, 1983; see also chapter 9). If one considers the original version of the λ-model based on the black box approach, both mechanisms of inhibition are taken into consideration automatically as far as they participate in the observed behavior of the motor system. In this case, λ still remains the only variable describing central control of a muscle, and separation of any part of the spinal circuitry into a separately controlled mechanism seems an unnecessary complication. We should, however, realize that in certain experimental conditions or pathological changes of the peripheral apparatus or spinal circuitry the hypothetical tonic stretch reflex can change its properties or cease to function (e.g., following total spinal cord transection). Then, it seems reasonable to shift description of motor processes to a different level and to consider separately different morphological and functional subsystems. In a normal body, these subsystems behave like a good orchestra with an outstanding conductor. Testing the normally functioning motor control system for a muscle reveals a monoparametric melody. My personal view is that we still lack the necessary information about the individual instruments to try to break the melody down into scores for individual musicians of the orchestra.

How Do the Electromyograms Emerge?

The present approach is based on a general idea that EMG patterns observed during changes in length of muscle fibers, which inevitably happen during both isotonic movements and "isometric"[1] contractions, are consequences of two factors: central motor commands and reflex actions of

---

[1] We are going to use the term *isometric* for muscle contractions performed without apparent changes in the external load. This definition corresponds to the common usage of the term. Note, however, that isometric contractions always induce changes in length of the muscle fibers due to elasticity of the tissues and the virtual impossibility of preventing at least minor changes in the joint angle. These length changes, as small as they are, can nevertheless occur at high speeds. They are sensed by the muscle spindles and therefore exert their effects upon the hypothetical length-sensitive reflex. So, in fact, isometric contraction is a very short movement against an increasing load. In chapter 5 we will assess characteristic times of these very short movements and incorporate them in the analysis of the EMG patterns.

peripheral receptors. Only in cases of deafferentation do these factors become discernible, and one can observe actions of "pure" central motor commands (Bizzi et al., 1976; Bizzi, 1980), which, by the way, may be considerably different from those used for controlling the intact system.

If we accept the view that motor control in an intact body is based on central regulation of parameters of muscle reflexes, which may, for example, include gain and threshold (Sechenov, 1863/1952; Lundberg, 1975; Berkinblit et al., 1986a), classification of inputs to the α-motoneurons into "central" and "reflex" becomes rather artificial. The α-motoneurons represent "output" elements (or the final common path, Sherrington, 1906) of the motor control system. This does not mean, however, that they receive separate signals from central and peripheral sources. Descending commands and signals from peripheral receptors converge on both α-motoneurons and different interneurons (Eccles, 1969; Hultborn et al., 1979; Fournier et al., 1983; Katz & Pierrot-Desseilligny, 1984) which in turn project to α-motoneurons. On the other hand, there is a possibility of strong presynaptic modulation of both central and reflex inputs to the α-motoneurons (Eccles, 1964; Lundberg, 1964; Granit, 1970). Therefore, the output elements (α-motoneurons) are likely to receive volleys from different sources already conveying "mixed" information that cannot easily be divided into central and peripheral.

Let us start with qualitative description of emergence of phasic EMG patterns (cf. Adamovitch et al., 1984; Abdusamatov & Feldman, 1986; Abdusamatov et al., 1987; Latash, 1989; Latash & Gottlieb, 1991a). It illustrates the approach and already leads to prediction of certain non-trivial qualities of the EMG patterns.

Control patterns of single-joint movement start with a shift of a centrally controlled variable (λ_{ag}) at a constant rate leading to a shift of the Ω-line and activation of the agonist muscle. This activation leads to shortening of the agonist and a joint movement. A phase-plane trajectory for a fast movement is shown in Figure 3.9. The Ω-line is assumed to shift at a much higher rate as compared with the actual joint movement. Therefore, at the first stage, muscle length is not changing much (it stays at point a in Figure 3.9), and the Ω-shift primarily defines patterns of α-motoneuron recruitment and, consequently, EMG patterns.

Later, the phase-plane trajectory starts playing its role. It follows the Ω-line and, at some moment, crosses it, leading to cessation of the agonist muscle activity (point b in Figure 3.9). Approximately at the same time, antagonist muscle activation starts. According to Feldman's version there are no delayed shifts in centrally controlled variables, and, therefore, later EMG events during a fast movement, including the antagonist burst, are defined exclusively by the reflex effects of muscle length changes. The alternative version considers the antagonist burst a consequence of two equally important factors: reflex effects of muscle length changes and a

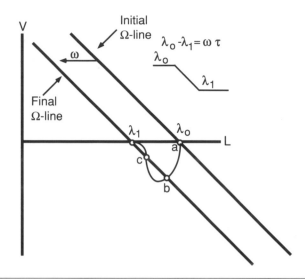

Figure 3.9 A single-joint movement is initiated by a shift of the agonist λ from an initial value λ_0 to a final value λ_1 at a constant rate ω. The Ω-line shifts and the muscle initial current state (a) appears in the activation zone. Activation of the agonist muscle leads to its shortening (thin line on the V-L plane). At some moment, the muscle current state can enter the zone of silence (point b), and its activation will seize. Later, the current state can once again enter the zone of activation (point c) which can be the cause of the second burst of the agonist EMG activity.

delayed central shift of λ_{ant}. These two approaches and their predictive abilities will be compared later (chapter 5). Irrespective of the causes of the antagonist burst, it leads to a decrease in the movement speed, and the phase-plane trajectory can once again enter the activation zone on the agonist V-L plane (point c in Figure 3.9). This may be the cause of the second agonist EMG burst that is frequently observed during very fast movements (see chapter 4).

Agonist EMG Changes

A motor command initiating a movement shifts λ_{ag} at a constant rate. The movement itself starts with a delay that is due to several factors, including signal transmission delays and inertial properties of the limb. This means that during a brief time, the current state of the agonist muscle on the V-L plane is stationary while the Ω-line is shifting to the left (Figure 3.9). The agonist EMG level is assumed to be proportional to the distance from the point on the V-L plane to the Ω-line. Since the point is stationary, the changes in EMG will depend only on the central control process (i.e., speed of Ω-line shift). This leads to an important conclusion: *Motor programs* *characterized by initial shifts of the Ω-line at the same rate should demonstrate identical initial slopes of the agonist EMG changes.*

This conclusion will be analyzed quantitatively later. However, note that it applies to both isotonic and isometric contractions because during this period there is no visible change in muscle length in either.

After the movement has started (in kinematic terms), the EMGs will be defined by two factors: central (shift of the Ω-line) and peripheral (shift of the point on the V-L plane). Note that signals from peripheral receptors during "matched" isotonic and isometric contractions will not be different until kinematic changes in the two cases start to differ. It will also take time for these changes to affect the α-motoneuron firing and, consequently, the EMG patterns due to delays in the reflex pathways (see detailed analysis later). Until that time, one may expect identical EMG patterns in isotonic and isometric contractions controlled by identical patterns of central variables.

For fast movements, the duration of λ_{ag} shift τ is assumed to be small as compared with the movement time. Therefore, the process underlying emergence of the first agonist burst may be divided into two phases. During the first phase, the EMG changes are defined primarily by the central motor command changes (shift in λ_{ag}). After λ_{ag} reaches its final value, EMG is affected only by changes in muscle length (i.e., by movement of the point corresponding to instantaneous muscle length and speed values on the V-L plane). Since the agonist is shortening, the EMG level should start to decrease until the point on the V-L plane crosses the Ω-line, leading to the agonist EMG disappearance. Therefore, for fast movements, the moment of termination of central λ_{ag} shift corresponds to the maximum agonist EMG level, thus giving a simple quantitative estimate of τ.

Antagonist EMG Changes

The initial "coactivation component" of the antagonist EMG before the "active braking" phase (antagonist burst) will also depend on two factors: a central shift in λ_{ant}, which is considered to be standard and short-lasting (c in Figure 5.1, chapter 5), and an increase in the antagonist muscle length. The latter factor will bring about an increase in the antagonist EMG level even in the absence of changes in its central command.

One can also qualitatively predict that changes in the antagonist EMG burst with changes in movement parameters might be more complex (e.g., nonmonotonic) than similar changes in the first agonist burst. During the antagonist burst, the EMG will also be defined by two factors: central (λ_{ant} shift), which is considered standard, and peripheral (lengthening of the antagonist muscle). For relatively short movements, an increase in movement amplitude is likely to lead to an increase in the speed of antagonist lengthening at the moment of λ_{ant} shift, thus leading to an increase of the antagonist EMG via the hypothetical tonic stretch reflex. However, for relatively large movements, the period of deceleration due to viscous properties of the limb will lead to a decrease in speed. Therefore, the antagonist EMG changes may not be monotonic with movement amplitude. The importance of proprioceptive feedback for the antagonist EMG changes has been corroborated by observations in functionally deafferented patients by Forget and Lamarre (1987).

Let us now move to description of the basic kinematic and electromyographic features of single-joint movements (chapter 4). In chapter 5, we will describe a simple model of the emergence of EMG patterns based on the original version of the λ-model.

Chapter 4

Patterns of Single-Joint Movements

Observations of kinematic and electromyographic patterns accompanying single-joint movements in conditions of minimal changes in muscle length (isometric) or minimal changes in external load (isotonic) have led to a variety of models attempting to account for the findings. Most of the physiological models are based on control signals that directly instruct the muscles when to be active and to what level. As such, these models are reformulations of the experimental findings. These models suggest different control patterns for isotonic and isometric contractions based on the differences in the observed patterns of muscle activation. An alternative is to consider muscle activation patterns as results of interaction of two equally important factors, descending control signals and reflexes.

Kinematic and Electromyographic Patterns of Single-Joint Isotonic Movements

In this and the following sections we are going to review a large body of experimental data and related models that study motor control during artificially simplified motor tasks. Although the whole ideology of such an approach has been seriously criticized (see the following section and the beginning of "Kinetic and Electromyographic Characteristics of Single-Joint Isometric Contractions," p. 116), it is still likely to provide interpretable information that may not, however, be directly generalizable to more natural unrestrained movements. In the next chapter, we will try to provide a link between these data and the λ-model and describe an example of how the equilibrium-point hypothesis can handle such data.

Do Single-Joint Isotonic Movements Exist?

Let us start with an attempt to systematically present an extensive body of experimental findings that resulted when the experimenter tried to confine the movements to a single joint, one degree of freedom, and, frequently, constant external load. Unfortunately for the experimenters, most of the commonly studied joints of human limbs (e.g., shoulder, wrist, and ankle) have more than one degree of freedom and are controlled by more than two muscles, some of which are biarticular (i.e., their contractions directly induce changes in joint torque and/or movements in two adjacent joints). These complications make the language commonly used for describing experimental results and models ambiguous.

For example, the presence of several degrees of freedom for a joint makes classification of muscles into "agonists" and "antagonists" or "flexors" and "extensors" quite artificial. It seems that most human limb muscles are multifunctional, as has been suggested by Buchanan et al. (1986, 1989) based on experiments measuring degree of muscle involvement in isometric contractions in different directions (see also van Zuylen et al., 1988; Flanders & Soechting, 1990; Tax et al., 1990). A muscle can exert agonistic or antagonistic action (i.e., assist or counteract a movement) depending on particular features of the task. Biarticular muscles are attracting more and more attention (Nichols, 1989; Zajac & Gordon, 1989; Gielen et al., 1990; see also "A Special Role for the Biarticular Muscles" in chapter 7), and it is now obvious that their potential role can no longer be ignored. Even when they are not actively recruited during a movement, they still provide a mechanical spring-like link between the joints, which affects the movement and should be taken into account by the control system. Therefore, it seems that movements are never confined to a single joint. The situation is also complicated by the fact that any joint movement leads to a change in geometry of muscle attachment to bones, thus changing the relation between muscle force and induced joint torque. This leads to a change in perceived external load for the muscle.

The bottom line is this: Single-joint movements against a constant external load (isotonic) do not exist.

Even if single-joint movement existed and were obtainable in experimental conditions, many authors would question their relation to "real-life" movements and to problems of motor control of voluntary movements in general. In a recent discussion in *Behavioral and Brain Sciences* (Gottlieb et al., 1989a), many commentators expressed their doubts about the applicability to real-life motor control of the data and conclusions derived from single-joint experiments. A number of studies have provided data suggesting that control of multijoint movements cannot be reduced to control of involved individual joints (Bernstein, 1967; Flash & Hogan, 1985; Berkinblit et al., 1986a,b,c; Georgopoulos, 1986; see also chapter 7).

In other words, multijoint motor programs are likely to be formulated in a different language.

So, what is studied in those numerous articles whose titles include "single-joint isotonic movements"? And why do people care to study these nonexistent phenomena?

Let us first suggest the following definition for this class of movements: *Movements in a joint that are not accompanied by movements of comparable amplitudes in other joints performed against an apparently constant external load will be termed single-joint isotonic.* Note that this definition does not impose limitations on muscle torque changes in other joints. However, it implies lack of changes in a component of the external load defined by the force of gravity. This requirement is usually met by confining movements to a horizontal plane. In humans, such movements are rela tively easily obtained by studying elbow or wrist movements in a horizontal plane when both distal and proximal segments of the limb are fixed on horizontal manipulanda. In most of the experiments cited below, the movements were indeed performed on one of these two joints. We will specifically point out if a different joint was studied.

There exist a number of considerations justifying spending time on single-joint motor studies while their results are likely not to be directly transferable to the field of multijoint movements.

First, the progress of science goes from simple to complex, and studies at a certain advanced level usually require understanding of the lower functional levels. It does not seem reasonable to discourage babies from learning how to crawl although in their future life they are quite unlikely to use this motor program on an everyday basis.

Second, although results of single-joint studies cannot be directly generalized for multijoint movements, they provide frameworks (e.g., the language of the equilibrium-point hypothesis) and approaches (e.g., analysis of a system's compliance, perturbation techniques, etc.) that can be helpful for understanding general principles of motor control irrespective of the number of joints and muscles involved. In this connection, one can consider single-joint movements as a simplified model for more complex motor system studies.

Third, investigations of single-joint movements have proven to be useful in clinical studies (see chapter 9) for detecting pathological or drug-induced changes in voluntary motor control. In some patients, voluntary motor control is restricted to single muscles or single joints; therefore, understanding basic principles of control of such movements is likely to be useful.

And fourth, experimental studies require reproducible conditions of performing experiments and recording, or otherwise taking into account, all the factors that are likely to influence a subject's performance. Even for human single-joint movements, it is, strictly speaking, virtually impossible to take into account all the factors, including multifunctionality of muscles, specific roles of biarticular muscles, changes in external load due to changes in limb geometry, and others. They are sometimes alluded to, rarely assessed, and most frequently just ignored. For multijoint movements, the situation is likely to be even more complicated because interjoint forces come into play, including centripetal and Coriolis forces; biarticular muscles are likely to start playing even more ambiguous roles, the force of gravity changes its direction with respect to orientation of the limb segments for most of the movements, and the number of involved variables increases dramatically. It seems that the likelihood of misinterpreting the data and performing inadequately controlled experiments increases for multijoint movement studies. Therefore, single-joint movements become more attractive as a paradigm for testing hypotheses that may, in future, appear helpful for analysis of more complex and more natural movements.

Are single-joint movements worth studying? Yes. This view is supported by many authors either explicitly (Gottlieb et al., 1989a) or implicitly, that is, by performing these studies (Lestienne, 1979; Wallace, 1981; Bizzi et al., 1982; Sanes & Jennings, 1984; Ghez & Gordon, 1987; Mustard & Lee, 1987). Otherwise, this chapter would have never been written!

Task and Performance Parameters

In most motor control experiments, the relations between task parameters (what is required from the subject) and performance parameters (what the subject is doing) are studied. For single-joint movement studies, the task parameters may include movement amplitude, movement time (or speed), external load, accuracy requirements, and other specific instructions to the subjects (e.g., "make a smooth movement," "avoid oscillation at the end of the movement," "do not correct the final position if you miss the target," etc.). Performance parameters include kinematic variables (joint position, speed, and acceleration), kinetic variables (muscle torque and its derivatives), electromyograms, accuracy indices (e.g., variability of the final position or the percentage of trials hitting the target), and other recorded or calculated indices.

Both task and performance parameters are usually chosen by the experimenter based on the objectives of a particular study. Diversity of these parameters makes comparison of data from different laboratories a hard job even when the apparent tasks and recorded variables are identical. Such factors as particular joint studied, inertial load (weight of the manipulandum), initial joint position, amount of relevant practice, level of motor fitness of the subject, and others are frequently considered secondary and relatively unimportant. There is also a diversity in equipment and methods of data processing that can sometimes lead to different conclusions based on identical sets of data.

For example, let us imagine that an investigator is interested in the presence or absence of differences between certain EMG or kinematic patterns in two experimental conditions. Since there is an inherent variability in any kind of human behavior, one cannot expect absolutely identical signals even when the subject is seemingly performing the same movement in the same conditions. There is also usually a good deal of excessive degrees of freedom for performing the task, such as relative participation of different agonist and antagonist muscles and amount of torque in planes different from the movement plane. So the question should be formulated as: "Is the difference between certain parameters of a movement in different conditions significantly bigger than characteristic values observed during reproduction of the same motor task in identical conditions?"

The null hypothesis is usually "There is no difference" and the problem is to reject it with a sufficient number of observations. Appropriate methods of parametric and nonparametric statistics can be found in Winer (1971). If an experimenter is searching for a difference, the task is relatively simple and the solution is relatively straightforward after choosing an appropriate statistical procedure: to collect as much data as necessary for showing statistically significant differences. However, even in this case, the conclusions will depend on how the signals were recorded, rectified, filtered, and averaged. For example, results of different filtering of the same EMG records can look quite different (see Figure 4.1a). Deriving acceleration by double-differentiating joint angle measured with a potentiometer and simultaneous recording of acceleration with an accelerometer can also bring about considerably different patterns (Figure 4.1b). Velocity signal can be obtained by integrating acceleration or differentiating angle (Figure 4.1c). It is now up to the experimenter to decide if the pairs of traces in Figure 4.1 are similar or different.

One of the favorite indices in motor control studies is movement time, although it would be hard to find a less reliable measure. Look at Figure 4.2, which shows typical kinematic traces for a fast elbow flexion. The beginning of the movement is relatively easy to define, although even here one can pick a moment of beginning of the agonist EMG burst or deviation of one of the kinematic traces above a certain threshold (cf. T_{acc}, T_{vel}, and T_{ang}). This choice may considerably affect the measured value. As far as the moment of movement termination is concerned, the problem becomes much worse, since fast movements are frequently accompanied by oscillations at the final position. One can pick the time when position trace first hits the target, or when velocity trace crosses zero, or when acceleration trace crosses zero for the second time, or some other measure (cf. T_1, T_2, and T_3 in Figure 4.2). Each index has its own pluses and minuses and can be used as "movement time."

Let me now describe general features of kinematic and EMG patterns accompanying single-joint isotonic movements. I am going to stress similarities of findings by different authors rather than differences because

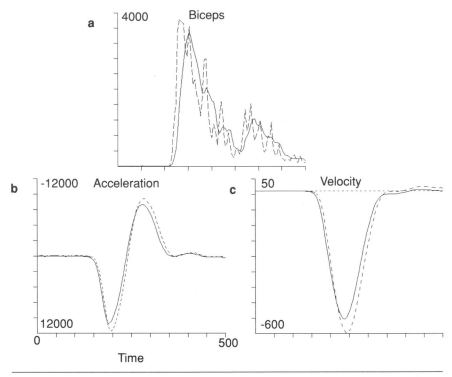

Figure 4.1 Examples of how data processing can create misleading impressions. Averages over 6 trials at 54° elbow flexion data are presented. (a) Biceps EMG plotted with 25 ms (solid trace) and 3 ms (dashed trace) moving averaging window. (b) Acceleration recorded with an accelerometer mounted on the distal part of the manipulandum (solid trace) and by double differentiating the angle signal recorded with a potentiometer mounted on the shaft of the manipulandum. (c) Velocity signals obtained by integrating acceleration (solid trace) and by differentiating angle (dashed trace). Note the considerably different peak amplitudes and slopes of the signals.

the latter can be due to poorly controlled experimental factors while the former are likely to represent experimental facts that are only strengthened by the same poorly controlled factors.

Kinematic Characteristics

If a subject is asked to perform a fast smooth movement to a target and stop there, the joint trajectory demonstrates a very typical pattern with a bell-shaped velocity and double-peaked acceleration (e.g., Bouisset & Lestienne, 1974; Freund & Budingen, 1978; Hoffmann & Strick, 1986; Milner, 1986; cf. Fig. 4.2). This kinematic profile is characterized by relatively symmetrical acceleration-deceleration phases, which have been

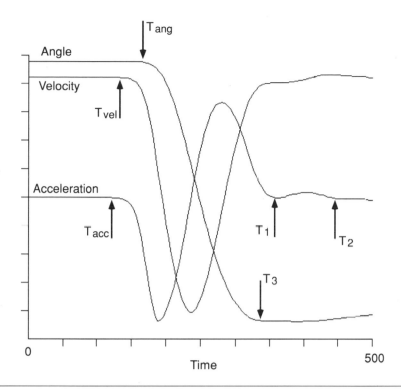

Figure 4.2   A typical kinematic pattern of a single-joint movement (6 elbow flexions over 54° were averaged). The beginning of the movement will be defined differently based on the first visible change in the acceleration (T_{acc}), velocity (T_{vel}), or angle (T_{ang}). It is even harder to define the end of the movement. One can use an instant when the acceleration reverses its direction (T_1), or when it comes to zero (T_2), or when the joint signal first hits the target (T_3), or some velocity-based criterion. Note that the figure shows a relatively smooth movement without major oscillations at the end that can only complicate the task.

shown to be independent of the inertial load (Flament et al., 1984). However, Nagasaki (1989) has reported asymmetric kinematic profiles in a wide range of movement amplitudes and speeds, and Brown and Cooke (1990) specifically studied movements with "symmetry ratios" (the ratio of the duration of the acceleration phase to the duration of the deceleration phase) varying in a wide range (from 0.4 to 2.0). An acceleration/deceleration asymmetry has also been reported by van der Meulen et al. (1990), who considered it a factor contributing to a decrease in the relative variability of the distance moved. Oscillations around the final position are sometimes observed with corresponding "bumps" in both velocity and acceleration. A decrease in movement speed is generally accompanied by an increase in variability of the trajectory (but not of the final position;

see chapter 6) and less smooth velocity curves. Slow movements demonstrate an initial increase in velocity, a period of relatively constant velocity, and its decrease while approaching the target (Mustard & Lee, 1987; Cheron & Godaux, 1986; Cooke & Brown, 1990). In this chapter, we are going to discuss a class of relatively fast movements that can be operationally defined as those having a bell-shaped velocity profile without a "flat" zone.

An increase in the amplitude of movements performed "as fast as possible" leads to a nearly linear increase in peak velocity and an increase in movement time (Fitts & Peterson, 1964; Hancock & Newell, 1985; for review see Gottlieb et al., 1989a,b). An increase in inertial load without changing movement amplitude leads to longer movement times and a decrease in velocity (Benecke et al., 1985; Gottlieb et al., 1989b). Expressions such as "movements with the same velocity" or "a decrease in velocity" are imprecise because velocity for any movement is a time function. So, when using these phrases, we imply a certain reasonable index of "movement velocity" such as peak velocity or average velocity (movement amplitude divided by movement time). Similar relations between movement velocity, time, and load can be observed with different instructions to the subjects, requiring moving at submaximal velocities (Gottlieb et al., 1990a,b; see later in this chapter).

Hoffman and Strick (1986) performed an extensive study of wrist movements and demonstrated linear relations between a variety of kinematic indices, including movement amplitude, peak speed, peak acceleration, and overshoot, which were observed under different instructions to the subjects. The subjects demonstrated a "preferred" movement time (see also Freund & Budingen, 1978) when they were given a chance to move at a comfortable speed but were able to modify it when requirements to movement amplitude and speed varied in parallel.

Adamovitch et al. (1984; see also Adamovitch & Feldman, 1984) studied elbow movements in a horizontal plane at a variety of velocities over a variety of amplitudes without changes in external load or specific accuracy requirements. All the trajectories observed by Adamovitch and Feldman can be classified into those having similar time courses during the first segment of the movement (class 1, cf. with the Speed-Insensitive strategy according to classification by Gottlieb et al. 1989a; see "Speed-Insensitive and Speed-Sensitive Strategies" later in this chapter) and those whose time courses diverge from the very beginning (class 2, cf. with the Speed-Sensitive strategy). Let us define a trajectory as a time function $S(t)$. Two observations by Adamovitch and Feldman seem particularly important:

• Compressing or expanding a natural trajectory (point by point) along the time axis leads to another natural trajectory of class 2. In other words, if one changes the argument in the trajectory function $S(t)$ to t/a ($a =$ constant) the resultant function will represent another natural trajectory performed at a different velocity (a times slower).

• If a subject performs two movements of different amplitudes pertaining to class 1 and the trajectories are aligned according to the beginning of the movements (let us accept, for the sake of discussion, that it can be somehow done), subtracting the shorter trajectory from the longer one yields another natural trajectory of class 1. In other words, a sum of two trajectory functions, one of which is shifted along the time axis by an appropriate value Δt, $S_1(t) + S_2(t + \Delta t)$, also represents a function of a naturally occurring trajectory.

These two sets of rules define a group of "naturally occurring trajectories" in which we can transform members of the group by multiplying (dividing) them by a constant, and/or adding (subtracting) them one to (from) another, and still have another member of the same group. Certainly, there are limits for these transformations, so that the resultant movement amplitudes and velocities do not exceed physiologically attainable values.

The studies and conclusions by Adamovitch and Feldman expanded a general idea that voluntary movements have a fixed timing structure and can be scaled (Glencross, 1973; Carter & Shapiro, 1984; Gielen et al., 1984; Milner, 1986; Shapiro & Walter, 1986; Ostry et al., 1987). They are in good agreement with Schmidt's notion of a generalized motor program (Schmidt, 1988; see chapters 1 and 6) that can be scaled with preservation of certain essential structural features. So, if a subject is asked to perform movements of different amplitudes, or at different velocities, or against different loads, he or she can use a standard general program and adjust a few parameters, thus fitting it to a particular task. This is an alternative to solving a complex mechanical problem each time some of the movement parameters need to be changed.

Electromyographic Characteristics

Recording muscle activity during fast isotonic single-joint movements reveals a typical EMG pattern usually termed the *triphasic pattern*. While discussing the features of this pattern, we are going to use a number of poorly defined terms. Unfortunately, this happens rather frequently when we deal with simplified descriptions of human behavior, and motor behavior is no exception. These poorly defined terms include *agonist, antagonist, EMG burst*, and several others.

A muscle whose active contraction accelerates the limb in the direction of movement will be addressed as an "agonist" while a muscle whose contraction decelerates the limb will be addressed as an "antagonist." These definitions can be criticized, but I am unaware of unambiguous alternatives. Since these definitions will be used only for isotonic single-joint movements, they avoid, at least partially, possible misinterpretation and confusion when dealing with the different levels of involvement for generating torques in different directions that is demonstrated for many human muscles (cf. the next section and Buchanan et al., 1986, 1989).

An "EMG burst" is something very much dependent on the method of recording and filtering (see Figure 4.1a). So, when discussing EMG bursts, let us keep in mind that the EMG signals have been "reasonably" filtered so that they show changes at times comparable with characteristic times for the kinematic variables (i.e., decimal fractions of the movement time) and do not show peculiarities of individual motor unit firing or very slow changes in the background EMG level.

The beginning of the agonist EMG burst is usually the first detectable event accompanying fast voluntary movement. It precedes first detectable kinematic changes by several tens of milliseconds. The shortest delays are reported for the EMG-acceleration delay (about 20 ms, Gottlieb et al., 1989a,b; Corcos et al., 1989), while EMG-velocity and EMG-angle delays may be higher (40-70 ms, Benecke et al., 1985; Hoffman & Strick, 1986). I have observed the shortest averaged EMG-acceleration delay of 11.4 ms when the accelerometer was taped to the wrist of a subject performing a fast elbow flexion.

The initial agonist burst is accompanied by a relatively low coactivation of the antagonist muscle and is followed by an antagonist EMG burst during which the agonist is relatively quiescent. Antagonist burst has been observed in deafferented monkeys (Polit & Bizzi, 1979) and in functionally deafferented patients (Hallett et al., 1975; Rothwell et al., 1982a; Cooke et al., 1985; Forget & Lamarre, 1987). When the subjects purposefully change the "symmetry ratio" (see previous discussion), antagonist burst latency is modified correspondingly (Brown & Cooke, 1990). The last two groups of observations have suggested that at least the beginning of the antagonist burst is centrally preprogrammed (see also Cheron & Godaux, 1986).

The antagonist burst can be followed by a second burst in the agonist. At the final position, there is usually a visible increase in both agonist and antagonist tonic muscle activity. A decrease in movement speed leads to less pronounced delayed EMG events, and eventually, slow movements are accompanied by sustained agonist activity with a low or absent activation of the antagonist. Cooke and Brown (1990) performed a series of experiments with single-joint movements performed over relatively large movement times (about 1 s) "at a constant speed." These movements were characterized by rather brief acceleration and deceleration separated by a long period of relatively constant velocity. Separating acceleration and deceleration periods has revealed a similar separation of agonist/antagonist burst pairs: The first pair (ag1/ant1) was observed during the acceleration, while the second pair (ant2/ag2) occurred during the deceleration. The authors concluded that the triphasic pattern may be the result of the merging of the two basic EMG burst pairs that occurs with a decrease in movement time.

Different measures have been used to characterize the EMG bursts and their changes with changes in movement parameters. They include rate of EMG rise, duration of the bursts, their relative timing, peak amplitudes, integrated activity, and others. Let us summarize the findings for an ideal experiment when the subject is making a fast smooth movement without corrections of the final position and without explicit accuracy requirements:

• An increase in velocity of movements over a constant amplitude against a constant load (Figure 4.3a) leads to an increase in the rate of EMG rise, peak value, and area of the first agonist burst (biceps and brachioradialis curves during elbow flexions illustrated in Figure 4.3), a decrease in the delay before the antagonist burst, and an increase in the antagonist burst amplitude and area (see the curves for the lateral and long heads of triceps in Figure 4.3) (Freund & Budingen, 1978; Lestienne, 1979; Shapiro & Walter, 1986; Mustard & Lee, 1987; Corcos et al., 1989). Some authors report increased duration of the first agonist burst with an increase in movement velocity (Brown & Cooke, 1984; Mustard & Lee, 1987; Corcos et al., 1989; Gottlieb et al., 1989a); others consider the duration to be constant (Hallett & Marsden, 1979; Brown & Cooke, 1981). These differences are likely to be due to different criteria used for assessing burst duration. The level of final coactivation of agonist and antagonist muscles increases with movement velocity.

• An increase in movement amplitude without changes in external load and instructions concerning movement velocity (e.g., "as fast as possible," although other instructions can be used; see Gottlieb et al., 1989a, 1990a,b, and later in this chapter) leads to relatively uniform rates of agonist EMG rise, a higher and longer first EMG burst with a corresponding increase in its area (biceps and brachioradialis curves in Figure 4.4), longer delays before the antagonist burst, and inconsistent changes in the antagonist burst amplitude and duration (lateral and long heads of triceps in Figure 4.4) (Marsden et al., 1983; Wadman et al., 1979; Brown & Cooke, 1984; Cheron & Godaux, 1986; Karst & Hasan, 1987; Mustard & Lee, 1987; Gottlieb et al., 1989b). It seems that an increase in movement amplitude for relatively small movements lead to an increase in the antagonist burst while further increase may lead to a decrease in the antagonist activity (Latash, 1989a,c; Latash & Gottlieb, 1991a).

• An increase in inertial load without changing movement amplitude and/or instructions concerning movement velocity (e.g., "as fast as possible," Figure 4.5) leads to a higher and longer agonist EMG activity, although without changes in the rate of the EMG rise, longer delay before the antagonist burst, and no apparent changes in the antagonist burst characteristics (Lestienne, 1979; Wadman et al., 1979), although an increase in the antagonist burst has also been reported (Mustard & Lee, 1987;

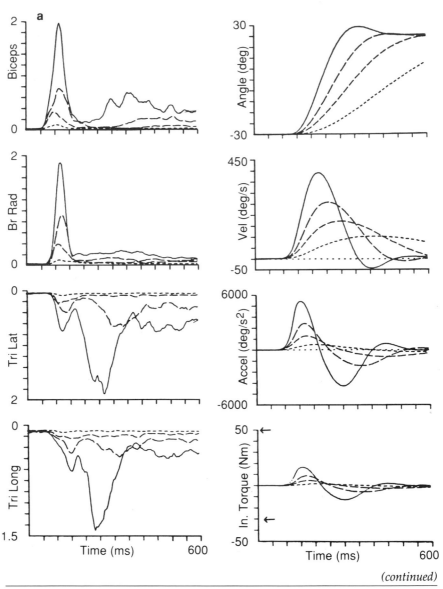

(continued)

Figure 4.3   EMG, kinematic, and inertial torque traces averaged across 10 trials for 54° elbow flexion movements at four subject-selected speeds (a) and to four different targets (3°, 6°, 9°, and 12°) with an instruction to be "both fast and accurate" (b, next page). Note different slopes of the agonist EMGs (biceps and brachioradialis), acceleration, and inertial torque starting from the very beginning of the movement.

Note. From "Organizing Principles for Single Joint Movements. II. A Speed-Sensitive Strategy" by D.M. Corcos et al., 1989, *Journal of Neurophysiology*, **62**, pp. 358-367. Copyright 1989 by The American Physiological Society. Reprinted by permission.

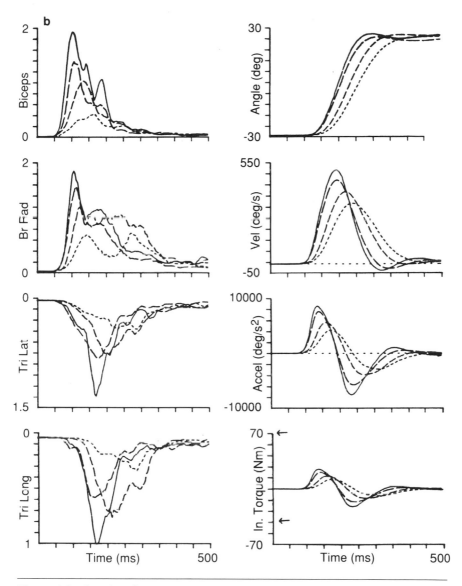

Figure 4.3   *(continued)*

Sherwood et al., 1988). Final coactivation of agonist and antagonist muscles increases with inertial load. Unexpected changes in the load lead to compensatory changes in both first agonist and antagonist burst at latencies comparable to spinal reflex loop delays (Lee et al., 1986; see also chapter 5, "Correspondence of the Model to the Data").

Variations in more than one of the three major task parameters (amplitude, velocity, and load) lead to combined effects on the EMG bursts, as in experiments when the subjects were required to perform movements of different amplitude within the same movement time (Lestienne, 1979; Mustard & Lee, 1987; Sherwood et al., 1988).

There have been several attempts to deduce more general rules for the EMG bursts. For example, Oguztoreli and Stein (1983) have shown a correlation of the antagonist burst area with the ratio of maximal kinetic energy during movement to movement amplitude. This correlation was shown for variations in movement amplitude, velocity, or load.

In general, it is assumed that the function of the first agonist burst is to accelerate the limb, the function of the antagonist is to decelerate it, and the function of the second agonist burst is somewhat obscure, probably dealing with terminating final limb oscillations, and/or correcting the final position (Marsden et al., 1983; Gottlieb et al., 1989a). It appeared possible to obtain reasonable movement trajectories by electrical stimulation of muscles with the parameters simulating naturally occurring EMG bursts (Wierzbicka et al., 1986). There is a high correlation between integrated accelerating torque, which is equal to peak velocity multiplied by inertial load, and area of the agonist burst (Gottlieb et al., 1990a). However, area of the antagonist burst does not demonstrate a similar correlation with the integrated decelerating torque (Gottlieb et al., 1990a,b), suggesting that other torque components (e.g., viscous torques) play a significant role in braking the movement. The last statement is more or less obvious, since slow movements do not demonstrate a visible antagonist burst and stop "by themselves," apparently because of the action of viscous torque components.

Kinetic and Electromyographic Characteristics of Single-Joint Isometric Contractions

The word *isometric* implies the lack of changes in muscle length that can never be secured even in animal experiments (Ritchie & Wilkie, 1958; Zajac, 1989; Griffiths, 1991). Any active contraction of muscle fibers leads to a decrease in their length. Although this decrease is relatively small, it can be realized at high velocities and lead to changes in activity of length- and velocity-dependent muscle afferents. In human studies, tendons and soft tissues surrounding the joints and the muscles cause inevitable small

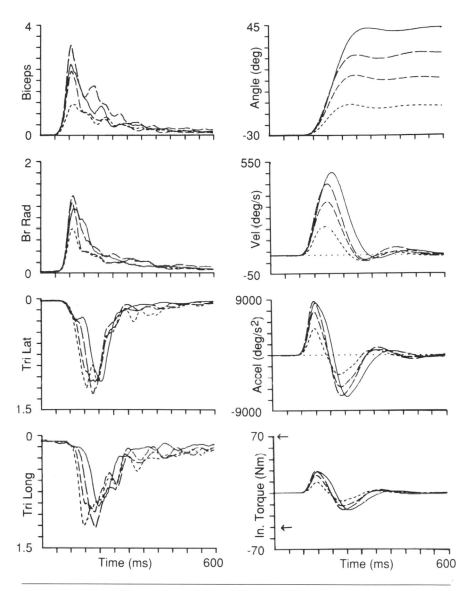

Figure 4.4   EMG, kinematic, and inertial torque traces averaged across 10 trials for elbow flexion movements performed "as fast as possible" over 18°, 36°, 54°, and 72° to a 9° target with a constant moment of inertia. Note coinciding initial fragments of the agonist EMGs, acceleration, and inertial torque. Note also a nonmonotonic change in the antagonist burst amplitude with an increase in the movement distance.

Note. From "Organizing Principles for Single Joint Movements. I. A Speed-Insensitive Strategy" by G.L. Gottlieb et al., 1989, *Journal of Neurophysiology*, **62**, pp. 342-357. Copyright 1989 by The American Physiological Society. Reprinted by permission.

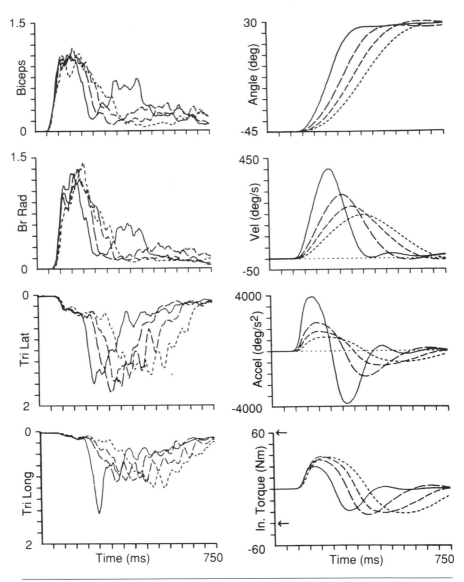

Figure 4.5 EMG, kinematic, and inertial torque traces averaged across 10
trials for elbow flexion movements performed "as fast as possible" over 54° to
a 9° target with different inertial loads. Note coinciding initial fragments of the
agonist EMGs and inertial torque (cf. Figure 4.4), while the kinematic traces resemble those typical for the Speed-Sensitive strategy (cf. Figure 4.3).
Note. From "Organizing Principles for Single Joint Movements. I. A Speed-Insensitive
Strategy" by G.L. Gottlieb et al., 1989, *Journal of Neurophysiology*, **62**, pp. 342-357. Copyright 1989 by The American Physiological Society. Reprinted by permission.

changes in joint position when muscle activation levels change. Small changes in joint angle and muscle length during isometric contractions have been suggested to play a major role in defining the triphasic EMG patterns accompanying rapid increase in joint torque (Feldman, 1986; Latash & Gottlieb, 1991b). In chapter 5, we will analyze the possible contribution of these small muscle length changes to EMG patterns mediated by hypothetical reflex mechanisms.

It has been shown that virtually all major human arm muscles are multifunctional (Buchanan et al., 1986, 1989; van Zuylen et al., 1988; Flanders & Soechting, 1990), that is, they are activated while exerting so-called isometric force in a wide range of directions. That's why it is even harder to introduce satisfactory definitions for agonists and antagonists in isometric contractions than in isotonic movements.

However, because we need to discuss the results of isometric experiments, some definitions should be agreed on. Let us define isometric contractions as those that do not lead to visible joint motion. If we are restricting the discussion to only contractions of a single degree of freedom, all the participating muscles can be classified into agonists (whose contraction leads to an increase in joint torque in a given direction) and antagonists (whose contraction leads to an increase in joint torque in the opposite direction). These definitions are bad and artificial, and I am sorry I have nothing better to suggest. In particular, most of the studied joints (e.g., wrist and ankle) have more than one degree of freedom.

Two types of isometric contractions have usually been studied: step and pulse. Step contractions require the subject to increase joint torque up to a certain level, whereas pulse contractions require returning to an initial level of joint torque. Additional instructions were used to specify time of the torque increase, accuracy constraints, and so on. Some authors (Meinck et al., 1984; Ghez & Gordon, 1987) consider isometric contractions simpler to analyze than isotonic movements; they argue that during isometric contractions the reflex contribution to the observed EMG and kinetic patterns is absent or minimal. Therefore, they consider isometric EMG patterns to be a more direct and undistorted reflection of central motor commands. But this point is not so obvious, because of the muscle length changes that can lead to corresponding changes in reflex effects on the α-motoneurons and EMGs (see chapter 5). These reflex effects are considered important in the framework of the equilibrium-point hypothesis (Feldman, 1986; Feldman et al., 1990a).

Fast isometric contractions are accompanied by triphasic EMG patterns similar to those observed during fast isotonic movements (Sanes & Jennings, 1984; Corcos et al., 1990a). However, the second delayed agonist burst is more frequently absent. Bursts of EMG activity degenerate with a decrease in the rate of torque increase, and slow increases in joint torque are accompanied by a tonic increase in the agonist EMG and a smaller increase in the antagonist EMG (Corcos et al., 1990a).

Similar measures were used for relating parameters of isometric contractions and isotonic movements to changes in the EMG patterns. Let us describe them starting, first, from the step experiments.

• An increase in the rate of torque rise keeping the final torque level constant leads to an increase in the rate of EMG rise, peak value, and area of the first agonist burst (see biceps and brachioradialis in Figure 4.6), no apparent changes in the delay before the antagonist burst, and an increase in the antagonist burst amplitude and area (lateral and long heads of triceps in Figure 4.6) (Corcos et al., 1990a). There are no obvious changes in the duration of either burst or the level of final coactivation of agonist and antagonist muscles with the rate of torque rise (Freund & Budingen, 1978; Ghez & Gordon, 1987; Corcos et al., 1990a).

• An increase in the final torque level without changes in instructions concerning the rate of torque rise (e.g., "as fast as possible," Figure 4.7) leads to relatively uniform rates of agonist EMG rise, higher agonist and antagonist burst amplitudes, higher integrated EMG activity for both muscles, higher final levels of both agonist and antagonist EMGs, and inconsistent changes in the duration of both bursts (Ghez & Gordon, 1987; Corcos et al., 1990a).

During pulse contractions, EMG patterns become more phasic, with better defined bursts, better pronounced second agonist burst, and lower level of muscle co-contraction at the final state. In this case, an increase in pulse torque amplitude without changes in the rate of torque increase leads to longer first agonist burst, a delay in the antagonist burst, and inconsistent changes in the antagonist burst amplitude and area (Ghez & Gordon, 1987; Corcos et al., 1990a). An increase in the rate of torque rise without changing pulse amplitude leads to faster rising first agonist and antagonist EMG bursts. Area of the antagonist EMG burst shows better relation to the rate of torque rise than to the pulse torque amplitude (Corcos et al., 1990a).

There is no controversy concerning the role of agonist muscles during isometric contractions. Certainly, they are used to induce torque increase. This is in fact the definition of their being agonists. The situation is not that clear with the antagonists. Their possible role is discussed in relation to truncating the rate of the agonist torque rise, stabilizing the torque trajectory, and returning the torque back to the original value (in cases of pulse contractions, Meinck et al., 1984; Sanes & Jennings, 1984; Corcos et al., 1990a).

Hypotheses and Models

In this section, we are going to briefly summarize two groups of models for single-joint motor behavior. Most of the models are driven by a desire

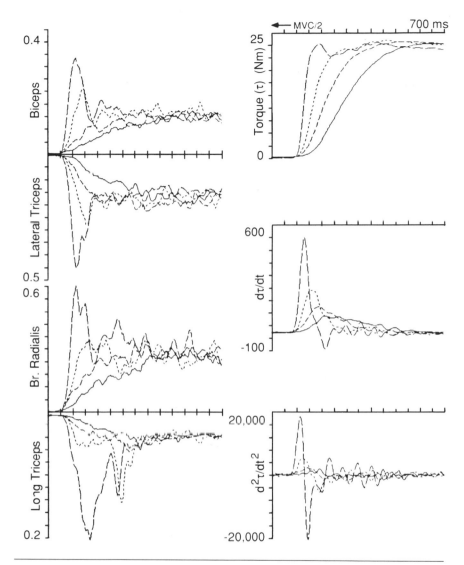

Figure 4.6 EMG, torque, and its time derivatives averaged across 10 trials during attempts at isometric elbow flexion contractions at four different rates of contraction. Note typical Speed-Sensitive patterns in both EMG and kinetic traces (cf. Figure 4.3).

Note. From "Organizing Principles for Single Joint Movements. IV. Implications for Isometric Contractions" by D.M. Corcos et al., 1990, *Journal of Neurophysiology,* **64,** pp. 1033-1042. Copyright 1990 by The American Physiological Society. Reprinted by permission.

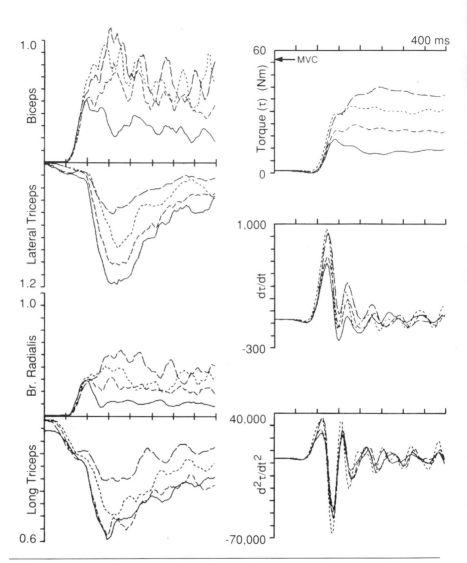

Figure 4.7   EMG, torque, and its time derivatives averaged across 10 trials during attempts at isometric elbow flexion contractions "as fast as possible" to four different final torque levels. Note typical Speed-Insensitive patterns in both EMG and kinetic traces (cf. Figure 4.4).

Note. From "Organizing Principles for Single Joint Movements. IV. Implications for Isometric Contractions" by D.M. Corcos et al., 1990, *Journal of Neurophysiology*, **64**, pp. 1033-1042. Copyright 1990 by The American Physiological Society. Reprinted by permission.

to account for or at least describe certain aspects of motor behavior. The models of the first group put emphasis on kinematic and/or kinetic characteristics of single-joint movements and deal with physical or engineering variables like joint torques, kinetic and potential energy, jerk (time derivative of acceleration), and others. The models of the second group are more neurophysiologically oriented and try to describe EMG patterns accompanying single-joint movements. Here, we shall purposefully restrain ourselves from discussing the λ-model and its predictions for the EMG patterns since the next chapter will be fully devoted to it.

Before starting to describe the models, let us once again recall some of the general principles of Bernstein (1967) that form the basis of the approach accepted in this book. In particular, Bernstein warned against drawing too close a relation between coordination and localization (see chapter 1). In other words, paying too much attention to anatomical structures and peculiarities of their connections can in fact distract us from discovering general principles of a system. We think that functioning of a complex system consisting of many relatively simple elements should not and cannot be derived from the properties of the elements. Let us illustrate this point by a quote from a recent review article on control of planned arm movements (Bullock & Grossberg, 1988, p. 49): "This raises a fundamental question for the theory of sensorimotor control and for neurosciences in general: How can the integrated activity of thousands of separate elements produce globally invariant properties?" This antithesis to our view implies that the function of a system represents a result of integrated activity of thousands of elements and has to be analyzed as such. This formulation of the problem seems to us as inappropriate as: "How can a car consisting of zillions of elementary particles move and even make turns?" Understanding a complex system does not require exhaustive information about all its elements and their connections. New variables should be introduced describing behavior of the system as a whole, similar to λ, which describes behavior of a muscle in general, without paying attention to particular subsets of α-motoneurons and muscle fibers subserving this behavior. Maybe λ is doomed to be deprived of its special role in future models of multijoint motor control. However, it might still possess value as a helpful parameter at a particular level of the motor system description, namely at the single-joint level.

"Engineering" Models

Many contemporary models of single-joint and multijoint motor control are based on the Bernstein approach (Flash & Mussa Ivaldi, 1984; Hogan, 1984, 1985; Berkinblit et al., 1986a,b,c; Flash, 1987; Mussa Ivaldi et al., 1989), that is, on analyzing global characteristics of the system's behavior and trying to model them with a relatively small number of controlled parameters. One of the prominent features of the motor control system is redundancy of the degrees of freedom (Bernstein, 1935), which leaves options for performing virtually any motor task in different ways. Bernstein formulated the main problem of the motor control as overcoming this redundancy. This can be done by imposing certain constraints upon the system, which would decrease the number of degrees of freedom (a detailed analysis of the Bernstein Problem can be found in chapter 7).

There is a group of models that are based on certain feasible considerations that would likely be used by engineers if they were asked to design an artificial moving device. These models will be considered in detail in chapter 8. They are based on certain "optimization principles" such as minimizing kinetic energy (Kathib, 1983), jerk (time derivative of acceleration) (Hogan, 1984, 1985), joint torques (Hollerbach & Suh, 1985), joint motion (Brooks, 1982), and others (see also Baillieul et al., 1984). Recently, a model has been introduced based on a so-called "passive motion paradigm" (Mussa Ivaldi et al., 1989; cf. a similar approach advanced by Bock, 1990) that implies simulating a passive limb motion from an initial to a final position with a field of elastic forces. The chosen trajectory is, generally speaking, defined by the law of minimizing potential energy. This particular model is less restrictive than the others. However, all the models of this group create a general impression that they are based on the assumption that the motor control system was either designed by a supreme engineer based on general engineering principles and physical laws known to us or that the system evolved independently based on those same laws and principles. This is not to say that the motor control system violates any general physical laws, just that the system might not care about minimizing anything that would likely be minimized by a 20th-century design engineer (cf. Stein et al., 1988).

A slightly different approach has been used by Hasan (1986). In his model, the kinematics is generated by a central motor command expressed in mass-spring terms, and the minimized parameter is the control system's "effort," that is, integrated changes in the control variables. In other words, the control system is associated with a lazy child who wants to get the task done with minimal thinking.

Another group of models is based on introducing constraints on the motor control system in a very abstract form without advancing a physical justification of why this particular constraint (or equation) is used for solving the problem of redundancy (Gutman & Gottlieb, 1989; see also Berkinblit et al., 1986a,b,c; Kay et al., 1987; Saltzmann & Kelso, 1987 for models of multijoint motor behavior). These models seem most promising in that they do not try to invent a general physical law based on experience with control in engineering and impose it on motor control in living beings. They try, instead, to find an "unbiased," consistent way of describing the behavior demonstrated by living systems (see chapters 7 and 8).

On the one hand, we do not have enough information to solve the problem of motor control directly and unambiguously based on physical principles. On the other hand, it seems that control strategies used by the central nervous system are reasonable from different engineering standpoints. The system is not using too much energy (although it may not be minimizing it), it is not exerting extreme stresses over the joints (although it may not be minimizing jerk in individual joints), and so on. The fact that many engineering principles are equally good (or bad) for describing naturally occurring movements suggests that we should search for specific features of the human motor control system that make it different from other, nonliving systems.

"Physiological" Models

Models of this group directly deal with the EMG patterns and their changes with changes in task parameters. It is tempting to describe reproducible EMG patterns as consequences of motor commands that directly specify the timing and amplitude of the EMGs of agonist and antagonist muscles or, equivalently, signals to the corresponding α-motoneuron pools (Hallett et al. 1975; Lestienne, 1979; Wallace, 1981; Bizzi et al., 1982; Enoka, 1983; Sanes & Jennings, 1984; Stein et al., 1988; Gottlieb et al., 1989a,b, 1990a,b). We shall call these *pattern-imposing* models. Note that pattern-imposing models are based on direct regulation of activity of α-motoneurons (cf. α-model, chapter 1). Bernstein (1935, 1967) stressed the theoretical impossibility of independently controlling the level of α-motoneuron activity during movement, and this view has been supported both experimentally and theoretically (e.g., Feldman, 1986; Latash & Gottlieb, 1990a). There are ways around Bernstein's objection, including "gating" of reflex feedback loops (suppressing their gain) during movements (Gottlieb & Agarwal, 1980b; Gottlieb et al., 1989a) and the possibility of predicting the reflex contribution in standardized experimental conditions (Gottlieb et al., 1990b). However, gating is unlikely to totally eliminate reflex inputs to the α-motoneuron pools. Relying on predictive abilities of the control system implies that even the simplest movements are regulated differently for different conditions of their execution. Even if it is not impossible, it is not very appealing.

An alternative is to consider the EMG patterns as consequences of an interaction between central motor commands and actual changes in muscle length, with the latter factor playing an equally important role. These models can be termed *pattern-generating*. They take advantage of the existent feedback loops and are based on controlling parameters of these loops rather than trying to get rid of them or compensate for their action. The next chapter will present one such model. The importance of the feedback signals from muscle spindles in coordination of single-joint movements has also been emphasized by Jongen et al. (1989). They, however, considered another aspect of the problem dealing with formation of synergist groups of muscles acting around a joint with more than one degree of freedom.

It may seem strange how different pattern-imposing models can coexist since they are, by definition, based on the experimentally observed EMG patterns. Possible effects of accuracy requirements upon kinematic and EMG patterns will be discussed in chapter 6. Some of the differences between EMG findings in different studies are likely to be due to different accuracy requirements, either explicit or implicit. These differences and also some ambiguities in the data (e.g., values of movement time or EMG burst durations that depend on the method of measurement) have led to somewhat different control patterns suggested for single-joint isotonic movements and isometric contractions.

Speed-Control Model.   Freund and Budingen (1978) studied movement time changes during very fast movements over different amplitudes and suggested that movements of different amplitudes were performed by regulating movement speed but not duration. Intensity of the agonist EMG was supposed to determine movement amplitude while its duration corresponded to movement time. Since movement time was considered constant, duration of the first agonist burst was constant too. So, we can call this a *pulse model* approach because movement amplitude is controlled only by intensity of an agonist EMG pulse that has a constant duration.

Pulse-Step Model.   Hoffman and Strick (1986) expanded the speed-control model by suggesting that movement amplitude can be regulated by controlling both movement velocity and time. They observed strong correlations between movement amplitude and peak values of different time derivatives of trajectory (velocity, acceleration, jerk) and also between movement velocity and duration of the same derivatives. These results are consistent with the pulse-step model of Ghez and Gordon (1987; see also Gordon & Ghez, 1987a,b), who stress the importance of the second derivative of angle for control of isotonic movements and second derivative of torque for control of isometric contractions. Addition of a step after an agonist pulse was intended to control final torque level, which was considered important for isometric contractions but not for isotonic movements. A difference in control patterns for isometric and isotonic tasks has also been suggested by Tax et al. (1989, 1990) based on their observations of the differences in recruitment of different elbow flexor muscles. A series of extensive studies of EMG patterns during isometric pulse contractions (Gordon & Ghez, 1987a,b; Corcos et al., 1989) provided data compatible with the pulse-step model.

Impulse-Timing Hypothesis.   A set of rules has been proposed by Wallace (1981) to describe how subjects control both movement amplitude and time (cf. Schmidt et al., 1979):

- An increase in movement time leads to an increase in duration of the first agonist burst and delay before the antagonist burst.
- An increase in movement velocity leads to an increase in intensity of both first agonist and antagonist bursts.
- An increase in inertial load while keeping movement time constant leads to an increase in agonist burst amplitude.

Wallace also suggested that the ratios of duration of the first agonist burst and of the delay before the antagonist burst to movement time are constant for different movement amplitudes, times, and loads.

In fact, all the rules used in different pattern-imposing models for describing EMG patterns are reformulations of experimental findings. As such, they may be helpful in providing more uniform description and classification of the data but are unlikely to contribute to understanding of the control mechanisms whose action we observe in the form of kinematic, kinetic, and EMG patterns. The most recent and elaborate model of this group is the "dual-strategy hypothesis," which in fact provides a link between pattern-imposing and pattern-generating approaches. That is, it can be described in terms of both controlling signals to (or from) the α-motoneurons and controlling parameters of reflex loops.

Excitation Pulse and the Dual-Strategy Hypothesis

Let us start with a quote from the original publication by Gottlieb et al. (1989a) describing the dual-strategy hypothesis: "To proceed further, we need an explicit definition of the variable which is controlled. Let us define it to be the neural input to the alpha motoneuron pool and refer to this as the *excitation pulse*. It represents the net descending presynaptic input, excitatory and inhibitory, which converges and summates in the alpha motoneuron pool" (p. 193). Further, the authors model the properties of the α-motoneuron pool as a low-pass filter, and therefore the "net descending presynaptic input" defines unambiguously the output of the pool, and, consequently, the EMG. This makes the model pattern imposing since all the features of the EMGs are already coded in the control signal. The model ignored possible contribution of reflex feedback loops to the experimentally observed muscle activation patterns. Let us return to this point later, and proceed with description of the dual-strategy hypothesis in its original form.

Excitation pulse was modeled as a rectangular signal whose height and duration could be modified by the central system separately or together (Figure 4.8). The authors have suggested that there exist two basic strategies for control of voluntary movements; the first one, termed the *Speed-Insensitive strategy*, was supposed to be controlled by changes only in duration of the initial excitation pulse to agonist α-motoneurons

(Figure 4.8b), while the other, termed the *Speed-Sensitive strategy*, was supposed to be controlled by changes in height of the excitation pulse to agonist α-motoneurons alone (Figure 4.8a) or together with its duration. Low-pass filtering of the excitation pulse by the α-motoneuron pool transforms height modulation into different slopes of the EMGs, and duration modulation into different times of EMG rise at the same rate (Figure 4.8b). The authors introduced also the possibility of EMG saturation when further increase in height or duration of the excitation pulse can lead only to prolongation of an already saturated EMG burst.

Kinematic predictions from the control exerted with the help of excitation pulses required another assumption, namely, that identical time-courses of excitation pulse bring about identical initial time-courses of contractile force. It, in turn, leads to identical acceleration profiles. Since

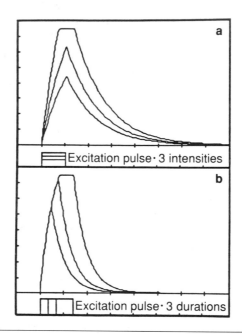

Figure 4.8 The dual-strategy hypothesis is based on control of the total presynaptic inputs to the α-motoneurons (excitation pulse). Two basic strategies correspond to controlling only the duration of the rectangular excitation pulse (Speed-Insensitive strategy, below) or its height alone or together with duration (Speed-Sensitive strategy, above). Low-pass filtering by the α-motoneurons transforms different heights of the excitation pulse into different EMG slopes and different durations of the pulse into the same EMG slopes over different times.

Note. From "Strategies for the Control of Voluntary Movements With One Mechanical Degree of Freedom" by G.L. Gottlieb et al., 1989, *Behavioral and Brain Sciences, 12*, pp. 189-250. Copyright 1989 by Cambridge University Press. Reprinted by permission.

the rules for the EMG emergence (low-pass filtering) also imply identical initial patterns for identical excitation pulses, the EMGs and acceleration show similar changes during the initial phases of movement, if the excitation pulse to agonist α-motoneurons is changed.

Speed-Insensitive and Speed-Sensitive Strategies

Gottlieb et al. (1989b, 1990a,b; Corcos et al., 1989) performed extensive studies of kinematic and EMG patterns during elbow flexion movements over different distances, at different speeds, against different loads, and with different accuracy requirements. Although not without exceptions (Gottlieb et al., 1989b), their results fit the dual-strategy hypothesis and let the authors formulate a number of simple rules specific for the two strategies.

The Speed-Insensitive strategy is used when there are no explicit or implicit constraints on movement time (or, more precisely, no *changes* in such constraints). For example, it is used when the subjects are asked to be both fast and accurate while moving over different distances or against different loads, and the size of the target is kept constant. Later, this strategy has been used when there are no explicit accuracy constraints and the subjects are asked to move "at a comfortable speed" or "at the same speed" (Gottlieb et al., 1990a).

The Speed-Sensitive strategy is used when the subjects need to adjust movement time either due to explicit task constraints (moving different distances at the same movement time) or due to changes in accuracy requirements (moving the same distance to targets of different size).

Rules for the Speed-Insensitive Strategy

- Intensity of the agonist excitation pulse is constant and duration is modified.
- EMG rises initially at the same rate irrespective of movement distance and load.
- Joint torque rises initially at the same rate. This corresponds to constant initial rates of acceleration if the inertial load is constant.
- Two components (two excitation pulses) are proposed for the antagonist. The first one defines initial "coactivation" phase of antagonist EMG, while the second generates the antagonist burst.

These rules were supported by observations of changes in different EMG indices (in particular, integrated agonist EMG during the first 30 ms, Q_{30}, and integrated agonist EMG during the acceleration period, Q_{acc}) with changes in accelerating and decelerating inertial torques in different tasks (Figure 4.9). That is, Q_{30} is independent of movement amplitude and inertial load reflecting identical initial slopes of the agonist EMG.

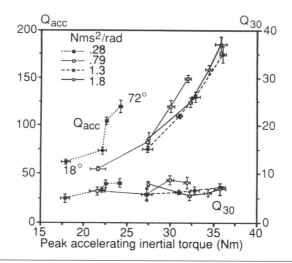

Figure 4.9   Integrated EMG activity over the first 30 ms (Q_{30}) and over the period of acceleration (Q_{acc}) for movements performed over four different distances and against four different loads "as fast as possible." Relatively constant Q_{30} values reflect the Speed-Insensitive strategy.

Note. From "Organizing Principles for Single Joint Movements. I. A Speed-Insensitive Strategy" by G.L. Gottlieb et al., 1989, *Journal of Neurophysiology, 62,* pp. 342-357. Copyright 1989 by The American Physiological Society. Reprinted by permission.

Q_{acc} increases with both amplitude and load reflecting longer times of the agonist EMG increase.

Gottlieb et al. have observed inconsistent changes in the antagonist EMG amplitude with changes in movement distance and inertial load (cf. also Marsden et al., 1983; Brown & Cooke, 1984; Cheron & Godaux, 1986; Mustard & Lee, 1987; Latash, 1989a,c; Latash & Gottlieb, 1991a), which do not easily fit into the excitation pulse control scheme (see previous discussion). These discrepancies are, nevertheless, quite predictable from the alternative view that afferent feedback signals play an important role in defining EMGs. Ignoring kinematic feedback contribution is likely to play a major role for delayed events, like the antagonist burst, and in comparing movements with considerably different kinematics, like those performed against different inertial loads.

Rules for the Speed-Sensitive Strategy

- Intensity of the excitation pulse is modified while its duration can be constant or changing.
- The initial slope of the agonist EMG rise is different for different tasks. Duration of the agonist burst is constant if the excitation pulse duration is constant.
- The slope of the initial joint torque rise is different for different tasks.

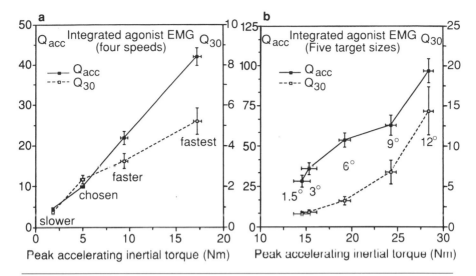

Figure 4.10 Integrated EMG activity over the first 30 ms (Q_{30}) and over the period of acceleration (Q_{acc}) for movements performed over four different distances at four speeds (a) and to four different targets (b). An increase in Q_{30} values reflects the Speed-Sensitive strategy.

Note. From "Organizing Principles for Single Joint Movements. II. A Speed-Sensitive Strategy" by D.M. Corcos et al., 1989, *Journal of Neurophysiology,* **62**, pp. 358-367. Copyright 1989 by The American Physiological Society. Reprinted by permission.

- Two components (two excitation pulses) are proposed for the antagonist as has been done for the Speed-Insensitive strategy. The first one defines the initial coactivation phase of antagonist EMG, while the second generates the antagonist burst.

Analysis of changes in EMG indices like Q_{30} and Q_{acc} with changes in task parameters was consistent with these rules (Figure 4.10). That is, both EMG parameters increased with movement speed, reflecting different initial rates of the EMG rise.

An attempt to apply the same sets of rules for the two strategies to isometric contractions (Corcos et al., 1990a) has led to predictable problems because the kinematics of the two regimes of muscle work is different, and therefore, the ignored contribution of afferent feedback signals is likely to be different too. The rules for isometric contractions are more complex. They involve, in particular, one more component in the centrally controlled excitatory input to the α-motoneurons, an "excitation step." It also appeared necessary to consider possibilities of ramp changes in excitatory signals to α-motoneurons. So, the control patterns have become nearly as diverse as the experimentally observed EMG patterns, which is expected from the pattern-imposing models since they do not imply any other source of changes in the EMG patterns aside from the central control signals.

Are There Strategies?

Initial discussion of the dual-strategy hypothesis revealed a number of controversial points, one of them being the authors' usage of the term *strategy*. As many commentators have pointed out (Bullock, 1989; Cordo et al., 1989; Hallett, 1989; Hasan & Karst, 1989; Heuer, 1989), strategy implies presence of choice. Most of the Speed-Insensitive patterns were illustrated with EMG and kinematic patterns obtained in experiments when the subjects were required to move "as fast as possible." This created an impression that the Speed-Insensitive strategy is a reflection of saturation when the highest possible rates of EMG and acceleration are observed. Therefore, its patterns do not reflect any volitional choice by the subjects. The Speed-Sensitive strategy has in fact been defined as "everything that is not the Speed-Insensitive strategy." So, if the first strategy does not imply choice, the second one does not imply it either.

If one considers the movement of a thrown rock, all its possible trajectories can be classified into those having a constant acceleration of gravity and those having any other acceleration (e.g., due to different external force fields). Is it possible to say that the rock uses two strategies for controlling its trajectory? The answer is probably "no" because the rock does not have a choice. Therefore, one of the crucial questions for the dual-strategy hypothesis has been "do the subjects have a choice?"

A series of experiments was performed by Gottlieb et al. (1990a) trying to answer this question. In this series, elbow joint flexions over different distances were studied. Specific instructions were used for the subjects moving at both maximal and submaximal velocities in order to induce Speed-Insensitive patterns. In particular, the subjects were asked to move "as fast as possible," "at a comfortable speed," "at the same speed," and to be "both fast and accurate." In the first three cases accuracy was not stressed; in the last case, the target size was kept constant for different movement amplitudes. For the three latter tasks, the subjects were free to choose their own kinematic and EMG profiles since the movements were performed at speeds much lower than the fastest possible.

The subjects demonstrated typical Speed-Insensitive patterns, including coinciding initial fragments of acceleration and agonist EMGs while moving over different distances at a wide variety of speeds. This finding suggests that the Speed-Insensitive strategy is not a "rock strategy" and there is freedom of choice for the system in certain conditions. Gottlieb et al. have suggested that the Speed-Insensitive strategy is a default used by the control system when there are no explicit or implicit requirements for movement time. Such requirements can be introduced by asking a subject to move over a certain distance in a certain movement time. If the distance is changed, and the time is kept constant, Speed-Sensitive patterns are likely to be observed. Another example is movements over the same distance with different accuracy requirements (Fitts & Petersen, 1964; Corcos et al., 1988), which force different movement times according to the speed-accuracy trade-offs (see chapter 6).

Which Variable Is Controlled?

Another controversial point in the dual-strategy hypothesis is the authors' definition of the control variable as the "net presynaptic input to α-motoneurons" (Gottlieb et al., 1989a, 1990b). We have already discussed some of the problems arising from direct regulation of the α-motoneuron activity. These problems are likely to be relatively small for the first several tens of milliseconds of a movement, since reflex feedback loops require time to start affecting the EMGs. This may be the cause of the relative success of such models in explaining EMG and kinematic patterns during initial fragments of fast movements, for example properties of the first agonist burst. The problems become obvious when the same approach is used for the delayed events, like the antagonist burst, or muscle contractions with significantly different kinematics (e.g., comparing isotonic movements and isometric contractions).

Movements performed under the Speed-Insensitive strategy frequently demonstrate nonmonotonic changes in the antagonist burst amplitude and area with an increase in movement distance (Marsden et al., 1983; Brown & Cooke, 1981; Cheron & Godaux, 1986; Mustard & Lee, 1987; Latash, 1989a,c). Gottlieb et al. (1989b) tried to resolve this problem by referring to changing relations between muscle torque and EMG for different joint angles. If the movements start from the same initial position, the antagonist burst appears later for longer movements, thus corresponding to different joint angles. So, Gottlieb et al. argued that, due to these effects, antagonist torque did in fact change monotonically while antagonist EMG did not. There are a number of arguments against this explanation. First, the dual-strategy hypothesis is based on regulation of not muscle torques, but excitation pulse, which is a prototype of the EMG. So, since duration of the excitation pulse to antagonist α-motoneurons is supposed to increase with movement distance, it should lead to an increase in the EMG, and there is still a mismatch between the hypothesized control signal and recorded antagonist EMG changes. Second, in another series of experiments (Latash, 1989c), it has been shown that similar changes in antagonist EMG burst with distance can be observed when the movements start from different positions and terminate at the same joint angle.

Analysis of isometric contractions in the same framework presents another challenge to the dual-strategy hypothesis. In order to match the model to the experimental findings, the authors were forced to introduce excitation steps along with excitation pulses and even a possibility of ramp changes in the excitation signal to the α-motoneurons (Corcos et al., 1990a, Figure 4.11). As a result, central control programs for isotonic movements and isometric contractions became different (compare similar programs for these two regimes of muscle contraction described in chapter 5 based on the equilibrium-point hypothesis). Different motor programs for isotonic movements and isometric contractions have also been suggested by other authors (Meinck et al., 1984; Sanes & Jennings, 1984; Ghez and Gordon, 1987; Tax et al., 1990) based on the experimentally observed differences in the EMG patterns.

The dual-strategy hypothesis does not require regulation at the α-motoneuron level. Another control variable, different from the excitation pulse, can be used for expressing the same ideas and introducing the same notions. It was in fact done, although not that explicitly, several years earlier by Adamovitch and Feldman (1984, see "Kinematic Characteristics" earlier in this chapter) who used the language of the equilibrium-point hypothesis for describing how single-joint movements are controlled. Adamovitch and Feldman have suggested that single-joint movements are controlled by ramp shifts in the reciprocal variable r and a simultaneous shift in the coactivation variable c (see chapters 1 and 2). Final value of r defines the final equilibrium state of the joint while final value of c defines the level of coactivation. Shifting r creates a disparity between current joint position and centrally specified equilibrium position. This disparity leads to muscle activation, generation of joint torques,

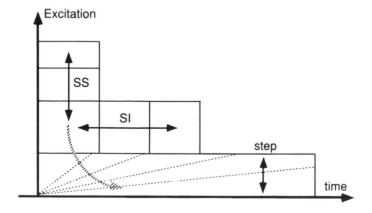

Figure 4.11 Excitation pulses, steps, and ramps for agonist excitation during isometric contractions performed under different instructions.
Note. From "Organizing Principles for Single Joint Movements. IV. Implications for Isometric Contractions" by D.M. Corcos et al., 1990, *Journal of Neurophysiology*, **64**, pp. 1033-1042. Copyright 1990 by The American Physiological Society. Reprinted by permission.

and eventually, joint movement to the new equilibrium position. The movement leads to changes in activity of proprioceptors that affect activity of α-motoneurons through the reflex feedback loops. Therefore, in this framework, the EMGs are not imposed by the control signals but rather emerge as results of dynamic interaction between control signals and peripheral conditions of movement execution.

Adamovitch and Feldman suggested the following simple scheme that can be used for realizing single-joint motor control. Figure 4.12a shows a chain of identical neurons whose activity specifies value of r (a similar consideration can be used for the c variable). At an initial state, all the neurons from the left up to a certain neuron a are active, thus specifying a value r_a. In order to make a single-joint movement, a wave of excitation is generated by a central controller, runs along the neuron chain with a certain speed ω, and stops at a certain neuron b. In isotonic conditions, the new value r_b corresponds to a new equilibrium state of the joint (movement amplitude). The speed of the wave of excitation controls the movement velocity while the final excited neuron defines movement amplitude (or final torque value, if the movement is blocked).

Figure 4.12b illustrates r patterns for movements performed at different speeds and over different amplitudes. Shifts in r at a constant rate (ω)

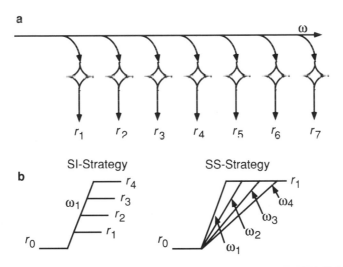

Figure 4.12   (a) A wave of excitation running along a hypothetical chain of neurons at a constant speed ω. Each neuron corresponds to a certain value of the central parameter r. Central command specifies ω and a point where the wave of excitation stops (i.e., final r). (b) Two patterns of r changes corresponding to two basic strategies: Speed-Insensitive (SI) and Speed-Sensitive (SS). In the SI strategy, r changes at the same rate ω but to different final values, while in the SS strategy, r changes at different rates.

over different time periods can be associated with the Speed-Insensitive strategy, while shifts at varying rates correspond to the Speed-Sensitive strategy. In the next chapter, we will demonstrate that these simple control patterns lead to reasonable EMG patterns for both isotonic movements and isometric contractions.

In chapter 2, we described a method for reconstructing virtual trajectories for single-joint movements. The virtual trajectories were associated with time functions of one of the basic central control variables $r(t)$. This method was applied to analysis of movements performed under the two basic strategies. In order to induce the Speed-Insensitive strategy, the subjects were instructed to move "as fast as possible" over different distances. In another series, they were asked to move over the same distance "with different speeds."

Figure 4.13 illustrates some of the results of these experiments. Virtual trajectories of fast movements were characterized by a typical nonmonotonic shape (N-shaped curves, cf. Figure 2.11 in chapter 2). Changes in movement distance did not lead to visible changes in the slope of the first arm of the N-curve, but its amplitude was scaled with movement distance. On the other hand, changes in movement speed led to pronounced changes in the general shape of the virtual trajectory, leading eventually to almost total disappearance of the N-shape.

According to my view (cf. Figure 2.12 in chapter 2), single-joint movements are controlled by an initial shift in λ_{ag} at a constant rate followed by a delayed "braking" shift in λ_{ant}. The latter shift becomes smaller for slower movements and eventually disappears. In this framework, the Speed-Insensitive strategy is characterized by λ_{ag} shifts at a constant rate over different times, thus leading to movements of different amplitude. The first arm of the N-curve is defined exclusively by this λ_{ag} shift, and therefore its experimentally observed uniform slopes and different amplitudes for movements over different distances fit the model. On the other hand, movements performed under a Speed-Sensitive strategy are controlled by λ_{ag} shifts at different rates (cf. with different slopes of the first arm of the N-curves in Figure 4.13).

As far as the second and third arms of the N-curve are concerned, less pronounced N-shapes for relatively slow movements correspond well to the idea that such movements are controlled with relatively small delayed shifts in λ_{ant}. It seems that the lowest point of the second arm of the N-curves for movements performed "as fast as possible" (Speed-Insensitive strategy) reverses its direction approximately at the same value of r. This would be a potentially interesting finding, especially if this value corresponds to the initial joint position (as it seems to show in Figure 4.13). However, the accuracy of the method does not allow such strong conclusions to be drawn.

Note that the method used for reconstructing virtual trajectories has been based on a simplified model originating from the equilibrium-point

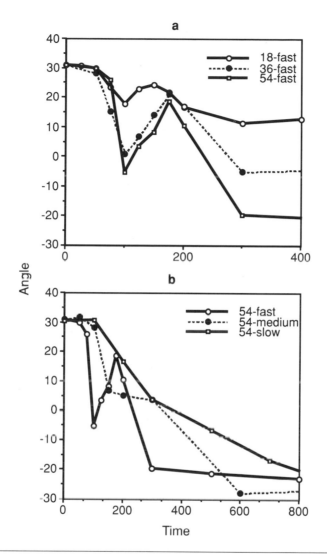

Figure 4.13   Virtual trajectories for single-joint elbow flexion movements performed under the Speed-Insensitive (a) and Speed-Sensitive (b) strategies. Time scale is in ms; angle scale is in degrees.
Note. Reproduced with permission from Latash & Gottlieb, 1992.

hypothesis. The findings represent nice illustrations for the two basic strategies of Gottlieb et al. and show that the dual-strategy hypothesis can quite naturally be based on the λ-model.

Now let us move to chapter 5, which describes a model of EMG emergence during isotonic movements and isometric contractions based on a universal central program of changes in λ for the participating muscles.

Chapter 5

Emergence of Electromyographic Patterns

Standardized time patterns of λs for agonist and antagonist muscles are used as control signals irrespective of the external load, in particular for both isotonic and isometric contractions. Analysis of trajectories of the muscles' current state on the phase plane leads to predictions of electromyographic patterns that are in good agreement with the experimental observations. In particular, qualitative changes in the patterns are described for movements and contractions of different amplitude and at different rates, and also of unexpectedly blocked movements.

In this chapter, a simple model is going to be based on the ideas of the equilibrium-point hypothesis that is able to account for most of the observed features of the EMG patterns accompanying fast single-joint isotonic movements and isometric contractions. This model, as any of its kind, is based on a number of assumptions and simplifications. Two possible extremes may then occur: the model can be oversimplified and made trivial and unable to account for specific experimental data; or the model can be overcomplicated by introducing everything one knows about the system, thus making it impossible to analyze.

I decided, first, to make the model as simple as possible, so that the resulting equations are analytically solvable. Then, if the model was unable to explain an important group of experimental data, additions and refinements could be introduced, complicating the model and improving its predictive power. Interestingly enough, this was not necessary. In other words, we are going to start from the simple pole of the spectrum and move into the complex area only if necessary. Note that the main objective was not to create the ultimate model of EMG emergence, but rather to demonstrate that the equilibrium-point hypothesis provides an adequate framework for analyzing EMG phenomena during fast single-joint movements.

Emergence of the EMG patterns in the framework of the equilibrium-point hypothesis has been analyzed by the group of Feldman (Abdusamatov & Feldman, 1986; Abdusamatov et al., 1987; Feldman et al., 1990a; see also Schieber & Thach, 1985). My approach shares a number of features with that of Feldman, including some of the basic notions and assumptions; however, it also has certain important distinctions that will now be discussed.

Basic Assumptions and Notions of the Model

1. The same central motor command will be used to generate an isotonic movement, an isometric contraction, or an intermediate regime of muscle work (against an elastic load) depending on external load characteristic. An isotonic movement and an isometric contraction produced by identical central commands in different external conditions will be termed *matched*. For example, if an isotonic movement is unexpectedly blocked at the starting position, there will be an increase in force without joint movement (isometric contraction) while the central command remains presumably the same (at least, for several tens of milliseconds). Movements in a pin joint will be analyzed, controlled by two muscles that will be addressed as agonist and antagonist.

2. For each muscle there is only one independently controlled variable, the threshold (λ) of the tonic stretch reflex (i.e., the muscle length at which recruitment of α-motoneurons starts under quasi-static conditions). Muscle length, movement amplitude, and λ will be measured in the same units. The thresholds for the agonist and antagonist muscles will be designated λ_{ag} and λ_{ant} respectively.

3. The central command changes λ at a constant rate (ω) that will be measured in units of length/time. We shall analyze only "fast" movements, that is, those for which the time during which λ_{ag} is changing (τ_{ag}) is small as compared with the total movement time. Since τ_{ag} is always greater than zero, changes in λ provide for a kind of internal or virtual trajectory of an intended movement (see chapter 2; cf. Flash, 1987; van Sonderen et al., 1988).

Until the experimental reconstruction of joint compliant characteristics *during* changes in λ (see chapter 2 and Latash & Gottlieb, 1990b, 1991c), it had not been obvious that the very concept of invariant or compliant characteristics can be applied to analysis of movement dynamics. Since the method used for reconstructing virtual trajectories during voluntary movements has been based on a number of controversial simplifications, skeptics may still consider λ an abstraction rather than a measurable parameter. However, we are going to demonstrate that even in this case such an abstraction is helpful in understanding how a standardized central command can be translated by a dynamic system into reproducible peripheral patterns of muscle activation, depending on existent conditions of movement execution, instruction, and/or intention of the subject. This would provide a uniform description of how a broad class of single-joint voluntary movements is controlled.

4. The control of fast, single-joint isotonic movements or isometric contractions can be partitioned into four phases (see Figure 5.1). Phase 1 (OA in Figure 5.1) consists of a shift of the agonist λ (λ_{ag}) from its starting value λ_{in} to the final value λ_{fin} at a rate ω_{ag} in τ_{ag} and an initial coactivation of antagonist muscles (c) by a small shift in λ_{ant} starting at the same moment as the λ_{ag} shift. We are going to analyze only relatively fast movements defined as those for which phase 1 is a small fraction of the total movement time. This assumption approximately corresponds to a movement time of under 500 ms. Phase 2 (AB in Figure 5.1) has constant values of both λ_{ag} and λ_{ant}. The acceleration of the limb is still positive but is decreasing. Phase 3 (BC in Figure 5.1) is active braking of the movement. It begins with a later shift of the λ_{ant} at a rate ω_{ant} lasting τ_{ant}. For an isometric contraction, this slows the rate of torque increase. This shift is delayed by t^* after τ_{ag} based on the planned movement time and approaches zero for short isotonic movements and isometric contractions. In phase 4, λ_{ant} shifts back in the opposite direction (CD, broken line in Figure 5.1), so that in the final state, agonist and antagonist levels of activation oppose each other and balance the external load. We will not analyze this part of the movement. The assumed four movement phases correspond to "cartoon" kinematic patterns (Figure 5.1, acc, V, and L), which are an approximation of the experimentally observed smooth sigmoid trajectories and bell-shaped velocities (see the first section of chapter 4). This approximation is mathematically tractable and leads to sufficiently simple equations through which some insight about the influence of kinematics on EMG can emerge.

5. All possible values of muscle length and velocity are represented by points on a phase plane (velocity-length or V-L plane; Figure 5.2). At any instant in time, the current values of muscle length and velocity will be referred to as the muscle's "current state." Following Feldman (1986), we divide this plane by a straight line (Ω-line) into two parts. When the state of muscle length and velocity places it to the right of the Ω-line, its α-motoneurons are active (points 1, 2, and 3 in Figure 5.2). When it lies to the left, the α-motoneurons are silent (points 4, 5, and 6 in Figure 5.2). The intersection of the Ω-line with the abscissa axis defines λ. Shifts of λ imply translation of the Ω-line. We will assume that the slope of the Ω-line does not change.

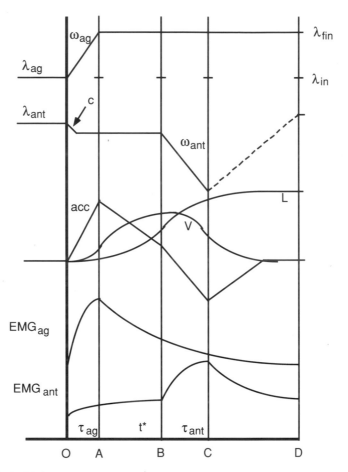

Figure 5.1   A fast single-joint movement is assumed to be controlled by a shift in λ_{ag} from an initial value (λ_{in}) to a final value (λ_{fin}) at a rate ω_{ag} over τ_{ag}; a simultaneous coactivation (c) of the antagonist by a shift in λ_{ant}; a delayed shift in λ_{ant} at a rate ω_{ant} over τ_{ant}; a "backward" shift in λ_{ant} (broken line; not analyzed in the model) leading to desired final joint position and muscle coactivation. Acceleration (acc) has been modeled as a combination of constant-jerk segments. Schematic drawings of velocity (V), muscle length (L), agonist (EMG_{ag}) and antagonist (EMG_{ant}) EMGs are also shown.

Note. From "An Equilibrium-Point Model for Fast Single-Joint Movements: I. Emergence of Strategy-Dependent EMG Patterns" by M.L. Latash and G.L. Gottlieb, 1991, *Journal of Motor Behavior*, **23**, pp. 163-177. Reprinted with permission of the Helen Dwight Reid Educational Foundation. Published by Heldref Publications, 4000 Albemarle St., N.W., Washington, D.C. 20016. Copyright © 1991.

6. The instantaneous EMG level (E) is directly proportional to the shortest distance from a point corresponding to the muscle's current state to the Ω-line.[1] This assumption ignores conduction delays in the neural pathways. This definition of E differs from Feldman's definition, which uses only angular disparity between the current position and the activation border (Abdusamatov & Feldman, 1986; Abdusamatov et al., 1987) but the assumed constant slope of the Ω-line makes the two definitions equivalent (see the first section in chapter 3).

Actual EMG depends on the method of recording and processing signals from the muscle fibers. It reflects intramuscular processes that define the development of muscle force and, therefore, limb kinematics. However, in the framework of this model, we are going to break the causal circle (α-motoneurons $\rightarrow$ EMG & torque $\rightarrow$ kinematics $\rightarrow$ receptors $\rightarrow$ reflex loops $\rightarrow$ α-motoneurons) by assuming a kinematic trajectory. As such, the EMG is a function of formally introduced central commands and assumed kinematics. Therefore, the model remains "open." This simplifying assumption is an alternative to those that must be made to dynamically "close" the model with quantitative assumptions relating EMG to torque and kinematics. Our kinematic assumption is based on a large body of observed kinematic patterns (see reviews in Gottlieb et al., 1989a,b, 1990a,b; Corcos et al., 1989, 1990a) and leads to analytically solvable equations that let us account for many of the experimentally observed EMG patterns. This approach differs from that of Feldman, whose modeling of the EMGs is based on direct incorporation of muscle torques, their relation to kinematics, and their dependence upon joint angle.

The EMGs will be calculated in arbitrary units, thus leaving room for only qualitative comparisons with the experimental data.

The four phases of the movement are defined in terms of λ. In order to compute a closed form solution for the EMG patterns, we will assume that the four phases can be described kinematically as intervals of constant jerk (the third derivative of angle). This assumption leads to accelerations that vary linearly with time (Figure 5.1). This assumption is a crude approximation of the actual limb acceleration profile but allows us to

---

[1]This assumption seemingly leads to a problem because the axes on the V-L plane have different units of measurement (length and length/time). However, since we are going to "measure" EMG in arbitrary units, it doesn't matter which units are used for the axes as long as the scales are kept constant. Formally speaking, there must be a coefficient that transforms length/time units into length units and another coefficient transforming length units into units of EMG (e.g., microvolts), just as for the muscle spindle we can speak of commensurable length and velocity sensitivities, both of which produce changes in spindle firing rate. Both these coefficients are considered here to be constant and equal to 1. However, later we assume, for simulation purposes, a different value of a coefficient transforming length/time units into length units.

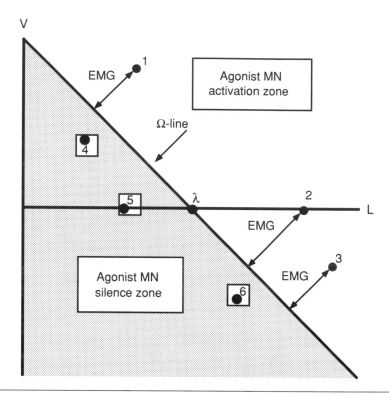

Figure 5.2 The phase plane for a muscle (V-L plane) is divided by a straight line (Ω-line) into zones of agonist motoneuron (MN) activation and silence. Intersection of the Ω-line with the L axis corresponds to the current value of λ. When muscle current state (combination of instantaneous values of its length L and speed V) is in the silence zone, the EMG is absent. When the current state is in the activation zone (points 1, 2, 3), the EMG is considered to be proportional to the distance to the Ω-line.

analytically solve the resulting equations that predict EMG.[2] Note that a direct relation of central commands to the acceleration timing was also proposed by Hemami and Stokes (1983).

Phase one of the movement ($0 < t < \tau_{ag}$) can be characterized with a linearly increasing acceleration, $a = A_{ag}t$, where A_{ag} is proportional to ω_{ag}

---

[2]One may suggest the following tentative justification for the introduced linear shifts in λ_{ag} and parallel linear changes in the acceleration. According to the model, the time of an initial linear shift in λ_{ag} is much smaller than the movement time. During this initial period, EMG changes will be defined mostly by this shift in a central variable. Since EMG is assumed to be proportional to distance from the muscle current state to the Ω-line, one may expect approximately linear changes in the agonist EMG as well. On the other hand, EMG has been shown to be proportional to muscle torque, at least for the static conditions (Bigland & Lippold, 1954; Buchanan et al., 1986), which is, in turn, approximately proportional to the acceleration.

($A_{ag} = \omega_{ag}/c_{\omega}$; c_{ω} is a constant characterizing inertial properties of the "limb + load"). We won't discuss an apparent mismatch in the units due to the introduction of an inertial term in the kinematic description, because we are interested only in qualitative description of the EMG patterns.

During the second phase (t^*), λ_{ag} is constant and feedback of muscle state causes the acceleration to decline at a constant rate (A^*). The viscoelastic properties of the limb (AB in Figure 5.1) (Gurfinkel et al. 1974; Lestienne 1979) will also contribute to this decline.

The third phase, active deceleration, can similarly be described by a linear fall of the acceleration with a coefficient A_{ant} proportional to ω_{ant} ($A_{ant} = \omega_{ant}/c_{\omega}$).

In the fourth and final phase, the deceleration declines to zero as λ_{ant} changes to a new equilibrium value at which agonist and antagonist torques are in balance.

In all four phases, the acceleration changes linearly with time and is continuous at the points of transition. This assumption is not fundamental to the model but is made because it allows analytical solution of the resulting equations that predict EMG.

In discussing movements of different distances and inertial loads, we shall divide the movements into two classes according to the dual-strategy hypothesis (Gottlieb et al., 1989a; see chapter 4, "Excitation Pulse and the Dual-Strategy Hypothesis"). Those sets of movements in which λ_{ag} shifts at the same constant rate of ω_{ag} (Speed-Insensitive strategy) are characterized by identical trajectories during phase one (OA interval in Fig. 5.1) (cf. with Adamovich & Feldman, 1984; Adamovich et al., 1984). Movements with different ω_{ag} (Speed-Sensitive strategy) are characterized by different phase one trajectories.

Initiation of an Isotonic Movement

Agonist EMG Changes

Assume that the threshold of the tonic stretch reflex (λ_{in}, the point X of intersection of the Ω-line with the abscissa axis at $t = 0$, Figure 5.3) is equal to the initial length of the agonist. This implies that the muscle is initially inactive but any increase in the muscle length (or decrease in λ) will start to recruit its α-motoneurons. (Consideration of a shorter muscle would only add a poorly defined constant to the λ changes in the equations.) At $t = 0$, λ starts to shift at a constant rate ω_{ag} ($\lambda = \lambda_{in} - \omega_{ag}t$). The Ω-line shifts to the left through point O towards λ_{fin}, the final value of λ, reached at $t = \tau_{ag}$. This gives rise to agonist muscle activation and its subsequent shortening.

The need to accelerate the inertial load of the limb assures that the changes in muscle length will proceed more slowly than the changes in

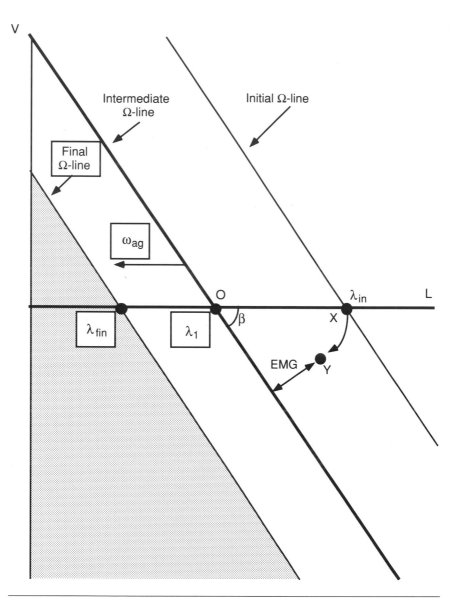

Figure 5.3   Originally, muscle length corresponded to λ_{in} (point X). Shifting of the Ω-line leads to muscle contraction (point Y). The EMG is proportional to the shortest distance from the current state (Y) to the current Ω-line.

Note. From "An Equilibrium-Point Model for Fast Single-Joint Movements: I. Emergence of Strategy-Dependent EMG Patterns" by M.L. Latash and G.L. Gottlieb, 1991, *Journal of Motor Behavior*, **23**, pp. 163-177. Reprinted with permission of the Helen Dwight Reid Educational Foundation. Published by Heldref Publications, 4000 Albemarle St., N.W., Washington, D.C. 20016. Copyright © 1991.

λ and at some later time, the muscle state reaches a point Y (Figure 5.3). At this point, the EMG level (E) is equal to the shortest distance from the point Y to the Ω-line (assumption 6). Acceleration $a(t) = -A_{ag}t$ (assumption 7) allows us to calculate the instantaneous values of velocity (V) and distance moved (S) by simple integration (initial conditions: $V_0 = 0$ and $S_0 = 0$):

$$V = -\frac{A_{ag}t^2}{2}; \quad S = -\frac{A_{ag}t^3}{6}. \tag{5.1}$$

Geometrical analysis leads to the following formula for the level of the agonist EMG (E_{ag}) as the muscle state moves from X to Y:

$$E_{ag} = t\omega_{ag}\sin\beta - \frac{t^3 A_{ag}\sin\beta}{6} - \frac{t^2 A_{ag}\cos\beta}{2}. \tag{5.2}$$

The first term can be regarded as the direct effect of the central command signal. The latter two terms are the effects of afference (although certainly affected by the central command). To calculate the rate at which E_{ag} changes, we differentiate Equation 5.2:

$$\frac{dE_{ag}}{dt} = \omega_{ag}\sin\beta - \frac{t^2 A_{ag}\sin\beta}{2} - tA_{ag}\cos\beta. \tag{5.3}$$

Let us now consider the implications of this equation for movements with different amplitudes, speeds, and moments of inertia.

Movements of Equal Amplitudes and Different ω_{ag} (Speed-Sensitive Strategy). To define time-to-peak of the first agonist EMG burst, let us make the right side of Equation 5.3 zero and substitute a constant c_ω for the ratio ω_{ag}/A_{ag} (assumption 7):

$$t^2 + 2t\cot\beta - 2c_\omega = 0. \tag{5.4}$$

The positive root of Equation 5.4 is

$$t_{max} = -\cot\beta + \sqrt{\cot^2\beta + 2c_\omega}. \tag{5.5}$$

Since t_{max} is only dependent on two constants (β and c_ω) the model predicts that time-to-peak of the first agonist EMG burst will be independent of ω_{ag} and A_{ag}, that is, independent of the movement speed (or movement time, for movements with a fixed amplitude). The above equations only imply this if $t_{max} < \tau_{ag}$ but we will show that this is more generally true.

During the time interval $0 < t < \tau_{ag}$, the changes in muscle length and velocity are small compared with λ changes (assumption 4), and since t is small, the values of E_{ag} and dE_{ag}/dt will be defined mainly by the first elements in Equations 5.2 and 5.3 correspondingly. Equation 5.2 implies that the peak value of E_{ag} (E_{max}) is positive and increases with ω_{ag}. Equation 5.3 implies that the rate of the EMG rise also increases with ω_{ag}. Thus, an increase in the movement speed will be associated with faster rise of the EMG and higher peak EMG values. These predictions are consistent with Figure 5.4a, which compares simulated and measured agonist EMG bursts for movements at four different speeds. The simulation was performed by solving Equation 5.2 and Equation 5.8 (upcoming). The lower part of Figure 5.4a shows averaged recorded EMGs from biceps for 54° movements at four movement speeds. Time-to-peak of these EMGs is slightly different but in most studies this time has been shown to be independent of movement speed (see later in this chapter and Corcos et al., 1989). The key feature is that both the modeled and the measured EMGs rise at rates and to peaks proportional to movement speed. The second derivatives

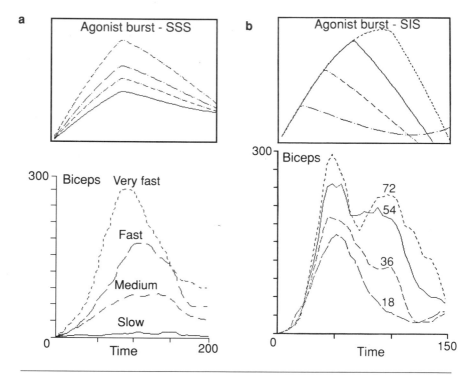

Figure 5.4 Computer simulations of the agonist burst (upper) and EMG recordings of the biceps (lower) during elbow flexion movements performed (a) at different speeds over 54° (Speed-Sensitive strategy [SSS]), and (b) over 18°, 36°, 54°, and 72° "as fast as possible" (Speed-Insensitive strategy [SIS]).

Note. From "An Equilibrium-Point Model for Fast Single-Joint Movements: I. Emergence of Strategy-Dependent EMG Patterns" by M.L. Latash and G.L. Gottlieb, 1991, *Journal of Motor Behavior, 23*, pp. 163-177. Reprinted with permission of the Helen Dwight Reid Educational Foundation. Published by Heldref Publications, 4000 Albemarle St., N.W., Washington, D.C. 20016. Copyright © 1991.

Table 5.1   Some EMG Measures for Movements Under the Speed-Sensitive and Speed-Insensitive Strategies

SSS (54°)	Q_{30}	Q_{acc}	$E_{ag}(max)$	Q_{dec}	$E_{ant}(max)$
Slow	0.9	5	0.5	8	1
Medium	2.1	11	1.5	9	1.5
Fast	3.2	22	4	20	3
Very fast	5.2	43	10	40	9.5

SIS (very fast)	Q_{30}	Q_{acc}	$E_{ag}(max)$	Q_{dec}	$E_{ant}(max)$
18°	9.5	80	2.8	148	8
36°	6	120	5.8	150	8.5
54°	7	155	6.3	168	7.8
72°	7.5	170	8.5	160	7.7

Note. For the Speed-Sensitive strategy (SSS), the data for elbow flexion movements over 54° are presented. For the Speed-Insensitive strategy (SIS), the data for movements performed "as fast as possible" over different distances are presented. Q_{30} is the integrated agonist (biceps) EMG over the first 30 ms after the beginning of a visible increase in the biceps EMG (t_0). Q_{acc} is the integrated biceps EMG during the period of limb acceleration (it involved the whole first agonist EMG burst). Q_{dec} is the integrated antagonist EMG from the moment t_0 up to the second crossing of zero line by the acceleration trace. $E_{ag}(max)$ and $E_{ant}(max)$ are maximal levels of the EMGs. All the data are in arbitrary units. The data in the upper and lower parts of the table are from different articles and should not be compared. (Adapted by permission from Figures 2, 4, 7 of Gottlieb et al., 1989b; and Figures 1, 2, 7 of Corcos et al., 1989.)

of the modeled and measured EMGs do not agree; this is probably due to the model's assumption of constant jerk. Table 5.1 presents some EMG measures presented in Gottlieb et al. (1989b) and Corcos et al. (1989) for movements performed under the two basic strategies. An index Q_{30} in Table 5.1 reflects the initial rate of the EMG rise while Q_{acc} reflects the total integrated agonist EMG. Note an increase in both indices with an increase in movement speed.

I would like to stress that my simulations are based on cartoon acceleration patterns and a number of simplifications and, therefore, lead to cartoon EMG patterns. This model is not designed to account for all the peculiarities of the actually observed EMG patterns, which are frequently irreproducible and subject-dependent. I will point out only some of the general characteristic features that are seen in both simulated and averaged actual EMGs. This comment applies both to Figure 5.4 and Figure 5.7.

Movements With Different Amplitudes and Fixed ω_{ag} *(Speed-Insensitive Strategy).* For movements of different amplitudes with the same constant ω_{ag}, the first agonist burst is also described by Equation 5.2 and Equation 5.3 above. Equation 5.3 shows that the rate of the EMG rise is independent of the movement amplitude. Figure 5.4b illustrates simulated and recorded EMGs for movements of different distances performed under this strategy. The model's EMGs all rise at identical rates and successively diverge at times that increase with movement distance (cf. Q_{30} and Q_{acc} in Table 5.1). The measured EMGs are also identical for about the first 30 ms for all distances and then diverge one at a time, but more gradually than in the model. The relative similarity of these initial EMG patterns when moving different distances "as fast as possible" (which means that peak movement speed increases with distance) should be compared with the differences in initial patterns when moving at intentionally different speeds. Although the measured EMG patterns are more complex than the model predicts, there is a good deal of experimental individual variability (see also Gottlieb et al., 1989b, 1990a). The model and the data agree however in the relative uniformity of onsets, distance dependent peaks, and distance dependent total activity (e.g., burst area; see Gottlieb et al., 1989a, 1990a; see also Table 5.1).

Changes in Moment of Inertia. The constant c_ω is proportional to inertial load, so an increase in the moment of inertia must lead to a decrease in the rate of acceleration A_{ag} for a constant ω_{ag}. This implies (a) an increase in time-to-peak of the first agonist burst, Equation 5.4, and (b) an increase in the peak of the EMG, Equations 5.2 and 5.3, for movements of constant amplitude and increasing inertial load. I do not illustrate this but will compare the prediction with published data later.

Antagonist EMG Changes (Early Component)

Assumption 4 assumes a low level of antagonist coactivation in the model that begins near the onset of agonist activation. This ensures that the antagonist λ value is already in its activation zone of the V-L plane (point X in Figure 5.5a). We assume no further central changes in the antagonist λ (λ_{ant} = constant; point l_0 in Figure 5.5a) until active braking of the movement starts. Constant λ_{ant} does not mean constant antagonist EMG since the muscle is being lengthened and excited by afferent input (point Y in Figure 5.5a).

The level of antagonist EMG (E_{ant}) can be calculated as

$$E_{ant} = (\lambda_{ant} + \frac{t^3 A_{ag}}{6}) \sin v + \frac{t^2 A_{ag} \cos v}{2}; \quad t < \tau_{ag}, \quad (5.6)$$

where v is an angle between the antagonist's Ω-line and the abscissa axis on the antagonist V-L plane. E_{ant} is an increasing function of t. For

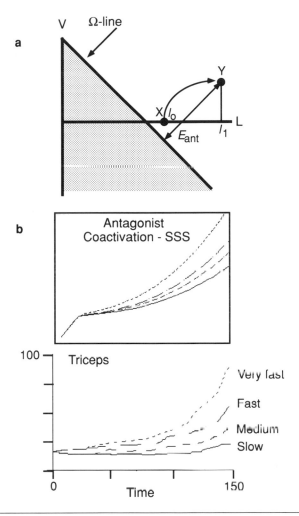

Figure 5.5   (a) Initial coactivation (c in Figure 5.1) leads to a shift of the Ω-line on the antagonist V-L plane so that the antagonist current state (X) is in the zone of muscle activation. Stronger activation of the agonist leads to antagonist stretching (point Y; cf. l_0 and l_1). The EMG (E_{ant}) is proportional to the distance from the antagonist current state to the Ω-line. (b) Computer simulations of the antagonist (lateral head of triceps) EMG changes before the active braking and actual EMG recordings for the elbow flexion movements over 54° performed at different speeds (Speed-Sensitive strategy).

movements of fixed amplitudes performed at different speeds (Speed-Sensitive strategy), an increase in ω_{ag} will bring about an increase in A_{ag}, giving rise to an increase in E_{ant}. Figure 5.5b shows the results of computer simulation (above) and experimental recordings of the antagonist EMG (below) for the Speed-Sensitive strategy. During the experiments, the subject was asked to perform elbow flexion movements over 54° at different speeds. Movements of different amplitudes performed with the "same speed" (Speed-Insensitive strategy, ω_{ag} = constant) will have time courses for E_{ant} that are similar and independent of movement amplitude.

Termination of an Isotonic Movement

Passive Braking

To simplify the analysis of phase 2 we have assumed a linearly falling acceleration (AB in Figure 5.1). The following equations describe acceleration, velocity, and distance starting from the time τ_{ag}:

$$a = -A_{ag}\tau_{ag} + A^*t',$$

$$V = -A_{ag}\tau_{ag}t' + \frac{A^*t'^2}{2} - \frac{A_{ag}\tau^2}{2}, \tag{5.7}$$

$$S = -\frac{A_{ag}\tau_{ag}t'^2}{2} + \frac{A^*t'^3}{6} - \frac{A_{ag}\tau_{ag}^3}{6} - \frac{A_{ag}\tau_{ag}^2 t'}{2},$$

where $t' = t - \tau_{ag} > 0$; A^* = constant > 0.

Since τ_{ag} is small (assumption 4), we can ignore the members in Equations 5.7 with τ^2_{ag} and τ^3_{ag}. Geometrical considerations give the following expression for the agonist EMG (E_{ag}) changes:

$$E_{ag} = t'^3 A^*\sin\beta - t'^2\left(\frac{A_{ag}\tau_{ag}\sin\beta}{2} - \frac{A^*\cos\beta}{2}\right) + \omega_{ag}\tau_{ag}\sin\beta \tag{5.8}$$
$$- t'A_{ag}\tau_{ag}\cos\beta.$$

To assess the duration of the first agonist burst, we need to compute the time at which E_{ag} goes to zero. Since our analysis is confined to relatively fast movements with movement time of under 500 ms (assumption 4), t' will not be over 100-200 ms (cf. Figures 1 and 3 in Gottlieb et al., 1989b; Figure 1 in Corcos et al., 1989). For such values of t', the cubic term is under 5% to 10% of the second (quadratic) term. The cubic term of Equation 5.8 also does not depend on ω_{ag} or τ_{ag}. Therefore we are going to omit this term and analyze a simpler expression:

$$t'^2\left(\frac{A_{ag}\tau_{ag}\sin\beta}{2} - \frac{A^*\cos\beta}{2}\right) + t'A_{ag}\tau_{ag}\cos\beta - \omega_{ag}\tau_{ag}\sin\beta = 0. \tag{5.9}$$

The positive root of Equation 5.9 is

$$t_1' = \frac{A_{ag}\cos\beta}{\left(A_{ag}\sin\beta - \dfrac{A^*\cos\beta}{\tau_{ag}}\right)} \tag{5.10}$$

$$+ \sqrt{\left\{\frac{A_{ag}\cos\beta}{\left(A_{ag}\sin\beta - \dfrac{A^*\cos\beta}{\tau_{ag}}\right)}\right\}^2 + \frac{2\omega_{ag}\sin\beta}{\left(A_{ag}\sin\beta - \dfrac{A^*\cos\beta}{\tau_{ag}}\right)}} \; .$$

Note that t_1' will be proportional to the duration of the agonist EMG burst and is an increasing function of τ_{ag}. In order to analyze movements of the same amplitude performed at different speeds, let us transform the right side of Equation 5.10 taking into account that $\omega_{ag} = c_\omega A_{ag}$:

$$t_1' = \frac{\cos\beta}{\left(\sin\beta - \dfrac{c_\omega A^*\cos\beta}{\omega_{ag}\tau_{ag}}\right)} \tag{5.10'}$$

$$+ \sqrt{\left\{\frac{\cos\beta}{\sin\beta - \dfrac{c_\omega A^*\cos\beta}{\omega_{ag}\tau_{ag}}}\right\}^2 + \frac{2c_\omega\sin\beta}{\sin\beta - \dfrac{c_\omega A^*\cos\beta}{\omega_{ag}\tau_{ag}}}} \; .$$

One can see from Equation 5.10' that t_1' is dependent only on $\omega_{ag}\tau_{ag}$ and certain constants. Therefore, since $\omega_{ag}\tau_{ag}$ is equal to movement distance, one can expect similar durations of the first agonist bursts for movements of equal distance performed with "different speed" (i.e., different values of ω_{ag}) (Speed-Sensitive strategy).

Getting back to Equation 5.10, note that movements with increasing amplitudes and equal values of ω_{ag} have increasing values of τ_{ag}. This leads to longer EMG bursts (Speed-Insensitive strategy). The omitted first member of Equation 5.8 would not qualitatively change the dependence of t_1' upon ω_{ag} and τ_{ag} because it does not depend on them. An increase in the moment of inertia would lead to an increase in c_ω and a decrease in A_{ag} if ω_{ag} is constant, which would increase t_1' (i.e., increase the duration of the agonist burst). Note that this analysis is based on Equation 5.9 rather than Equation 5.8 and, therefore, is applicable only to rather fast movements, when ignoring the cubic term in Equation 5.8 is likely not to lead to significant changes in the predictions.

Figure 5.4a,b illustrates the subsidence of the agonist EMG burst for both simulated EMGs (above) and averaged biceps EMGs (below) for the two strategies. Although the simulated patterns cannot reproduce the details of the measured EMG data, they capture the important property of burst duration and illustrate that speed-sensitive patterns tend to have similar burst durations while speed-insensitive patterns tend to have clearly different burst durations (cf. Table 5.1).

Active Braking

We now assume that phase 3 active braking of the movement is initiated by changes in λ_{ant} at $t = \tau_{ag} + t^*$, when the acceleration, velocity, and position are a^*, V^*, and S^* (point B in Figure 5.1). These values are described by expressions similar to Equation 5.7, transformed to account for changes of signs (i.e., shortening of the agonist is accompanied by an equal lengthening of the antagonist):

$$a^* = A_{ag}\tau_{ag} - A^*t^*,$$
$$V^* = \frac{A_{ag}\tau_{ag}^2}{2} + A_{ag}\tau_{ag}t^* - \frac{A^*t^{*2}}{2}, \tag{5.11}$$
$$S^* = \frac{A_{ag}\tau_{ag}^3}{6} + \frac{A_{ag}\tau_{ag}^2 t^*}{2} + \frac{A_{ag}\tau_{ag}t^{*2}}{2} - \frac{A^*t^{*3}}{6},$$

where τ_{ag} is the total time for active acceleration at a constant rate $A_{ag} > 0$, and t^* is the total time of phase 2 at a constant rate $A^* > 0$.

One might expect that movement dependent changes in the antagonist EMG burst might be more complex (e.g., nonmonotonic) than similar changes in the first agonist burst. The antagonist EMG burst is also defined by two factors: central (λ_{ant} shift), which is considered standard, and peripheral (lengthening of the antagonist muscle). For short movements, t^* approaches 0, and a^*, V^*, and S^* are increasing functions of τ_{ag}. Therefore, an increase in movement amplitude is likely to lead to an increase in the speed of antagonist lengthening up to the start of the λ_{ant} shift, leading to an increase of the antagonist EMG burst. However, for relatively large movements the period of decreasing acceleration in phase 2 may lead to negative accelerations and a decrease in speed. Then a^*, V^*, and S^* are decreasing functions of t^* (note assumption 4, $\tau_{ag} < t^*$). Because an increase in movement amplitude leads to an increase in both τ_{ag} and t^*, a^*, V^*, and S^* may be nonmonotonic functions of movement amplitude, and might lead to a nonmonotonic dependence of the antagonist EMG burst on movement distance.

Let us denote the rate of the antagonist λ changes as ω_{ant}. The antagonist EMG (E_{ant}) is described by the following expression (see Figure 5.6):

$$E_{ant} = \left(\omega_{ant}t'' + S^* + V^*t'' + \frac{A_{ant}t''^2}{2} - \frac{A^*t''^3}{6}\right)\sin v \tag{5.12}$$
$$+ \left(a^*t'' + V^* - \frac{A^*t''^2}{2}\right)\cos v,$$

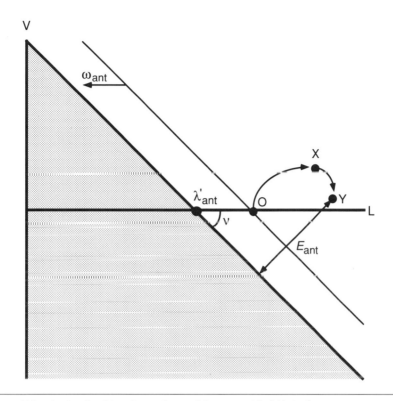

Figure 5.6   Active braking is performed by a rapid shift in λ_{ant} at a rate ω_{ant} leading to a corresponding shift of the Ω-line on the antagonist V-L plane. The speed of muscle lengthening will start to drop (cf. points X and Y) but the EMG will increase.

Note. From "An Equilibrium-Point Model for Fast Single-Joint Movements: I. Emergence of Strategy-Dependent EMG Patterns" by M.L. Latash and G.L. Gottlieb, 1991, *Journal of Motor Behavior*, **23**, pp. 163-177. Reprinted with permission of the Helen Dwight Reid Educational Foundation. Published by Heldref Publications, 4000 Albemarle St., N.W., Washington, D.C. 20016. Copyright © 1991.

where $A_{ant} = \omega_{ant}/c_\omega$ is a constant rate of acceleration change defined in the same way as A_{ag} for the agonist (see previous discussion), v is an angle between the Ω-line and the abscissa axis, and $t'' = t - \tau_{ag} - t^* > 0$.

For short t'' (in cases of large movements when $t^* \gg t''$), E_{ant} is an increasing function of t^*. We are going to ignore the cubic term of Equation 5.12, as has been done during the analysis of Equation 5.8. The justification is, once again, relatively short movement times for which t'' is not going to be over 100-200 ms (cf. Figures 1 and 3 in Gottlieb et al., 1989b; Figure 1 in Corcos et al., 1989). Then the cubic term is under 5% to 10% of the quadratic term. After a simple transformation of Equation 5.12 we get

$$E_{ant} = t''^2 \left(\frac{A_{ant}\text{sinv}}{2} - \frac{A^*\text{cosv}}{2} \right) + t'' \left(\omega_{ant}\text{sinv} + V^*\text{sinv} + a^*\text{cosv} \right) \quad (5.13)$$
$$+ S^*\text{sinv} + V^*\text{cosv}.$$

To find the maximum value of E_{ant}, let us differentiate Equation 5.13 by t'' and equate the derivative to zero,

$$\frac{dE_{ant}}{dt''} = t'' \left(A_{ant}\text{sinv} - A^*\text{cosv} \right) + \omega_{ant}\text{sinv} + V^*\text{sinv} + a^*\text{cosv} = 0,$$

leading to

$$t_{max}'' = \frac{\omega_{ant} + V^* + a^*\text{cotv}}{A^*\text{cotv} - A_{ant}}, \quad (5.14)$$

$$E_{ant}(\text{max}) = S^* + V^*\text{cotv} + \frac{(\omega_{ant} + V^* + a^*\text{cotv})^2}{2(A^*\text{cotv} - A_{ant})}. \quad (5.15)$$

To estimate the duration of the antagonist EMG burst and its dependence on a^*, V^*, and S^*, make the right half of Equation 5.13 zero. Solving for the positive root gives:

$$t_1'' = \frac{\omega_{ant} + V^* + a^*\text{cotv}}{A^*\text{cotv} - A_{ant}} \quad (5.16)$$
$$+ \sqrt{\frac{\omega_{ant} + V^* + a^*\text{cotv} + (S^* + V^*\text{cotv})(A^*\text{cotv} - A_{ant})}{(A^*\text{cotv} - A_{ant})^2}}.$$

Because, by definition, $t'' = 0$ at the beginning of the antagonist burst, Equation 5.16 describes the duration of the antagonist burst.

For movements of constant amplitude, an increase in the movement speed (an increase in ω_{ag}; Speed-Sensitive strategy) would lead to an increase in A_{ant} and a decrease in both t^* and the movement time (see previous discussion). This would (see Equation 5.11) bring about an increase in a^*, V^*, and S^*. Thus, an increase in ω_{ag} would lead to an increase in both antagonist EMG maximal amplitude and burst duration (see Equations 5.15 and 5.16).

$E_{ant}(\text{max})$ is an increasing function of a^*, V^*, and S^*. Returning to Equation 5.11, for large movements, $E_{ant}(\text{max})$ is also a decreasing function of t^* or of the movement time. So, in the case of movements performed at the same speed (same ω_{ag}, Speed-Insensitive strategy), the peak level of the antagonist burst will increase for small movements but decrease for movements of large amplitude. Therefore, $E_{ant}(\text{max})$ is an "inverted U" function of movement amplitude. Table 5.1 presents peak levels and integrals (Q_{dec}) for the antagonist EMG burst. Note a nonmonotonic change in both indices.

For short movements, a^*, V^*, and S^* are increasing functions of τ_{ag} (or movement amplitude). Therefore, similar analysis demonstrates an increase of the antagonist burst amplitude and duration with the movement amplitude for the Speed-Insensitive strategy. This might be the cause for inconsistent experimental data on the antagonist burst amplitude changes (see chapter 4).

Note that the right side of Equation 5.16 is an increasing function of a^*, V^*, and S^* and, therefore, all the above considerations concerning $E_{ant}(\max)$ apply to that expression as well. In particular, duration of the antagonist burst will increase for small movements and decrease for movements of sufficiently large amplitude.

Simulations of the antagonist EMG burst using Equation 5.12 and averaged EMGs of the lateral head of the triceps for the two strategies are shown in Figure 5.7a,b. Slower movements are characterized by a later occurring smaller antagonist burst. Note the nonmonotonic behavior of the antagonist burst amplitude for movements of different amplitude performed under the Speed-Insensitive strategy (as fast as possible). This behavior is evident in both recorded and simulated EMGs.

Strictly speaking, Equation 5.13 can be applied to the analysis of the antagonist burst amplitude but not duration because it is effective only for the time period $0 < t'' < \tau_{ant}$. The analysis of the antagonist burst duration should be performed similar to the one in "Passive Braking" earlier in this chapter. For brevity, I have omitted these calculations. However, the conclusions are similar to those drawn from Equation 5.16. The present results are compatible with the mathematical model proposed by Karst and Hasan (1987).

Let us briefly summarize some of the predictions from the model and compare them to the experimental data (see also "Electromyographic Characteristics" in chapter 4).

For movements of equal amplitudes performed using a Speed-Sensitive strategy (at different speeds, different ω), an increase in the movement speed would *not* change the time to the maximal agonist EMG level or the total duration of the first agonist EMG burst, but the amplitude of the EMG burst would increase (Freund & Budingen, 1978; Lestienne, 1979; Shapiro & Walter, 1986; Mustard & Lee, 1987; Corcos et al., 1989). Therefore, the initial EMG slope would increase (Corcos et al., 1989; Gottlieb et al., 1989a).

The level of antagonist coactivation would increase with an increase in speed (Figure 1 in Lestienne, 1979; Figure 1 in Mustard & Lee, 1987). The antagonist burst would increase in both amplitude and duration (Freund & Budingen, 1978; Mustard & Lee, 1987; Corcos et al., 1989).

In cases of movements of different amplitudes performed with a constant ω (Speed-Insensitive strategy), an increase in the amplitude would bring about similar time courses of the agonist EMG rise although for

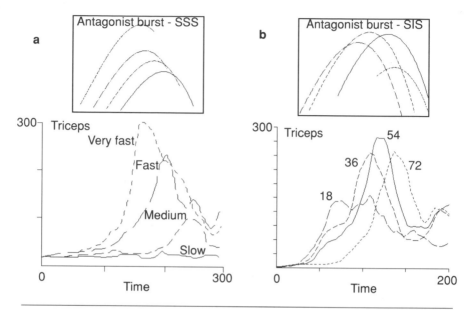

Figure 5.7 Computer simulations of the antagonist burst (above) and EMG recordings of the lateral head of the triceps during elbow flexion movements performed at different speeds over 54° (a: Speed-Sensitive strategy—SSS) and over 18°, 36°, 54°, and 72° "as fast as possible" (b: Speed-Insensitive strategy—SIS). *Note.* From "An Equilibrium-Point Model for Fast Single-Joint Movements: I. Emergence of Strategy-Dependent EMG Patterns" by M.L. Latash and G.L. Gottlieb, 1991, *Journal of Motor Behavior, **23**,* pp. 163-177. Reprinted with permission of the Helen Dwight Reid Educational Foundation. Published by Heldref Publications, 4000 Albemarle St., N.W., Washington, D.C. 20016. Copyright © 1991.

longer time periods, an increase in the peak agonist EMG, in the time-to-peak of the EMG, and in the total agonist first EMG burst duration (Wadman et al., 1979; Cheron & Godaux, 1986; Mustard & Lee, 1987; Gottlieb et al., 1989b).

Time course of the coactivation component of the antagonist EMG would not change. The maximal level of the antagonist EMG burst during the braking phase and the total duration of the burst would decrease with an increase in the movement amplitude for large movements and increase for short movements (Marsden et al., 1973; Brown & Cooke, 1984; Cheron & Godaux, 1986; Karst & Hasan, 1987; Mustard & Lee, 1987).

An increase in the moment of inertia associated with the Speed-Insensitive strategy would lead to an increase in movement time. This would be accompanied by an increase in the first agonist peak EMG level and total duration of the first agonist burst (Lestienne, 1979; Wadman et al., 1979; although Mustard & Lee, 1987 and Sherwood et al., 1988 report changes in the EMG amplitudes without visible differences in the timing).

Note that an increase in moment of inertia accompanied by instruction to the subject to maintain a constant movement time (Lestienne, 1979; Mustard & Lee, 1987; Sherwood et al., 1988) leads to necessary changes in ω (ω/c_ω = constant), that is, to the Speed-Sensitive strategy.

Initiation of an Isometric Contraction

Earlier, for isotonic movements, it was shown that agonist EMG (E) during the initial phase of the first agonist burst can be described by an equation:

$$E = t\omega_{ag}\sin\beta - \frac{t^3 A_0 \sin\beta}{6} - \frac{t^2 A_0 \cos\beta}{2}, \; 0 < t < \tau_{ag}, \tag{5.17}$$

where β is an angle between the Ω-line and the abscissa axis on the agonist V-L plane. Because matched isotonic and isometric contractions are kinematically identical during this time (assumption 6), Equation 5.17 applies to isometric contractions as well.

Measurement of joint angle, velocity, and acceleration shows that the kinematic trajectories of isotonic movements start to deviate from those recorded during isometric conditions approximately 40-50 ms after the beginning of the first agonist burst (see Figures 5.9 and 5.10 on pages 163 and 165 and compare with similar assessments by Lee et al., 1986). It will take about 50 ms before these changes can affect the EMGs due to the time delays in a hypothetical reflex arc mediating the tonic stretch reflex (Matthews, 1959; Houk, 1976, 1979; Feldman, 1979; Sinkjaer et al., 1988; Smeets et al., 1990).

EMG changes result from two factors: Central commands shifting λ_{ag} and proprioceptive feedback reflecting the current state of the muscle (cf. assumptions 5 and 6). The first, central, factor is considered to be identical for matched isotonic and isometric contractions (cf. assumption 1). Let us consider that in cases of identical kinematic trajectories, the proprioceptive signals are also identical, until there is a visible divergence in the kinematics. This is likely to be true for the signals from spindle afferents, especially if one takes into account that we consider the case of identical kinematics and identical central commands. The central variable λ describes general behavior of a muscle and is assumed to define descending drive to the α-motoneurons, γ-motoneurons, and interneurons in the reflex loops. Therefore, identical motor commands (cf. assumption 1) imply, in particular, identical descending γ-drives. As far as the Golgi tendon organs are concerned, one might presume that isometric force during this interval is matched by inertial force, and the elastic force component is low. Thus, both factors affecting the agonist EMG are considered identical during the first 40 ms. If one takes into account the delay in the hypothetical tonic stretch reflex arc, the EMG patterns of isotonic movements and matched isometric contractions should not differ significantly during the first 90-100 ms.

Therefore, since Equation 5.17 applies when $t < \tau$, that is, during approximately the first 40-50 ms, all the conclusions drawn from Equation 5.17 for isotonic movements can be applied to isometric contractions. In particular, for contractions performed under Speed-Sensitive strategy, time-to-peak of the first agonist EMG burst is independent of ω, that is, independent of the rate of force production; an increase in the rate of force production will be associated with a faster rise of the EMG and higher peak EMG levels (cf. Corcos et al., 1989, 1990a).

For contractions performed under Speed-Insensitive strategy (i.e., when ω is fixed), the rate of the EMG rise is independent of the contraction amplitude (cf. Sanes & Jennings, 1984; Ghez & Gordon, 1987; Gottlieb et al., 1989b; Corcos et al., 1990a).

Divergence of Isometric and Isotonic Patterns

Time-to-peak of the first agonist burst for short (in biceps for 36° elbow flexion) and fast movements (performed "as fast as possible") is approximately 40-50 ms (see Figure 6 in Gottlieb et al., 1989a; see also Figure 5.9), which in the model corresponds to τ. According to assumption 3, at this moment λ_{ant} starts to shift at a rate of ω_{ant}.

Up to the moment τ_{ag}, matched isotonic and isometric movements are kinematically identical and, by definition, have identical central commands. Only for time $t > \tau_{ag} + \Delta t$, where Δt is a delay due to the reflex loop signal conduction, will their EMG traces diverge.

During isotonic contractions, shortening of the agonist muscle leads to a decrease in spindle afferent firing rate, while lengthening of the antagonist muscle leads to an increase in spindle activity (Lennerstrand, 1968; Houk et al., 1981; Matthews, 1981). In the terms of the model, the current state for a shortening muscle on the V-L plane follows the Ω-line, thus leading to a decrease in its EMG level. For a lengthening muscle, the current state moves away from the Ω-line, thus increasing its EMG. During an isometric contraction, for $t \geq \tau_{ag}$, the current states for both agonist and antagonist muscles are fixed corresponding to a certain agonist muscle length and zero velocity (points B, Figure 5.8). In isotonic conditions, the agonist state corresponds to a shorter muscle with a non-zero speed of shortening (point F, left part of Figure 5.8) thus leading to a lower EMG level (compare E_F and E_B). For the antagonist, similar considerations show higher EMG level in isotonic conditions (compare points B and F and EMGs E_F and E_B in the right part of Figure 5.8).

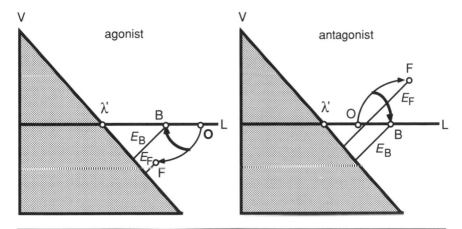

Figure 5.8   Imagine that a movement starts from a point O corresponding to certain values of agonist and antagonist muscle lengths. If at some moment of time t_0 the movement is unexpectedly blocked, the changes in muscle length will quickly cease. That is, at any time after t_0, the agonist EMG will be bigger and the antagonist EMG will be smaller in the blocked (B) trials as compared with the free (F) trials (cf. E_B and E_F).

Note. From "An Equilibrium-Point Model of Dynamic Regulation for Fast Single-Joint Movements: II. Similarity of Isometric and Isotonic Programs" by M.L. Latash and G.L. Gottlieb, 1991, *Journal of Motor Behavior*, **23**, pp. 179-191. Reprinted with permission of the Helen Dwight Reid Educational Foundation. Published by Heldref Publications, 4000 Albemarle St., N.W., Washington, D.C. 20016. Copyright © 1991.

Failure of the agonist muscle to shorten and of the antagonist muscle to lengthen under isometric conditions will lead to a relative increase in the agonist and a decrease in the antagonist EMG levels when compared with matched isotonic patterns for $t \geq \tau + \Delta t$.

Introducing Reciprocal Inhibition

Without taking into account possible effects of reciprocal inhibition, the agonist and antagonist EMGs for $t \geq \tau + \Delta t$ will depend on a number of factors:

- peak value of the EMG at $t = \tau$, the first members in Equations 5.18;
- muscle length changes due to the active acceleration phase. See F_1 and F_1^* in Equations 5.18;
- muscle length changes due to the active deceleration phase. See F_2 and F_2^* in Equations 5.18; and
- the shift of the antagonist Ω-line on the antagonist EMG due to the active deceleration. See F_2^* in Equations 5.18.

These factors can be summarized in the following equations:

$$E(t') = E_p - F_1 [A_0\tau, t'] + F_2 [A^*(t' + \Delta t)],$$
$$E^*(t') = E_p^* + F_1^* [A_0\tau, t'] - F_2^* [A^*(t' + \Delta t) + \omega_{ant}(t' + \Delta t)], \quad (5.18)$$
$$t' = t - \tau - \Delta t > 0,$$

where E_p is the peak value of agonist EMG, E_p^* is the peak value of antagonist EMG at $t = \tau$; F_1, F_2, F_1^*, F_2^* are monotonically increasing functions; $A_0 = \dfrac{\omega_{ag}}{c_\omega}$; $A^* = \dfrac{\omega_{ant}}{c_\omega^*}$ are constants of acceleration rise and fall.

Note that for isometric contractions, after the joint is fixed (point A, Figure 5.1), the agonist EMG will stay constant because both factors, central command and kinematics, are constant. This certainly does not correspond to experimental observations. This apparent discrepancy suggested introducing effects of reciprocal inhibition in the most simple and formal way. These effects have not been introduced earlier for reasons of simplification and because their absence did not lead to any obvious discrepancies with the experimental data.

If one assumes a monotonic relation between signals from α-motoneurons and EMGs, the reciprocal inhibition between the antagonist motoneuron pools will lead to:

$$E(t) \sim F[E^*(t - \partial t)],$$
$$E^*(t) \sim F^*[E(t - \partial t)],$$

where F and F^* are monotonically decreasing functions; E is the agonist EMG; E^* is the antagonist EMG; ∂t is a time delay in the reciprocal inhibition arc, which is close to the monosynaptic reflex delay; and $(t - \partial t) > 0$. For the purposes of simulation (see the next section), we are going to model the F and F^* functions as:

$$F = \frac{k}{k + E^*(t)} \text{ and } F^* = \frac{k}{k + E(t)}, \text{ where } k = \text{constant.}$$

Equations 5.18 will now look like this:

$$E(t') = F[E^*(t' + \Delta t - \partial t)] \{E_p - F_1[A_0\tau, t'] + F_2[A^*(t' + \Delta t)]\},$$
$$E^*(t') = F^*[E(t' + \Delta t - \partial t)] \{E_c^* + F_1^*[A_0\tau, t'] \quad (5.19)$$
$$- F_2^*[A^* (t' + \Delta t) + \omega_{ant} (t' + \Delta t)]\}.$$

Similar considerations lead to the following equations for the EMGs during an isometric contraction:

$$E(t') = F[E^*(t' + \Delta t - \partial t)] [E_p - F_3(A_0\tau, \Delta t) + \Delta E],$$
$$E^*(t') = F^*[E(t' + \Delta t - \partial t)] [E_c^* + F_3^*(A_0\tau, \Delta t) - F_4^*(A^*\Delta t) \quad (5.20)$$
$$+ \omega_{ant}(t' + \Delta t) - \Delta E^*],$$

where ΔE is a change in the agonist EMG due to the movement blocking; ΔE^* is a change in the antagonist EMG due to the movement blocking; and F_3, F_3^*, and F_4^* are monotonically increasing functions.

The results of the simulation are shown in Figure 5.9. Note that in both cases—(a) without reciprocal inhibition and (b) with reciprocal inhibition—there is an increase in the agonist EMG and a decrease in the antagonist EMG when the movements are blocked (thick traces) as compared with the unblocked isotonic movements (thin traces). This result corresponds to a qualitative prediction made earlier. Introducing reciprocal inhibition makes the EMG patterns more phasic, especially that of the agonist EMG after the movement is blocked (cf. upper thick traces) without any other qualitative changes. Note also the beginning of the second agonist burst apparent in B.

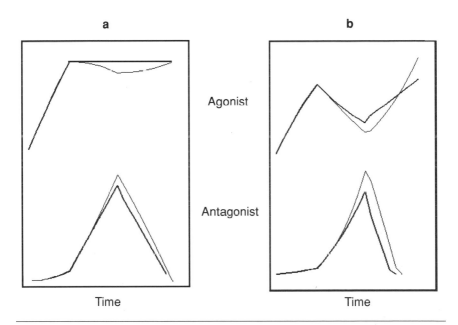

Figure 5.9 Computer simulations of the agonist and antagonist EMGs during blocked (bold lines) and free (thin lines) trials. The left graph shows the EMGs without taking into account the effects of reciprocal inhibition, while the right side shows the graphs incorporating the effects of reciprocal inhibition. Note more realistic agonist EMGs in the right graph.

Note. From "An Equilibrium-Point Model of Dynamic Regulation for Fast Single-Joint Movements: II. Similarity of Isometric and Isotonic Programs" by M.L. Latash and G.L. Gottlieb, 1991, *Journal of Motor Behavior*, **23**, pp. 179-191. Reprinted with permission of the Helen Dwight Reid Educational Foundation. Published by Heldref Publications, 4000 Albemarle St., N.W., Washington, D.C. 20016. Copyright © 1991.

Correspondence of the Model to the Data

Initiation of an Isometric Contraction

According to the model, there will be no differences between isometric and isotonic EMG patterns during the first 90 to 110 ms. The first half of this interval is for peripheral mechanical variables to diverge (cf. Lee et al., 1986), and the second is for reflex loops to affect muscle activity. We cannot be certain about the anatomical components of a hypothetical reflex arc that is not the monosynaptic reflex. If this is an equivalent to the tonic stretch reflex, EMG changes are unlikely to appear with a latency of less than 50 ms (Matthews, 1959; Feldman, 1979; Sinkjaer et al., 1988).

In our experiments with unexpected blocking of elbow flexion movements at the starting position (Latash & Gottlieb, 1991b), we have observed practically coinciding EMG trajectories during the first 100 ms of unperturbed and unexpectedly blocked isotonic movements. Similar data have been reported in other studies of unexpectedly blocked movements (Denier van der Gon & Wadman, 1977; Wadman et al., 1979; Lee et al., 1986) and of movements with unexpectedly increased inertial load (Brown & Cooke, 1981; Lee et al., 1986; Smeets et al., 1990).

EMG Divergence in Unexpectedly Blocked Movements

Figure 5.10 shows the EMG and kinematic traces for isotonic and unexpectedly blocked at the initial position (isometric) elbow flexions. Visible differences in the EMG patterns occur at times (t_d) about 100 ms after the beginning of the first biceps burst. Generally speaking, they represent an increase in the agonist (biceps and brachioradialis) and a decrease in the antagonist (lateral and long heads of the triceps) EMG (cf. Wadman et al., 1979; Lee et al., 1986; Smeets et al., 1990) as predicted from the model based on action of a hypothetical reflex loop mediating the tonic stretch reflex.

The tonic stretch reflex is not the only possible mechanism for these responses. Alternative explanations include long-loop reflexes or preprogrammed reactions (Desmedt & Godaux, 1978; Brown & Cooke, 1981; Bonnet, 1983; Marsden et al., 1983). Volitional modulation and even reversal of the preprogrammed reactions have been demonstrated following changes in the instruction (Evarts & Granit, 1976; Strick, 1978; Chan & Kearney, 1982; Bonnet, 1983). To dissociate between these two explanations, we trained our subjects to reverse movement direction from elbow flexion into extension when the movement was unexpectedly blocked. The earliest increase in the extensor EMG could be detected at 150 ms after the first visible increase in the agonist (flexor) EMG (Figure 5.11). However, even in these cases, an initial decrease in the extensor EMG in blocked trials (lateral and long heads of triceps, bold lines in Figure 5.11) is seen preceding the surge reversing the movement.

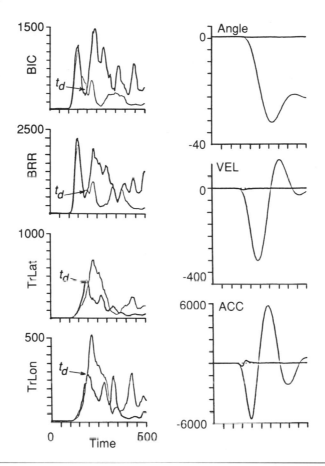

Figure 5.10 EMG and kinematic traces during elbow flexion movements "as fast as possible" over 36°. Part of the trials were blocked (bold traces). Note an increase in the agonist (BIC—biceps and BRR—brachioradialis) activity and a decrease in the antagonist (TrLat—lateral and TrLon—long heads of triceps) activity in the blocked as compared with the free trials. For quantitative assessment of the EMGs, their level was measured 50 ms after the time of their apparent divergence (t_d). EMG scales are in μV; kinematic scales are in degrees, °/s and °/s²; time scale is in ms.

Note. From "An Equilibrium-Point Model of Dynamic Regulation for Fast Single-Joint Movements: II. Similarity of Isometric and Isotonic Programs" by M.L. Latash and G.L. Gottlieb, 1991, *Journal of Motor Behavior*, **23**, pp. 179-191. Reprinted with permission of the Helen Dwight Reid Educational Foundation. Published by Heldref Publications, 4000 Albemarle St., N.W., Washington, D.C. 20016. Copyright © 1991.

The two explanations for the EMG changes are basically different. Preprogrammed reactions are a modification of or addition to the central motor command (see also "Preprogramming in Motor Control" in chapter 1). In our model, reactions arise from sensory-based (reflex) responses to changes in trajectory while the central motor command stays the same.

Termination of a Contraction

One of the interesting features of the EMG patterns in the latter part of unexpectedly blocked (isometric) contractions is pronounced reciprocal oscillation of the EMGs observed in several of the subjects (e.g., Figure 5.12). The frequency of these oscillations could be close to 10 Hz, which is close to the frequency of the physiological tremor (Lippold, 1971), or to 20 Hz.

Late in a movement ($t \geq t + t^* + \Delta t$), when the quantity ω_e in the second equation in Equations 5.20 has become constant, the EMG changes for isometric contractions will depend only on the final state of coactivation and reciprocal relations between the antagonist motoneuron pools. For isotonic movements, the EMGs will also depend on ongoing muscle length changes. This difference leads to a peculiar consequence.

Note that reciprocal relations like Equations 5.18 and 5.20 imply two possible final types of states: stable, in which both variables (E or E^*) converge to constants, which is common for reciprocal relations; and unstable, with oscillations. Precise analysis of systems like Equations 5.18 and 5.20 requires information about all the F functions, which are inaccessible. However, speaking qualitatively, the existence of considerable coactivation at the latter part of isometric contractions (cf. Ghez & Gordon, 1987; Corcos et al., 1990a) without any other external influence favors emergence of oscillations with a period characteristic for reciprocal reflex connections. In cases of isotonic movements, this possibility seems less likely since an "external" influence provided by changing muscle lengths (functions F_1, F_2, F_1^*, and F_2^*) plays a role in damping the oscillations.

One theory of physiological tremor is oscillation in the monosynaptic stretch reflex arc. Our assumption of imperfect fixation in isometric contractions allows slight changes in muscle lengths at the final contraction level and might be the cause of the observed EMG oscillations. Slight oscillation in the acceleration trace in the blocked trials can be seen in Figure 5.12 and is evident if plotted with a higher resolution.

Standard Motor Programs Can Lead to Different Peripheral Patterns

The reproducibility of peripheral patterns (kinematic and EMG) during the execution of standardized motor tasks has led to models in which

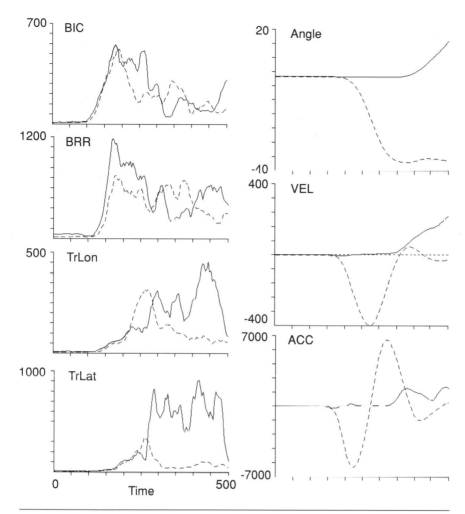

Figure 5.11   The subject was instructed to reverse the direction of elbow flexion movements only in blocked trials. This particular subject demonstrated the shortest time to the inversion. However, even in this case, an initial drop in the triceps EMG is apparent, preceding the burst that reversed the direction of the movement.

Note. From "An Equilibrium-Point Model of Dynamic Regulation for Fast Single-Joint Movements: II. Similarity of Isometric and Isotonic Programs" by M.L. Latash and G.L. Gottlieb, 1991, *Journal of Motor Behavior,* **23**, pp. 179-191. Reprinted with permission of the Helen Dwight Reid Educational Foundation. Published by Heldref Publications, 4000 Albemarle St., N.W., Washington, D.C. 20016. Copyright © 1991.

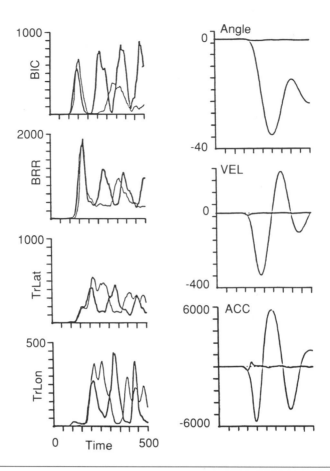

Figure 5.12   In some subjects, unexpected blocking of a voluntary fast elbow flexion resulted in pronounced EMG oscillations (bold traces). Scales and abbreviations are the same as in Figure 5.10.

Note. From "An Equilibrium-Point Model of Dynamic Regulation for Fast Single-Joint Movements: II. Similarity of Isometric and Isotonic Programs" by M.L. Latash and G.L. Gottlieb, 1991, *Journal of Motor Behavior*, **23**, pp. 179-191. Reprinted with permission of the Helen Dwight Reid Educational Foundation. Published by Heldref Publications, 4000 Albemarle St., N.W., Washington, D.C. 20016. Copyright © 1991.

central commands directly specify changes in signals to α-motoneuron pools (pattern-imposing models, see "'Physiological' Models" in chapter 4). However, the theoretical impossibility of absolute regulation of all inputs to and, therefore, signals from α-motoneurons has been stressed since the classical works of Bernstein (1935, 1967). An alternative approach is based on standard shifts of independently regulated variables (e.g., λ) which are specific for the subject's intentions but not for the peripheral conditions.

These two approaches lead to different interpretations of the same findings. For example, the decrease and eventual disappearance of the antagonist burst with a decrease in movement speed (Lestienne, 1979; Mustard & Lee, 1987) can be regarded as (a) "smart" action of the regulating system, which can predict that slow movement requires less braking torque for its termination or may even stop by itself, or (b) a result of a decreased speed of antagonist lengthening leading to a reflex decrease in the EMG for the same central program. Forget and Lamarre (1987) observed the antagonist burst in patients who lacked peripheral feedback (see also Hallett et al., 1975; Rothwell et al., 1982; Sanes et al., 1984). This burst was, however, considerably diminished, which could be a consequence of absent reflex actions while the patients still used the same motor programs.

Isotonic Movements and Isometric Contractions

From the pattern-imposing perspective, central commands are qualitatively different for isotonic movements and isometric contractions (Meinck et al., 1984; Ghez & Gordon, 1987; Tax et al., 1989, 1990); from the equilibrium-point perspective, fast isotonic movements and isometric contractions are regulated by similar shifts of the same central variables (Feldman, 1979, 1986; Latash & Gottlieb, 1991b). EMG patterns in isotonic and isometric conditions are indistinguishable until peripheral receptors can detect differences in trajectories and reflex loops can affect the EMGs. This pattern of the EMG divergence can be explained by hypothetical tonic stretch reflex arc action. Such observations are consistent with the idea that identical motor programs can be used for isotonic and isometric muscle contractions.

Tax et al. (1989, 1990) studied muscle activity during isotonic movements and isometric contractions at times when muscle forces were similar during both tasks. They described certain differences between the electromyographic patterns in these two regimes of muscle work and attributed them to differences in central motor programs. However, as originally suggested by Bernstein (1947, 1967; see also The Problem of Postural Stability in chapter 7), each motor task consists of two components, postural and related to movement itself. Most of the muscles studied by Tax et al. were likely to participate in both components. Therefore, the electromyographic differences might reflect the differences in requirements to maintain appropriate arm posture during isometric contractions and isotonic movements rather than the differences in central motor programs.

Muscle Coactivation

Coactivation of antagonist muscles from almost the beginning of fast contractions is well-known (Wadman et al., 1979; Lee et al., 1986). For isotonic movements, it is thought to stabilize movement (Karst & Hasan, 1987). However, it is also present in isometric contractions when it does not seem necessary to stabilize a trajectory. According to the equilibrium-point hypothesis, coactivation reflects one of the two basic commands (c) used for controlling both single-joint isotonic movements and isometric contractions. This view was indirectly supported by observations in monkeys of motor cortex cells whose activity was related to coactivation of antagonist muscles (Humphrey, 1982). Ghez and Gordon (1987; see also Gordon & Ghez, 1987a,b) favor specific control mechanisms for isometric contractions; however, in discussing the coactivation phenomenon, they propose lack of specificity of descending control mechanisms. I agree with this statement but think that this lack of specificity applies not only to the coactivation phase but to the whole motor program.

Comparison to Feldman's Approach

The model my colleagues and I have developed has many features in common with that of Feldman and his colleagues (Adamovitch & Feldman, 1984; Adamovitch et al., 1984; Abdusamatov & Feldman, 1986; Abdusamatov et al., 1987; Feldman et al., 1990a), but it is not identical. Both models describe the emergence of the EMG patterns during fast movements as results of muscle torque changes due to a centrally controlled equilibrium muscle length and actual length and velocity of its changes. One difference is that Feldman and his colleagues chose to use a different pair of control variables for describing single-joint motor control and the EMG patterns, the reciprocal and coactivation commands. However, r and c are defined as:

$$r = \frac{\lambda_{ag} + \lambda_{ant}}{2}, \qquad (5.21)$$

$$c = \frac{\lambda_{ag} - \lambda_{ant}}{2},$$

so any theory in terms of r/c is equally defined in terms of $\lambda_{ag}/\lambda_{ant}$.

The choice between r/c and $\lambda_{ag}/\lambda_{ant}$ and an attempt to make a parsimonious model have led Feldman and his colleagues to postulate a central control pattern that includes an early quick change in both control variables (r and c) without any explicit programming of the antagonist burst (cf. Figure 2.12a; Abdusamatov & Feldman, 1986; Feldman et al., 1990a). The antagonist burst occurs only as a result of antagonist muscle length changes. While such a scheme works, at least qualitatively, for many of the movement conditions we have considered, it predicts a smaller antagonist burst due to lower movement speed when the inertial load is increased. It also predicts that the latency of the antagonist burst will be sensitive to the trajectory of the movement. These predictions are inconsistent with experimental observations (Lestienne, 1979; Mustard & Lee, 1987; Gottlieb et al., 1989a). The possibility of independent central programming of the antagonist burst (by a delayed shift in λ_{ant}) *and* its dependence on the peripheral feedback in our model is supported by a variety of experimental and clinical observations (Hallett et al., 1975; Rothwell et al., 1982a; Cooke et al., 1985; Cheron & Godaux, 1986; Forget & Lamarre, 1987; Ghez & Gordon, 1987; Mustard & Lee, 1987; Brown & Cooke, 1990; Cooke & Brown, 1990).

Our approach here (cf. Figure 2.12b) is more straightforwardly described in terms of the original pair of controlled variables (λ_{ag}, λ_{ant}) and leads to nonmonotonic changes in both r and c variables. The experiments with reconstruction of virtual trajectories during single-joint fast movements (chapter 2) suggest that r and c do in fact change nonmonotonically during fast movements. Predictions of nonmonotonic virtual trajectories have also been made from theoretical models by Hogan (1984) and Hasan (1986a).

In our model, preserving a more independent control over antagonist muscles permits a large antagonist burst for slow movement arising from large loads and a small antagonist burst for kinematically similar but intentionally slow movements against a small load. Note, however, that the system's dynamics exerts its effects on all the EMG components, including the antagonist burst, through the feedback loops from the length- and speed-sensitive muscle receptors. In this respect, our approach is basically different from the pattern-imposing models described in the previous chapter and can be labeled pattern-generating.

Limitations of the Model

The model describes EMG patterns for movements with different speeds, over different amplitudes, and against different loads. However, it is obvious that it must have a limited range of applicability. For example, very slow movements (Freund & Budingen, 1978; Corcos et al., 1989) and movements with an unexpected stop (Hoffman & Strick, 1986) do not demonstrate an antagonist EMG burst. One should not expect the model, at least at the present stage, to be able to account for all of the experimentally observed EMG characteristics, taking into account the assumed cartoon kinematics and introduced assumptions and simplifications.

Two of our assumptions impose limitations on the range of movement speeds: (a) small values of τ as compared with total movement time, and (b) standard shift in λ_{ant}. The latter assumption implies that braking due to viscous properties of the effectors is insufficient for effectively decelerating the limb to rest. These limitations are not implicit in the λ-model but result from simplifications introduced in order to make the equations analytically solvable.

A general problem of motor control is whether the motor control system finds a unique program every time it is required to perform a coordinated movement, or whether it uses one of a few standardized subprograms with a limited number of variables to match it to the specific task (cf. Stein et al., 1988). In the first case, it is necessary to continuously monitor peripheral information and adjust the motor command accordingly. In the second case, the motor program is formulated in variables that take peripheral information into account automatically. For single-joint movements, the threshold of the tonic stretch reflex is such a variable. For multijoint movements, certain characteristics of the "working point" are likely to represent these variables (Bernstein, 1935, 1967; Berkinblit et al., 1986a,b,c; Flash, 1987; Mussa Ivaldi et al., 1989; see chapter 7). The present model is based on the second approach; one of its goals is to unify isotonic and isometric contractions as consequences of a standard central motor command.

In the next chapter, we are going to discuss the issues of motor variability and motor learning. This analysis will force us to reconsider and go more deeply into some of the general issues of motor control, in particular the notion of "motor program." We are going to start with the most general question: "Is there such a thing as motor program?"

Chapter 6

Issues of Variability and Motor Learning

Two views on motor program coexist. According to the first one, there is an abstract representation for a future movement that includes the order of events, their temporal structure, and level of performance ("strong" or "weak"). According to the second one, motor program does not exist, and movements are manifestations of self-organizing processes within a nonequilibrium system that are induced by imposing a new order upon the motor system. Different aspects of motor variability and motor learning can be ascribed to processes at different levels of a hypothetical system for movement production. Experimental manipulations can lead to different kinds of speed-accuracy trade-offs by suggesting certain strategies to the subject.

Is There a Motor Program?

In the previous chapters, I have used such expressions as "descending control signals," "motor command," "central program," and so on without really defining any of them. These expressions sound like lay terms and create a false feeling of understanding. In fact, it seems more or less obvious that there is some sequence of movement-related signals ("motor command") coming from some brain structures ("descending") giving rise to a voluntary movement and in some way coding its features ("central program"). However, there is still no consensus on what "motor program" is, how and where it is created, how it is implemented by the so-called lower structures, and even if it exists at all. Until now, we have mostly been dealing with the question: "How is the motor program implemented?" In this chapter, I will start discussing (and will continue this discussion in chapter 10) the first half of the problem: "What is motor program?"

Let us return to the scheme used in the first chapter (Figure 6.1, cf. Figure 1.2) that shows a homunculus, sitting somewhere in the brain and generating motor programs that are then correctly understood and implemented by a lower level (level 2 in Figure 6.1) and eventually lead to desirable behavior of the peripheral motor apparatus (level 3). Even this scheme can already be a point of controversy. There are at least two basically different approaches. One of them discusses where and how the motor program is generated while the other rejects the notion of motor program as misleading.

Engrams and Synergies

The first approach originates from the classic works of Bernstein (1935, 1947, 1967) and is based on his original concepts of synergies and engrams.

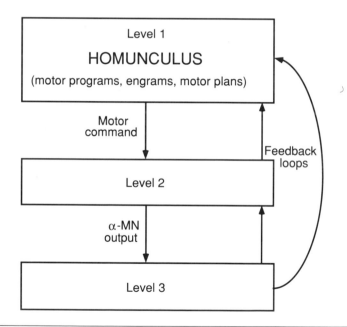

Figure 6.1 A tentative scheme of voluntary motor control. A smart homunculus knows a lot of motor programs (or engrams) from which it can generate appropriate motor plans (cf. generalized motor program). As a result, a motor command is generated to the intermediate level, which processes the descending signal, taking into account the current conditions of movement execution (peripheral information). The output of the second level is activity of motoneuron pools, which leads to changes in muscle activation patterns, muscle forces, and/or joint angles leading to a movement.

Synergies are time sequences of control signals to groups of muscles that lead to simple coordinated motor acts. These simple acts are used as bricks for building more complex movements. Engrams are prototypes of planned movements in abstract variables and abstract time, and their sequences can be considered motor programs. There are basic engrams corresponding to synergies that may be parts of more complex engrams for voluntary movements. Bernstein thought that the engrams were formed in certain brain structures, transmitted to lower (spinal and supraspinal) levels of the central nervous system, and eventually realized by the peripheral apparatus. The engrams were supposed to be continuously updated with the help of feedback loops from both peripheral receptors and intermediate levels of signal processing within the central nervous system (cf. Prablanc et al., 1986; van der Meulen et al., 1990).

Several solutions have been proposed for the hypothetical variables in which the engrams are formulated. There have been attempts to directly associate the notion of motor program with patterns of muscle activation (Hallett et al., 1975; Lestienne, 1979; Wallace, 1981; Bizzi et al., 1982; Enoka, 1983; Gottlieb et al., 1989a). In these studies, differences or resemblances in EMG patterns were equated with differences or resemblances in hypothetical motor programs. Such attempts originate from the α-model (chapter 1, "The α-Model"), which has been mostly based on studies of deafferented animals (Bizzi et al., 1976; Bizzi, 1980). As a result, the feedback components of the motor system were considered auxiliary to a primary motor program that directly specifies which muscles to activate, to what extent, and when. However, results of experiments on deafferented animals may not be directly related to control of voluntary movements in animals and humans with an intact nervous system. This simplified concept of motor program does not seem very practical because it requires separate programs for virtually every movement in a human's life because of the changing external conditions of movement execution (e.g., relative direction of the force of gravity). Many experiments studying multijoint movements (see Chapter 7) have demonstrated poor reproducibility of individual muscle forces or even joint angles during consecutive attempts at executing the same task (Bernstein, 1935, 1947; Georgopoulos et al., 1981; Flash & Hogan, 1985; Berkinblit et al., 1986a; Georgopoulos, 1986). In particular, Bernstein (1935), in his studies of highly automatic professional movements (e.g., hitting an anvil with a hammer) has noted much more reproducible trajectories of the "working point" (head of the hammer) as compared with the trajectories of individual joints. Patients with polyneuropathy or neuropathy of large sensory fibers (which can be considered a model of profound deafferentation) can perform coordinated movements, but even the simplest single-joint movements are accompanied by significantly changed EMG patterns (Sanes et al., 1985; Forget & Lamarre, 1987). These observations suggest that intact afferent loops are an important part of the system of motor control and contribute to the patterns of muscle activation in healthy subjects.

An alternative to the α-model approach, the λ-model, which is advocated in this book, originates from classic works by Sherrington (1908), who considered voluntary movements to be the consequences of modulation of parameters of muscle reflexes. Engrams for single-joint movements can be identified with functions $r(t)$ and $c(t)$ (see chapters 1 and 2). We have already discussed major differences of the equilibrium-point hypothesis from controlling patterns of muscle activation (chapter 5). Let me emphasize here that limb deafferentation leads, in this framework, to the impossibility of using the "normal" way of controlling movements with learned engrams and forces the animal to search for new ways to control movements in the absence of reflexes. In this case, there is no other choice than directly regulating activity of the α-motoneuron pools, and λ-control is reduced to α-control.

Generalized Motor Program

Schmidt (1980a,b, 1988) has developed a concept of generalized motor program that is formulated in a more abstract way. This concept is very close to the original ideas of Bernstein. It implies the existence of an abstract representation for a future movement that includes the order of events, their temporal structure, and level of performance ("strong" or "weak"). In particular, a motor program can be realized within different time periods.

An elegant series of experiments supporting this aspect of Schmidt's concept was performed by Viviani and Terzuolo (1980), who studied professional typists performing at different speeds. Typing a fixed sequence of characters several times revealed temporal patterns that became amazingly similar after a simple scaling procedure. This observation suggests that different temporal patterns were the result of playing back at different speeds an original, template pattern, formulated in some abstract form and in an abstract time. Later, the possibility of realization of a motor program at different speeds was corroborated in a number of studies (Soechting & Lacquaniti, 1981; Atkeson & Hollerbach, 1985; Ojakangas & Ebner, 1991).

Another aspect of the concept of generalized motor program based on impressive classic observations of Bernstein (1967) and supported later by Raibert (1977) was that similar patterns can be observed during cursive writing with different joints (wrist or elbow), limbs (arm or leg), or even with a pen in one's mouth. Those classic pictures of the Russian word "Coordination" and the phrase "Able was I ere I saw Elba" have been reproduced in a number of publications (Gelfand et al., 1971a; Keele, 1986). Bernstein and Raibert argued that the striking similarity of those examples of cursive writing was a strong argument for an abstract representation of a motor program that can be realized with different effectors. In other words, this is an example of a transfer of a motor program to a different effector apparatus.

Taking a risk of being accused of blasphemy, let us argue that those classic figures are not absolutely persuasive. Handwriting is certainly automatic and specific for a given individual. During acquisition of this skill, people usually practice it with their dominant upper limb using either wrist (writing on a desk) or elbow and shoulder (writing on a blackboard). So, it seems that demonstration of individual specific features of cursive writing with a pen attached to one's foot or in one's mouth is a very strong argument for an abstract central representation of this skill. However, let us dissociate cursive writing (which is certainly highly automatic) from copying a picture (which is presumably not). A person can draw with a pencil attached to the foot, although the results are generally poor. Performing a drawing task must be clearly dissociated from using "the same motor program" as during writing with the dominant hand. So, the two possibilities are that

- a subject is actually using the same unique, individual generalized motor program for writing with different effectors; or
- a subject can use this program only with the effectors that were used during its acquisition, and when asked to perform the same task with a different effector apparatus, she draws a picture similar to the one she has in her mind or in front of her eyes.

A simple experiment can be performed to illustrate these two possibilities. Figure 6.2 shows a short phrase written by a subject with the right (dominant) hand (sample 1), left hand, right leg, left leg, and mouth. Another subject (not a professional artist) was asked to copy the original phrase with the same set of effectors. Note that only the first subject was supposedly skilled at reproducing his own handwriting, while the second subject was just copying a picture. However, I don't think that it is easy to say which samples in Figure 6.2 were "written" by the first and which were "drawn" by the second subject. The members of our laboratory were able to correctly match the samples with the subjects in less than 50% of the cases. Such an identification may require quite a complicated graphological analysis. The bottom line is that similarity of cursive writing patterns may well be a consequence of drawing skill and not of transfer of a motor program to different effectors.

Dynamic Pattern Generation

An alternative approach has been developed during the last decade (Kelso, 1981, 1984; Kay et al., 1987; Saltzman & Kelso, 1987; Kelso & Schoner,

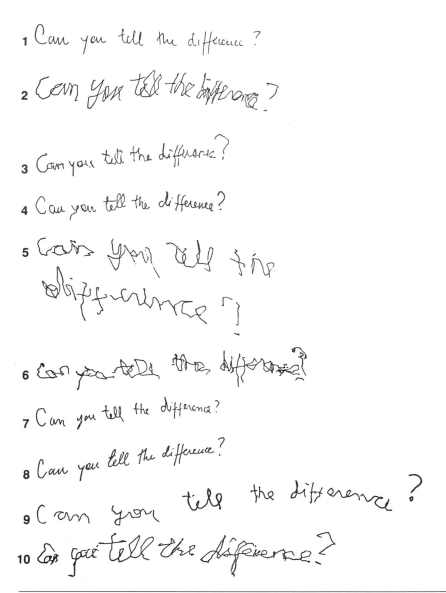

Figure 6.2 Sample 1 was written by subject A with the right (dominant) hand. Four more samples were written by the same subject using the left hand, right foot, left foot, and the mouth. Five other samples were drawn by subject B using the same set of effectors (right and left hand, right and left foot, and the mouth). Subject B was trying to copy the upper template sample. Members of our laboratory were able to identify which samples were written by which subject in less than 50% of cases. Try to beat their results!

Subject A wrote samples 2, 5, 7, and 9. Subject B drew samples 3, 4, 6, 8, and 10.

1988; Schoner & Kelso, 1988; Turvey et al., 1988; Kugler et al., 1990; Schmidt et al., 1990; Scholz & Kelso, 1990; Schoner, 1990; Turvey, 1990b; deGuzman & Kelso, 1991; Kelso et al., 1991). It is based on the ideas of dynamics of nonequilibrium systems and considers movements as manifestations of self-organizing processes leading to stable coordinative structures that demonstrate a limit cycle behavior. Motor behavior is considered to be implemented by imposing a new order on the system. As a result, external movement patterns depend on both motor task and initial state of the system. If a motor system (including the central control structures) needs to come to a certain new state defined by a motor task, it may demonstrate different trajectories both in its state space and in the external space, all converging on an "attractor." This process can be described and analyzed with the methods of stochastic nonlinear dynamics. Although the physical and mathematical apparatus of this approach is rather sophisticated, it can be illustrated with a simple example.

Imagine that a person sitting in a chair is given a task to pick up an object from a table. The person needs to stand up, make a few steps, and take the object with his dominant hand. Repetition of the same task in seemingly identical initial conditions will lead to different trajectories of all the joints and even the working point (the dominant hand). In the framework of a generalized motor program, there is no need for the subject to use different sequences of motor commands in different trials. So, the variability in the performance will probably be explained with the help of nonideal reproducibility of the external conditions, and some kind of noise inherent to the nervous system. The dynamic pattern generation approach has a straightforward explanation for this variability. The motor control system (in a general sense; i.e., including all the central neural structures) is considered to be essentially nonequilibrium, that is, continuously fluctuating. Even if the subject's initial position and external conditions are ideally reproduced from trial to trial and the subject is absolutely relaxed (note that both these conditions are experimentally unattainable), the state of his nonequilibrium motor control system fluctuates, although this can happen without visible external effects (e.g., changes in posture). Therefore, the initial conditions cannot be reproduced even theoretically. Trajectories of the motor control system to a final equilibrium state (attractor) defined by the motor task will be different corresponding to different initial conditions. This will be manifested by different external trajectories on the body and the limbs. This simple example is, however, not so simple to analyze experimentally because of too many variables that need to be monitored. Experimental support for the dynamic pattern generation is based mainly on studies of two-limb interactions (Kelso et al., 1979a,b, 1983; Kelso, 1984; Kay et al., 1987, 1991; Kelso & Schoner, 1988; Schoner & Kelso, 1988; Schmidt et al., 1990; Schoner, 1990; deGuzman & Kelso, 1991) and will be discussed in the next chapter.

This approach looks basically different from the notion of generalized motor program. It does not even have an analogue for the motor program. After the task is defined, the motor control system goes to the final state "by itself" according to the laws of stochastic nonlinear dynamics.

I accept, in this book, a dynamic approach to generation of voluntary movements, that is, consideration of movements as results of interaction between the body and the environment and between different subsystems within the body. However, I have also extensively used time functions of certain hypothetical central variables for describing how single-joint movements are controlled, which suggests a relation to the notion of generalized motor program. Let us try to reconcile the two approaches, although this attempt is likely to provoke a strong negative reaction from the adherents of both views. For this purpose, let us once again redraw the general scheme of motor control (Figure 6.3; cf. Figure 6.1; see also Latash & Gutman, 1992).

The first axiom is: There is always a motor task for a purposeful voluntary movement. This task is defined either by an experimenter or by the

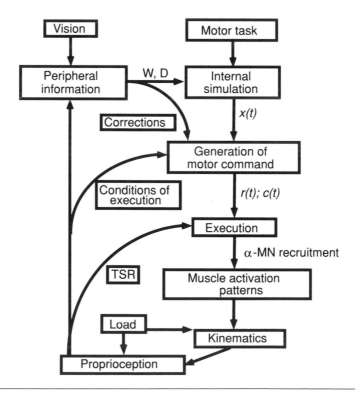

Figure 6.3 A general scheme of production of voluntary movement. Three major steps include internal simulation, generation of motor command, and execution.

subject's will. Bernstein was the first to suggest that planning voluntary movements is performed in terms of kinematics in the external Cartesian space (Bernstein, 1936; see also chapter 7). That is, the "goal" of the planned movement is expressed in terms of its trajectory. For this purpose, the system should be able to perform an "internal simulation" of a planned movement taking into account the conditions of movement execution, including predictable external load changes, and generate a function that would directly reflect the whole desired trajectory (and, consequently, amplitude and velocity) of the future movement. In a recent study, Simonetta et al. (1991) have shown that *Bereitschaftpotential* ("readiness potential"), which is seen in motor cortex prior to a relatively complex, sequential voluntary movement, depends not only upon the features of the first simple subcomponent of the movement but also upon the remainder of the motor task. This means that the whole planned movement is reflected in the brain's preparatory activity, tentatively corroborating the idea of internal simulation.

The second step is to translate the simulated trajectory into variables interpretable by the lower structures. For a single-joint movement this is equivalent to generation of time functions $r(t)$ and $c(t)$. These functions (or commands) will lead to changes in muscle activation and to a movement that, in the ideal case, should exactly follow the simulated trajectory.

Until now, the description better fits Bernstein's hierarchical scheme. Let us, however, consider $r(t)$ and $c(t)$ as defining a trajectory of an attractor. As a result, these functions do not directly encode properties of the movement, even when the external conditions of movement execution are ideally reproduced, but certainly are major factors in defining them (along with initial state of the system). So, on the one hand, $r(t)$ and $c(t)$ can be considered engrams of a single-joint movement or reflections of a generalized motor program (cf. with the notion of virtual trajectory; see chapter 2, "Static Experiments"). This view suggests that they partially encode certain properties of the movement, including patterns of transition from an initial to a final position. On the other hand, the same two functions can be considered as specifying only a trajectory of an attractor for the motor control system, and transitions from one state to another occur stochastically according to the rules of nonlinear dynamics.

Most of the studies within the dynamic pattern generation approach do not specify the level at which the formation of the patterns takes place. They rather consider the whole system of motor control as a one-step process of pattern generation based on motor task and external conditions of movement execution. However, a number of dynamic pattern generation systems can form a hierarchy so that output of an "upper" level is an input to a "lower" level. In a hypothetical hierarchical system for production of voluntary movement (Figure 6.3), each level may represent a nonequilibrium system generating output patterns processed by the lower levels that also represent nonequilibrium systems. Output patterns of each level will be only probabilistically defined by the input. For example, the output of the level of "internal simulation" will ambiguously depend on motor task, and the $[r(t), c(t)]$ command variables may be different for the same ideal trajectory and external loading conditions. These patterns, at a certain intermediate step, may be considered analogous to the Bernstein notion of motor program. At some level, the differences between the two approaches become intangible.

Variability of Single-Joint Movements

The notion of variability can be introduced in two ways. The first considers deviations of movement characteristics from those observed during a certain "ideal" or "average" performance. For example, this notion can be applied to locomotion for describing different patterns (kinematic, kinetic, and/or EMG) observed during consecutive steps, or to any other repetitive action. The second approach requires introduction of a new notion of "target" that imposes explicit or implicit restrictions on changes in a certain parameter of the performance (e.g., final position, movement time, or others). In this case, variability describes how successful (or, rather, unsuccessful) the subject is in complying with the restrictions imposed by the target. In general, variability is an antonym of reproducibility, and is a curse of motor control studies that do not investigate it explicitly. Reducing variability is frequently the goal of training procedures in both "real life" (including training athletes or rehabilitation of patients with motor disorders) and laboratory environments. On the other hand, variability by itself is a fascinating phenomenon that obeys its own laws and demonstrates consistent relations between task and performance parameters. Within the framework of dynamic pattern generation (see the previous section), variability is considered an inherent part of the motor control mechanism (Mpitsos, 1990).

Two Types of Variability

According to the scheme shown in Figure 6.3, the major steps in generation of voluntary motor command are "internal simulation," "generation of motor command," and "execution." Output of the execution step leads to activation of α-motoneuron pools and, eventually, to limb kinematics. Instruction to the subject is a component of the motor task. For example, if there is an explicitly presented target, its dimensions (W) and amplitude of the future movement (D) are supplied by the "peripheral information" and affect the step of internal simulation.

Let us consider only very fast movements that are supposedly preprogrammed by the subject and no corrections are introduced at the levels of internal simulation and generation of motor command. Target width

(W) acts only at the level of internal simulation where the future movement trajectory $x(t)$ is planned. Therefore, if there is a dependence of a certain parameter of performance (e.g., peak velocity (V) or movement time) on W (or a ratio W/D), its mechanism should be described at this level.

It is also conceivable that, at each level, there is some inherent noise leading to suboptimal processing of the incoming information. As a result the output of internal simulation $x(t)$ does not exactly correspond to the optimal planned trajectory, $r(t)$ and $c(t)$ do not ideally correspond to $x(t)$, the mechanism of execution involving the hypothetical tonic stretch reflex is not 100% reproducible, and, finally, the external conditions of movement execution vary slightly from trial to trial. Each of these factors will contribute to the observed variability, in particular to the actual variability in final position (movement amplitude). Since movement speed (V) can theoretically be reflected at each level, the dependence of variability in movement amplitude on movement speed (speed-accuracy trade-off) can be defined by many factors related to the mechanisms acting at each of the levels.

So, we have come to two conclusions that are in fact axioms (cf. Latash & Gutman, 1992):

- *The dependence of movement speed on target width (W/V trade-off) emerges at the level of internal simulation.*
- *The dependence of dispersion of the final position on movement speed (V/D trade-off) gets contribution from each of the levels shown in Figure 6.3.*

The mechanism of the W/V trade-off is just one of the factors influencing the V/D trade-off, and it does so after a long processing. Therefore, the results of the experiments that study changes in movement speed induced by manipulations with the target size correspond to the W/V trade-off. Experiments that study changes in the dispersion of final position as a function of movement speed correspond to the V/D trade-off. These two groups of experiments study different aspects of motor variability that occur at different levels of the hypothetical control scheme and, therefore, should not be directly compared.

Speed-Accuracy Trade-Offs

We are now going to briefly review certain aspects of motor variability. Extensive reviews on this topic have been published (Keele, 1986; Meyer et al., 1990).

The most famous general law of motor variability is probably Fitts' law, which is sometimes called the *speed-accuracy trade-off*. It was originally introduced for describing variability of repetitive (Fitts, 1954) and, later, discrete arm movements (Fitts & Peterson, 1964) performed under visual control with explicitly presented targets. Target size was assumed to restrict the variability and therefore be its measure. Based on some of the aspects of information theory (Shannon, 1948), Fitts suggested a logarithmic relation between movement time (T), movement distance (D), and target width (W):

$$T = a + b \log_2 \frac{2D}{W} \qquad (6.1)$$

where a and b are empirically defined constants. Note that Fitts' law is a typical W/V trade-off in the framework introduced in the previous section, although visual control implies possibility of corrections during the movements, which makes the previous analysis inapplicable.

Since the original works by Fitts, logarithmic relations like Equation 6.1 have been used for describing experimental findings in a variety of conditions with movements of different complexity involving different joints (Knight & Dagnall, 1967; McGovern, 1974; Flowers, 1975; Langolf et al., 1976; MacKenzie et al., 1987) including, in particular, single-joint movements (Crossman & Goodeve, 1983; Corcos et al., 1988; Meyer et al., 1988). Analysis of variability of movements performed by aged subjects (Welford et al., 1969) or under water (Kerr, 1978) has led to a more general formulation of Fitts' law separating its distance and target size components:

$$T = a + b_1 \log_2 D + b_2 \log_2 \frac{1}{W}, \qquad (6.2)$$

where b_1 and b_2 can be different constants.

Some of the experimental findings failed to obey the logarithmic relation (Ferrell, 1965; Kvalseth, 1980) and suggested another form of the trade-off:

$$T = a + b \left(\frac{D}{W}\right)^p, \qquad (6.3)$$

where $0 < p \le 1$.

The later interpretations of Fitts' law implied an important role of visual control for guiding the movements (cf. Keele, 1968, 1986). Therefore, it is supposed to be applicable only for those movements when there is enough time for the subjects to introduce corrections based on the visual feedback. Estimations of the shortest time necessary for a visual correction vary considerably in different experiments, from the lowest estimate of 135 ms (Carlton, 1981) up to 200 to 250 ms (Keele & Posner, 1968) or even 290 ms (Beggs & Howart, 1970). The lowest estimates are, however, likely to reflect the minimal time from detecting that a correction is necessary (decision-making moment) to actually implementing it, while the higher estimates include the time from movement initiation to decision making.

Studies of very fast force pulses or movements have revealed a different relation between movement time, movement distance, and variability assessed as the standard deviation (SD) of dispersion of final position or force level (Schmidt et al., 1978, 1979; Sherwood & Schmidt, 1980; Wright & Meyer, 1983):

$$SD = a + b\,\frac{D}{T}\,,\qquad(6.4)$$

where a and b are constants. Note that this is already an example of V/D trade-off and should not be compared to Fitts' logarithmic relation, Equation 6.1. Later, linear relations (Equation 6.4) have been demonstrated for movements at relatively slow speeds (Zelaznik et al., 1988). Relations of the Equation 6.4 type were observed in a wide range of muscle forces. However, at forces more than 60% to 70% of the maximal voluntary contraction force, the linear relation breaks in a rather unexpected way: An increase in force leads to an increase in accuracy (or a decrease in variability, Schmidt & Sherwood, 1982). We shall discuss this seemingly paradoxical finding in the next section.

Interpretations of Equation 6.4 have been mostly based on an assumed increase in variability of muscle forces with an increase in absolute values of muscle forces that are supposed to occur at higher movement speeds. Newell and Carlton (1985; see also Newell et al., 1984) have demonstrated that force variability does increase with muscle force in isometric conditions indirectly corroborating this assumption. However, during isotonic movements, a nonlinear relation between joint torque and its variability (Sherwood et al., 1988) and between variability of movement time with an increase in movement amplitude (Newell et al., 1984) has been reported. A variety of EMG and kinematic parameters have been shown to change nonlinearly with movement speed and amplitude by Carlton et al. (1985).

Recently, Meyer and his colleagues have advanced a "dual-submovement model," which tries to reconcile different expressions for the speed-accuracy trade-off (Meyer et al., 1988, 1990). In my opinion, such an attempt is doomed because different types of the trade-off (W/V and V/D) have basically different underlying mechanisms. The model of Meyer is based on a number of assumptions. The most important of them are these: (a) Aimed movement consists of several submovements; each consecutive submovement tries to improve accuracy of the whole movement. (b) There is "motor noise," which may lead to inaccurate submovements and thus require corrections. (c) Each submovement is organized in such a way that the total movement duration is minimized.

So, when a subject initiates an aimed movement, he or she preplans the first movement component (the first "submovement"). In the course of the movement, the subject assesses its possible outcome. If the movement is likely to be successful, no corrections (further submovements) are introduced. If, however, the target is likely to be missed, the subject introduces a correction (a second submovement). Generally speaking, any number of corrections can be used during one movement. Note that the dual-submovement model implies presence of continuous feedback (e.g., visual) to the subject during the course of the movement, which informs him or her of the possibility of success or failure. The importance of submovements in accuracy-constrained three-dimensional movements has recently been suggested by Milner and Ijaz (1990). They have shown that a relative increase in duration of the deceleration phase with an increase in the task difficulty (putting a peg into a hole) was accompanied by irregularities in the decelerative phase trajectory, which included small changes in the direction of the hand path, suggesting an increase in the number of submovements.

Analysis of variability in the framework of the model has led Meyer et al. (1990) to a relation similar to Equation 6.3 with p depending on the number of submovements for a particular task. An increase in the number of submovements leads to a decrease in p, so that for movements consisting of only one submovement $p = 1$, and a linear relation is observed similar to Equation 6.4. Presence of only one submovement means that visual feedback has not been used for movement corrections, which is likely to happen during very fast movements or force pulses (cf. Schmidt et al., 1978, 1979). For movements consisting of two submovements, a square-root relation was predicted and experimentally observed (Meyer et al., 1988):

$$T = a + b \sqrt{\frac{D}{W}}. \qquad (6.5)$$

Slow movements performed under continuous visual control are likely to consist of many submovements leading eventually to the nearly logarithmic relation of Fitts.

An idea that adjustments of a motor program based on efferent and/or afferent information are contributing to the accuracy of movements (and, therefore, affecting the speed-accuracy relationship) has been analyzed recently by van der Meulen et al. (1990). In their study, it has been shown that a large relative variability in the initial acceleration is accompanied and partially compensated by a strong negative correlation between the initial acceleration and the duration of the acceleration phase (cf. Newell et al., 1982; Sherwood, 1986). Additional decrease in the relative variability in the distance moved happened during the deceleration phase. Although van der Meulen et al. did not use the term *submovements* for the corrections, their results are compatible with the dual-submovement model.

The empirical success of the dual-submovement model and the possibility of accounting for a variety of data make the model attractive despite its flawed attempt to reconcile W/V and V/D trade-offs. Testing the dual-submovement model in various experimental conditions does not look easy because experimental detection of submovements is very ambiguous. Submovements are, by definition, consequences of central corrections and can be illustrated by separate shifts of a central variable (for example, r). One can use some kinematic index, for example, the number of positive or negative acceleration peaks, as a tentative measure of the number of submovements, but any such measure can be criticized (consider, as an extreme case, movements performed on the background of tremor).

Variability in a Kinematic Model

Let us discuss motor variability using a kinematic model suggested recently by Gutman and Gottlieb (1989; see also "Kinematic Solutions" in the next chapter and Latash & Gutman, 1992). The model of Gutman is based on an idea of nonlinear "internal time" and eventually leads to the following formula for the simulated movement trajectory:

$$x(t) = D \exp\left(-\frac{t^{\alpha}}{\tau_{mov}}\right), \tag{6.6}$$

where D is movement distance, α is a constant, and τ_{mov} is a "movement time constant" reflecting the planned movement time (MT). Analysis of simulated movement trajectories for different values of α has demonstrated a good fit to the experimental data at $\alpha = 3$.

According to Fitts' law, MT increases with an increase in the ratio of distance to target width (D/W). Assume that a subject originally plans to perform a movement within a MT corresponding to some value of $\tau_{mov} = \tau_{mov}{}^0$. Now let us introduce an axiom:

Visual information on the target width and movement amplitude leads to a change in τ_{mov} defined by the value of the D/W ratio.

In the first approximation, this assumption can be modeled as a linear relation between the corrected value of τ_{mov} and D/W:

$$\tau_{mov} = \tau_{mov}{}^0 \left(1 + \frac{D}{W}\right). \tag{6.7}$$

Experimental definition of MT is an ambiguous task, mostly because of the difficulties in determining the moment of movement termination. A common although arbitrary approach is based on an assumption that the movement terminates at some moment when the endpoint achieves some location close to the target. Using this definition, MT can be expressed from Equations 6.6 and 6.7, assuming $\alpha = 3$, in the form

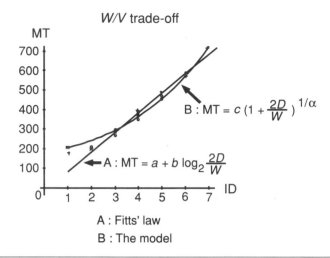

Figure 6.4 Original Fitts' data fitted with a straight line (Fitts' law) using two parameters, *a* and *b*. A simulated curve is also shown based on a kinematic model using only one parameter *c*. ID—index of difficulty, MT—movement time in ms.

$$\text{MT} = \sqrt[3]{-\tau_{\text{mov}}^{0} \ln \xi} \ \sqrt[3]{1 + \frac{D}{W}}, \tag{6.8}$$

where ξ is a small positive value.

Equation 6.8 is an expression of the W/V trade-off different from the logarithmic relation of Fitts. Figure 6.4 shows a curve corresponding to Equation 6.8 (bold line) superimposed on the original Fitts' data. Although Equation 6.8 is monoparametric (with the only parameter τ_{mov}^{0}) it fits the data at least not worse than the original two-parametric Fitts' curve (thin line in Figure 6.4). Fitting the Fitts' data with a power function (Ferrell, 1965; Kvalseth, 1980; Gan & Hoffman, 1988) has led to the best results for the index of the power in the range of $\frac{1}{2.6} \cdots \frac{1}{3.2}$. It is consistent with our value of $\frac{1}{3}$.

If a subject is reproducing an aimed movement several times, standard deviation of the instantaneous position of the endpoint at different times during the movement is due to a number of factors, including errors in setting the time constant. This is a V/D trade-off, according to our terminology. Let us denote standard deviation of instantaneous endpoint position defined by the error in the timing constant as $SD_{\text{tr.-off}}$. From Equation 6.6, one gets the following expression for the coefficient of variation of this component:

$$\frac{SD_{tr.-off}}{D} = \frac{SD\ (\tau_{mov})}{\tau_{mov}}\ \frac{t^3}{\tau_{mov}}\ \exp\left(-\frac{t^3}{\tau_{mov}}\right), \tag{6.9}$$

where SD (τ_{mov}) is standard deviation of τ_{mov}.

Figure 6.5a shows a trajectory of a simulated reaching movement $x(t)$ with two curves (dotted lines) corresponding to $x(t) + SD_{tr.-off}$ and $x(t) - SD_{tr.-off}$. An error curve corresponding to Equation 6.9 is shown in Figure 6.5b together with a simulated velocity profile. It has the following major features (see Latash & Gutman, 1992):

• The error curve starts with a lag after the velocity curve. This lag reflects the fact that the error is accumulated in the beginning of the movement. Note that a similar conclusion has been drawn by van der Meulen et al. (1990).

• The error curve has a maximum in the middle of the movement and then drops down. Van der Meulen et al. (1990) have reported a similar increase in the positional variability in the middle of the movement, which dropped during the deceleration period. They concluded that there was some compensatory mechanism correcting the errors accumulated during the first half of the movement by adjusting movement parameters during the deceleration period. Note, however, that in our model, this form of the error curve has emerged as a result of purely kinematic considerations without any compensatory mechanisms. Note also that "ellipses of variability" reported by Darling and Cooke (1987) are in good agreement with the positional error curve in Figure 6.5b.

• The value of this function at the beginning and at the end of the movement is zero. This means that the V/D trade-off due to the error in the timing constant exists only during the movement.

A "positional error" curve shown in Figure 6.5b illustrates that if one measures endpoint location at some distant moment of time (t_3), when the movement has presumably stopped, there will be no velocity-dependent positional errors (i.e., no V/D trade-off). If the measurement is made at some time before the movement has terminated (t_2), a V/D trade-off is likely to occur. It will be maximal at a time (t_1) somewhat after the moment of peak velocity. Note that in many experiments studying V/D trade-offs, movements were considered stopped and measurements were made at the moment of touching the target (Schmidt et al., 1978, 1979; Sherwood & Schmidt, 1980; Sherwood et al., 1988). The velocity of the endpoint at the moment of impact was presumably not zero (cf. with t_2 in Figure 6.5b). According to our model, this "minor" experimental detail could have led to a V/D trade-off that would otherwise be absent. The importance of the impact forces for the speed-accuracy trade-off has recently received support from a study of pointing movements by Teasdale and

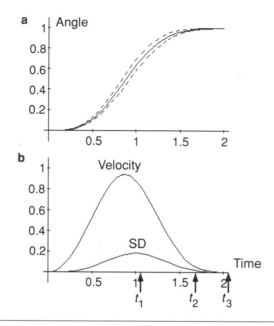

Figure 6.5   A simulated movement trajectory is shown with the velocity-dependent component of standard deviation (a). The lower graph (b) shows the standard deviation curve and velocity curve. Note that the velocity-dependent error exists only during the movement (times t_1 and t_2) but not after the movement (time t_3). All scales are in arbitrary units.

Schmidt (1991), who have shown that the kinematic profiles can be affected by the amount of impact force the subjects are allowed to use.

Until now, we have discussed the possiblity of emergence of two kinds of trade-offs at the level of internal simulation. Other kinds of trade-offs can emerge at later stages of generation of voluntary movement. The next sections are devoted to analysis of trade-offs that can emerge at the level of generation of motor command (Figure 6.3). For this purpose, I shall return to the framework of the equilibrium-point hypothesis.

Variability in the λ-Language

The equilibrium-point hypothesis has been criticized from a variety of standpoints, in particular for the failure to explain the observed variability in motor performance (Darling & Cooke, 1987). However, it provides a suitable framework for discussing the issues of variability and even for advancing certain testable hypotheses.

The general idea is simple. There are three basic parameters characterizing movement kinematics: distance, speed, and time. They are obviously interrelated so that only two of them can be chosen independently, and the third appears as a result of simple mathematics. We suggest that each of the three parameters has a counterpart (a control variable) at the control level. A subject can choose, at the control level, two of the three control variables with some level of variability that is inherent to the control process and cannot be reduced ("neural noise," cf. with "motor noise" in Meyer et al., 1988). Variability of the third, dependent variable will be defined not only by the variability of the first two, but also by their magnitude. This approach suggests three basic strategies corresponding to three pairs of independently chosen controlled variables leading to different kinds of trade-offs depending on the motor task or the subject's intention.

Basic Assumptions

According to the λ-model, regulation of each muscle can be described with time shifts of a variable λ, the threshold of a tonic stretch reflex. As described in chapter 1, a joint controlled by a pair of muscles (agonist and antagonist) can be described by a pair of λ variables or a pair of reciprocal and coactivation commands, $r = \dfrac{(\lambda_{ag} + \lambda_{ant})}{2}$ and $c = \dfrac{(\lambda_{ag} - \lambda_{ant})}{2}$ (Feldman, 1980a,b; Feldman & Latash, 1982a). Fast isotonic movements can be similarly described by time shifts of (r, c) or $(\lambda_{ag}, \lambda_{ant})$. In the case of isotonic movements, r defines equilibrium position and c corresponds to joint stiffness. In chapter 2, I described different patterns of r and c presumably used for controlling fast single-joint movements. One of the patterns (Figure 2.12a; Abdusamatov & Feldman, 1986; Feldman et al., 1990a) includes a shift of r at a constant rate. Another pattern implies a more complex nonmonotonic shift in r (N-curve, see Figure 2.12b), which has been supported by experimental observations (Latash & Gottlieb, 1991c; see "Shifting Joint Compliant Characteristics During Fast Movements" in chapter 2). However, even in this case, one may consider the first arm of the N-curve as being characteristic for movement speed and amplitude. Therefore, let us consider, for simplicity, that movement parameters are defined by a shift in r at a rate ω during a time τ from an initial value of r_{in} to a final level r_{fin} (Figure 6.6). We are not going to discuss different versions of the speed-accuracy trade-off but rather its nature in general.

Two of the three interrelated control variables (r_{fin}, τ, and ω; $r_{fin} - r_{in} = \omega\tau$) can easily be associated with parameters of an isotonic movement: r_{fin} with movement amplitude, and τ with movement time. To reduce the number of subscripts and superscripts, I will use just r for r_{fin}. We may associate ω with a loosely defined notion of "movement speed." Although any fast movement is performed with constantly changing limb speed, I

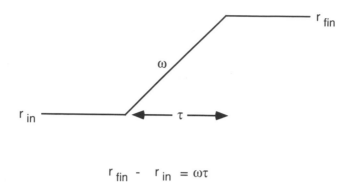

$$r_{fin} - r_{in} = \omega\tau$$

Figure 6.6 According to this simplified scheme, single-joint movements are controlled by a unidirectional shift in a controlled variable r from its initial value (r_{in}) to its final value (r_{fin}) at a rate ω over τ. If r_{in} is known, movement amplitude is related to r_{fin}, movement time to τ, and movement speed to ω. Only two of these three variables can be fixed independently, and the third emerges as a result of a simple equation (lower part of the figure). Fixing independently different pairs of the central variables defines three strategies leading to different kinds of trade-offs.

suggest that ω can in fact be considered a central variable. That is, when subjects think in terms of controlling movement speed, they act in terms of controlling ω. Naive subjects have no problems in understanding an instruction to perform movements of different amplitudes with "the same speed." Performing with the same speed (which translates to the same ω) can be associated with the Speed-Insensitive strategy while performing with different ω is associated with the Speed-Sensitive strategy (see chapter 4 and Corcos et al., 1989; Gottlieb et al., 1989a,b, 1990a). It has been demonstrated (Gottlieb et al., 1990a,b) that the instruction to move with the same speed leads to predictable kinematic and EMG patterns in a wide range of speeds (e.g., characterized by peak speeds).

Let us suppose that there is a neural noise that leads to irreducible errors in specifying values of r, ω, and τ. It is not a priori evident if a subject is able to choose any of the three variables with a higher accuracy than the others. Everyday experience suggests that "going to the same point" is easier than "going with the same speed," which, in turn, is easier than "going within the same time." This order of difficulty corresponds also to the amount of explanation and practice that is required to make a naive subject follow instruction. This assumption can be expressed in a more formal way, in terms of variability in the three variables, inherent to the central control process:

$$\frac{d\tau}{\tau} > \frac{d\omega}{\omega} > \frac{dr}{r} \, , \tag{6.10}$$

where $d\tau$, $d\omega$, and dr are hypothesized errors in specifying τ, ω, and r characterizing inherent variability of the nervous system. We will assume that these errors are mutually independent and also independent of the values of τ, ω, and r.

A series of experiments was performed trying to validate, at least partially, Equation 6.10 (Latash, 1989c). During nearly isotonic elbow flexion movements, the subjects were required in different series to reproduce one of the following three parameters: (a) a learned movement amplitude starting from different initial positions and moving "as fast as possible," (b) a learned final position starting from different initial positions and moving as fast as possible, and (c) a learned movement time for movements of different amplitudes. During a short training period, the subjects knew in advance which parameters they would be asked to reproduce in future series. The results have demonstrated a marked difference in the coefficients of variability for the three experimental conditions (Table 6.1). The lowest values of the coefficients of variation were observed during reproduction of final position, while the highest values corresponded to reproduction of movement time.

Better reproduction of location rather than movement distance has been demonstrated in a number of studies (Keele & Ells, 1972; Marteniuk et al., 1972; Marteniuk, 1973; Wrisberg & Winter, 1985), although the opposite

Table 6.1 Averaged Values of the Coefficients of Variation and Standard Deviations for the Experiments With Movement Time, Amplitude, and Final Position Reproduction

	MT-1	AMP-1	FP-1	MT-all	AMP-all	FP-all
Mean	9.4	6.1	3.5	10.8	7.0	4.1
SD	3.3	2.6	1.4	4.3	3.4	1.5

t Values for the Student's t Test

	First series		All the series	
	t	p	t	p
MT vs AMP	1.9	0.12	3.5	0.003
MT vs FP	3.6	0.015	6.9	0.0001
AMP vs FP	2.4	0.06	4.2	0.007

Note. The coefficients of variation (in %) are shown separately for the three instructions requiring movement time (MT), amplitude (AMP), and final position (FP) reproduction. Data for the first sets of trials (1) and for all the trials in all three series (all) are shown separately. The t values are based on Student's two-tailed paired t test.

result has also been reported (Bock & Eckmiller 1986). Better reproduction of location is a natural consequence of the equilibrium-point hypothesis, since, in isotonic movements, location is directly related to the value of one of the main control variables, r. So, in order to reproduce a final position, the subjects could "remember" a final value of r and move there at the highest available speed ω (as required by the instruction).

On the other hand, reproduction of movement amplitude requires control of two of the variables: speed of r changes (ω) and time of r changes (τ). Since the instructions required moving with the highest speed, the subjects were likely to choose the highest available value of ω and to remember and reproduce τ. Higher variability in the amplitude reproduction experiments suggests that a hypothesized irreducible error in τ is higher than in r ($d\tau/\tau > dr/r$; cf. with Equation 6.10).

Movement time as measured in the experiments (from the first deflection of the acceleration to the crossing of the zero line by a linear extrapolation of the descending deceleration curve) also depends on two control variables, τ and r or τ and ω. However, changing movement distance required the subjects not to reproduce but rather to adjust both variables. So, it is not surprising that the variability in these experiments was higher than in the first two series.

Let us return to Equation 6.10 and introduce the following assumption:

In different situations, humans have a choice of specifying any pair of the three variables (r, ω, τ). This choice depends on requirements particular to the movement task and will influence the interrelation between movement speed, distance, and accuracy.

Fitts' Strategy (ω/τ Strategy)

If a task is to move quickly to a visually presented target, an ω/τ strategy may be used (although not necessarily; see the next section). A visual target implies relying on visual information for planning and terminating the movement at an appropriate time; that is, by fixing ω in advance and adjusting τ based on the visual feedback (cf. with iterative correction models, Keele & Posner, 1968). This strategy will also be used for movement amplitude control (Bock & Eckmiller, 1986). The error in the final position (Δr) will depend on the errors in the regulated variables ($d\omega$ and $d\tau$). We will use d for inherent errors (see previous discussion) in the centrally regulated variables and Δ for the calculated error in the dependent variable.

Let us assume that a subject specifies τ with an error $d\tau$ and ω with an error $d\omega$. Then, the error in specifying r will be

$$\Delta r = \omega d\tau + \tau d\omega. \tag{6.11}$$

Equation 6.11 implies a close to linear relation between spatial variability

(Δr) and movement time. Note that we consider *smooth* movements that, in the framework of the dual-submovement model (Meyer et al., 1988), correspond to a single submovement, which, in turn, leads to Schmidt's relation (Equation 6.4). An increase in movement speed from ω to ω_1 without changing movement distance ($r_1 = r$) is associated with a decrease in the time variable from τ to τ_1 because $\tau_1 = r_1/\omega_1$.

$$\frac{\omega}{\omega_1} = \frac{\tau_1}{\tau} = \alpha,$$

where α is a constant. Because the errors $d\tau$ and $d\omega$ are supposed to be independent of τ and ω,

$$\Delta r_1 = \omega_1 d\tau + \tau_1 d\omega = \frac{\omega d\tau}{\alpha} + \alpha\tau d\omega. \qquad (6.12)$$

An increase in movement speed means that $\alpha < 1$. Then, changes in spatial error (Δr) will depend on relative speed ($d\omega$) and timing ($d\tau$) errors. From the assumption (Equation 6.10), it follows that $\omega d\tau > \tau d\omega$.
   Because $\alpha < 1$, then

$$\frac{\omega d\tau}{\alpha} > \tau d\omega,$$

and

$$\frac{\omega d\tau}{\alpha}(1 - \alpha) > \tau d\omega (1 - \alpha).$$

After a simple transformation, we get

$$\frac{\omega d\tau}{\alpha} + \alpha\tau d\omega > \omega d\tau + \tau d\omega. \qquad (6.13)$$

The right side of this inequality is Equation 6.11, and the left side is Equation 6.12. Therefore Equation 6.13 demonstrates that $\Delta r_1 > \Delta r$, which means that spatial variability increases with speed. This might be the basis of an aspect of the speed-accuracy trade-off that predicts increased variability of movements at higher speeds (lower movement times).

Closed-Eyes Strategy (r/ω Strategy)

Another strategy is used in tasks requiring preplanning a movement to a point (cf. Schmidt et al., 1979; Schmidt & McGown, 1980; Meyer et al., 1982) if, for example, visual feedback is not available. Let us suppose that a subject chooses values of r and ω with accuracy dr and $d\omega$. According to a previous assumption, these errors are independent of the values of r and ω. Then, timing errors can be found from:

$$\tau + \Delta\tau = \frac{r + dr}{\omega + d\omega}; \quad \frac{r}{\omega} + \Delta\tau = \frac{r + dr}{\omega + d\omega}; \quad \Delta\tau = \frac{\omega dr - r d\omega}{\omega^2}. \quad (6.14)$$

The original assumption of Equation 6.10 leads to $\omega dr < r d\omega$ and $\Delta\tau < 0$. Then, Equation 6.14 implies that an increase in movement speed, preserving distance, will lead to a decrease in absolute timing variability. Similarly, an increase in movement distance, preserving the speed parameter ω, will lead to an increase in absolute value of timing variability. Since dr and $d\omega$ are postulated to be independent of both r and ω, there is no reason to expect any kind of speed-accuracy trade-off when the subject uses this strategy.

We have modeled this strategy in a series of pilot experiments in which the subjects were required to make movements of different amplitude to a small target with their eyes closed (Latash, 1989b,c; Latash & Gottlieb, 1990c). The data were compared to performance of the same subjects in a more classical Fitts'-type paradigm when the eyes were open. Standard deviation of the final position was used as an index of variability, while peak speed was used as an index of movement speed. Two findings are of interest. First, movements with closed eyes were characterized by considerably higher peak speeds (Figure 6.7) without an increase in variability of the final position. Second, movements over the largest amplitude performed with open eyes had a higher index of variability (standard deviation of the final position) than movements over smaller amplitudes, while movements with closed eyes did not show an apparent increase in variability with movement amplitude (note that an increase in variability with movement amplitude was reported by Chapanis et al., 1949, and Zelaznik et al., 1983, for movements without visual feedback although in different experimental situations).

A similar lack of effect of movement speed on the variability of the location of working point (finger and thumb joint end-positions during grasping) has been reported by Darling et al. (1988) for both fast and relatively slow movements. The slower movements in these experiments demonstrated a number of submovements (two to four) that, according to the dual-submovement hypothesis of Meyer et al., should lead to different kinds of trade-offs. However, no detectable trade-off was observed. This result can be expected from our model if the subjects were using the closed-eyes strategy.

Fitts' paradigm implies that the subjects are trying their best to perform at the highest possible speed when some level of accuracy is imposed. However, our results suggest that changing the instruction can "force" a subject to perform better, that is, to increase speed without sacrificing accuracy. For example, the first blocks of movements in our experiments were performed by the subjects in both conditions in virtually identical situations after the same number of practice trials. The only difference was in open or closed eyes and somewhat different instructions. The

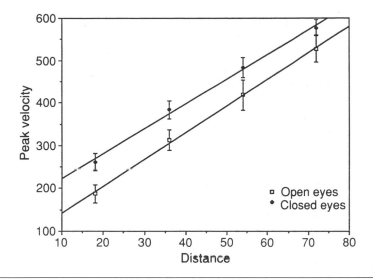

Figure 6.7   Values of the peak speed averaged across the subjects show a nearly linear increase with movement distance in both series of experiments, with open and with closed eyes. Absolute values of the peak speed were higher in the closed eyes experiments for all the distances. This difference was especially pronounced for the shortest distance (18°). Vertical bars show standard deviations; scales are in degrees and °/s.

instruction in the experiments with open eyes was typical for the Fitts-type experiments, that is, "be fast and accurate" (Fitts, 1954; Corcos et al., 1988). In the experiments with closed eyes, the instruction was "go as fast as possible to the final position; consider the final position as a point." Therefore, it is not surprising that the subjects in the latter series performed at higher peak velocities (and correspondingly shorter movement times). However, they did it without an increase in the final position variability. In fact, even the subjects with a lot of experience in similar experiments were quite surprised by the very high accuracy they demonstrated with their eyes closed. After becoming experienced in such experiments, the subjects realized that it was possible to use the closed-eyes strategy even with the eyes open, that is, they paid no attention to the target and went to a final position (middle of the target bar) as fast as possible, thus improving the performance in seeming violation of all the existing trade-off laws. This demonstrates that a seemingly clear and concise instruction (e.g., "go to the target as fast as you can and stop inside the target") may not be sufficient to ensure that a subject is doing what an experimenter wants, although the performance may not show obvious indications of misunderstanding. In fact, all these experiments originated from my own experience as a "bad subject" in a modification of the typical Fitts-type experiment, which required reproduction of the same movement hundreds of times and which, therefore, was extremely boring. Somewhere in the middle of the experiment, I started to invent different ways of "doing the same," in particular trying to move without looking at the screen. To my surprise (and, later, to the justified dismay of the experimenter!) my performance improved dramatically, and thus the whole set of data was trashed. The bottom line is that extensive, detailed instructions, specifying every particular aspect of the experiment and severely restricting the subject to the experimenter's requirements seem necessary.

Getting back to the results, they suggest that the central control structures are capable of being fast and accurate simultaneously. The problem is to choose the right strategy. There is another question: Why didn't the subjects use this strategy from the very beginning? Perhaps the reason is that from everyday experience we know that it is hard to hit a small and distant target under visual control. So, the subjects are reluctant to move fast to small targets, and a special instruction is required to overcome this reluctance.

Another finding that appears to violate Fitts' law is the seeming lack of dependence of the spatial variability indices on the movement amplitude in the experiments with closed eyes. The pilot nature of the data prevents certainty; however, this finding is quite compatible with the model predictions and, therefore, corroborates the equilibrium-point approach to analyzing variability.

One may suggest another explanation for these findings. We have already mentioned reports by Schmidt and Sherwood (1982; see also Sherwood & Schmidt, 1980) that an increase in force amplitude in force pulses over 60%-70% has led to a paradoxical drop in force variability. An increase in peak speed in our experiments suggests a considerable increase in inertial joint torques (and correspondingly, muscle forces). Inertial torques during very fast and large movements have been shown to be close to those observed during attempts at isometric maximal voluntary contraction (Gottlieb et al., 1989b). Therefore, the lack of increase in variability in our experiments and observations by Schmidt and Sherwood can share the same mystic nature.

Badminton Strategy (r/τ Strategy)

The third option is to fix r (final position) and τ. The error in ω is (cf. with Equation 6.14):

$$\Delta\omega = \frac{\tau dr - r d\tau}{\tau^2}.$$ (6.15)

Using the assumption (Equation 6.10), $\tau dr < r d\tau$ and $\Delta\omega < 0$. Taking Equation 6.15 into account, we find that variability in ω (or related speed parameters, e.g., peak speed) will increase with an increase in movement distance, and will decrease with an increase in movement time.

This strategy is likely to be used in tasks that require getting to a point at a certain moment of time; for example, striking a moving target (badminton, baseball, etc.). The prediction implies that hitting a shuttlecock with a desired force (speed before impact) will lead to higher variability in the striking force for longer and faster swing movements. I have not consulted professional badminton players, but my own amateur experience suggests that this is likely true.

Relation to Motor Learning

Schmidt, in his extensive monograph *Motor Control and Learning* (1988), defines motor learning as "a *process* of acquiring the *capability* for producing *skilled actions*" (p. 345). This definition contains terms that are understood and expressed differently in frameworks of different models of motor control. Most of the theories of motor learning are formulated in psychophysiological or behavioral terms and do not incorporate specific hypothetical models or mechanisms of motor control (Pew, 1966; Adams, 1971; Schmidt, 1975). Since we have accepted the black box approach to analysis of different aspects of motor behavior, prior to addressing issues of motor learning it seems necessary to define a framework for future analysis, and therefore, to answer these questions: What is controlled? What are the variables used by the central control system for motor programs? What are the central correlates of improved performance?

Chapter 1 was designed to persuade the reader that the equilibrium-point hypothesis (λ-model) is the best available framework for discussing problems of motor control, at least for single-joint movements. The next chapters analyzed different aspects of motor behavior from this viewpoint. Now we are approaching the limits of applicability of the equilibrium-point hypothesis, which has been introduced as an essentially single-joint model (although see Feldman et al., 1990b). Terms such as *skilled actions* generally imply multijoint or even multilimb movements in three-dimensional space. Therefore, any single-joint model of motor learning can be criticized as an apparent oversimplification that may have nothing to do with acquisition of "real" motor skills. However, I think that the general dynamic approach, represented for single-joint movements by the equilibrium-point hypothesis, looks promising for analysis of unrestricted multijoint movements as well and will discuss these issues in chapter 7. On the other hand, there is enough data on the effects of practice on single-joint movements (McGrain, 1980; Gottlieb et al., 1988; Corcos et al., 1990b) to justify its analysis with the methods of the λ-model. So, I am going to discuss in this chapter observations of changes in performance parameters during practice of simple single-joint movements and hope that at least some of the conclusions will not be misleading and can in future be generalized for three-dimensional multijoint movements. Some of the more general aspects of motor learning will be discussed in chapter 10.

Remember (Figure 5.1 in chapter 5) that fast single-joint movements are supposed to be controlled by the following pattern: (a) a shift in the agonist controlled variable (λ_{ag}) from an initial to a final value, (b) a simultaneous coactivation of the antagonist (small shift in λ_{ant}), (c) a delayed shift in λ_{ant} (active braking), and (d) a reversed change in λ_{ant} bringing the limb to a desired equilibrium with a certain final coactivation. In terms of the reciprocal and coactivation commands (r and c), this process implies a nonmonotonic N-shaped change in r with a phasic peak in c in the middle of the movement.

Let us analyze a simple case when a subject is performing a single-joint nearly isotonic movement in a horizontal plane and is asked to move "as fast as possible" from a fixed initial joint position to a certain final position. In order to simplify the analysis even more, let us instruct our subject not to pay attention to accuracy and to consider the target as a dimensionless point approximately showing the distance to be moved. The subject will still have some notion of required accuracy (how far from the target he or she can terminate the movements) but at least it will not be imposed upon him or her and, hopefully, will remain constant during the experiments. The main index of performance that is required to be optimized is movement time or a related speed parameter (e.g., peak movement speed).

What options does the subject have to improve the performance? In terms of the λ-model, the major possibilities are to adjust the rate of the initial λ_{ag} shift, the delay before the active braking starts, and the rate of the λ_{ant} shift. There are certainly more options that we have not discussed yet and that include, for example, changing the initial coactivation shift in λ_{ant}, changing the final level of coactivation, and changing the amplitude of the delayed λ_{ant} shift. But let's keep things as simple as possible for now.

Most of the studies of the effects of practice in single-joint movements are described in terms of *refining the patterns of muscle activation* (Gottlieb et al., 1988; Corcos et al., 1990b). In other words, they are analyzed in the framework of independent central control of the levels of muscle activity, which are usually directly associated with the EMG levels and patterns. This approach has been criticized in the previous chapters; it would be redundant to repeat all the reasons against it here. The alternative, dynamic approach, as far as I know, has never been used for this purpose.

Effects of Practice on Kinematic and EMG Patterns

An extensive study of the effects of short- and long-term practice on the speed of single-joint movements has been recently performed by Corcos et al. (1990b; see also Gottlieb et al., 1988). Naive subjects were presented with a small target (3°) and instructed to be either as fast as possible or both fast and accurate (a Fitts-type paradigm). They performed blocks of 20 trials at isotonic elbow flexion over 54° in the middle of the physiological range of joint angles. Ten blocks (200 trials) were performed each day during 10 consecutive days. Testings during the first and the last days included movements "as fast as possible" over different distances.

Two groups of findings are of particular interest. The first group describes changes in performance that occurred within one day and were not transferred to the next day. I will address these as "transient." The second group includes changes that were apparent when comparing performance at different days. I will call these "stable."

Not unexpectedly, training resulted in a considerable increase in peak movement speed and other related kinematic parameters. This improvement was reflected in both transient and stable changes. Simultaneously, the following major changes were observed in the EMG patterns (see also McGrain, 1980; Schmidtbleicher et al., 1988): (a) an increase in the rate of rise, peak values, and integrals of the first agonist EMG burst; and (b) considerably higher peak values and integrals of the antagonist burst, sometimes with a decrease in the delay before the antagonist burst. The authors have tried to explain the findings with refinement of motor programs using the language of the dual-strategy hypothesis (see "Excitation Pulse and the Dual-Strategy Hypothesis" in chapter 4), that is, in terms of changing parameters of the excitation pulse to the agonist and antagonist α-motoneuron pools. Remember that the excitation pulse, by definition, is a sum of all presynaptic inputs to an α-motoneuron pool and, therefore, is a direct precursor of the EMG, which unambiguously defines its pattern. So, the observed EMG changes have been interpreted as reflections of analogous changes at the control level (i.e., in duration and amplitude of the excitation pulse).

Let me make two points. First, as we have already discussed in chapter 4, it is necessary (at least for delayed events like the antagonist burst) to take into account the contribution of reflex feedback signals to the total level of the α-motoneuron activity. Therefore, changes in the central commands are only one factor leading to the observed EMG changes. Second, Corcos et al. (1990b) have noted that kinematic and EMG patterns observed at different stages of training were similar to those observed in experiments when the subjects used the Speed-Sensitive strategy, that is, adjusted movement speed to match the task. Figure 6.8 shows averaged kinematic and EMG patterns during series of trials at 36°, 54°, and 72° before (thin lines) and after (bold lines) the training. Note that movements over different amplitudes performed on the same day demonstrate patterns typical of the Speed-Insensitive strategy (overlapping initial fragments of the kinematic and EMG traces) while movements over the same

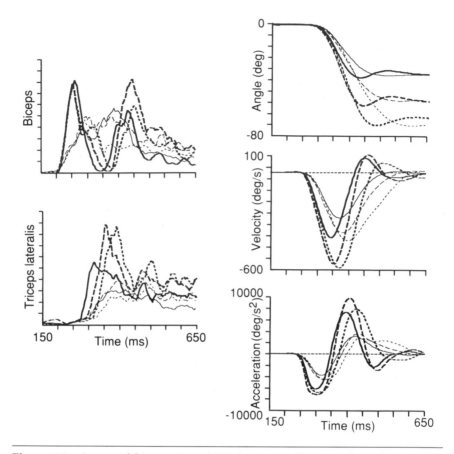

Figure 6.8 Averaged kinematic and EMG patterns for one subject during series of trials at 36°, 54°, and 72° before (thin lines) and after (bold lines) the training. Movements over different amplitudes performed on the same day demonstrated patterns typical for the Speed-Insensitive strategy (overlapping initial fragments of the kinematic and EMG traces) while movements over the same distance at different days show typical Speed-Sensitive patterns (kinematic and EMG traces diverge from the very beginning).

Note. From "Organizing Principles Underlying Motor Skill Acquisition" by D.M. Corcos et al. In *Multiple Muscle Systems. Biomechanics and Movement Organization* (pp. 251-267) by J.M. Winters and S.L.-Y. Woo (Eds.), 1990, New York: Springer-Verlag. Reprinted by permission of Springer-Verlag.

distance at different days show typical Speed-Sensitive patterns (kinematic and EMG traces diverge from the very beginning).

It seems only natural to assume that, in the course of training, central motor commands of the subjects changed qualitatively in the same way as during movements over the same distance performed at different speeds during one day. Using the λ-language, this means an increase in the rate of the initial shift in λ_{ag}. There is no need to postulate any other changes.

Let us look more closely at Figure 6.8. Note that the initial coactivation component of the antagonist EMG has approximately the same rate of rise in all the series (although the resolution is suboptimal), suggesting that the initial coactivation shift in λ_{ant} has not changed considerably. Note also that there are no apparent changes in the delay before the antagonist EMG burst (compare thick and thin traces over the same amplitude). This suggests a lack of change in the delay before the active braking λ_{ant} shift. The antagonist EMG rises much more steeply after the training, which may be attributed to either increase in movement speed leading to an increase in the contribution of the feedback component, or to an increase in the rate of the λ_{ant} shift, or to both. However, it seems that the most parsimonious explanation of the observed changes includes only one event (i.e., an increase in the rate of initial λ_{ag} shift).

If the subjects are asked to perform "as fast as possible," extensive training leads probably to an increase of the rate (ω) of λ_{ag} shift up to its maximal level. Another consequence of this may be a decrease in variability in ω which may, in turn, lead to a decrease in variability of various kinematic parameters as reported by Darling and Cooke (1987) and Corcos et al. (1990b), although Darling et al. (1988) reported a lack of changes in the variability of the location of fingertip contact on the thumb after practicing a simple grasping movement. A seemingly paradoxical increase in variability of certain EMG indices has been reported by Darling and Cooke (1987) in the process of training fast single-joint movements. However, their method of measurement (EMG integrals over short time bins) might well have reflected the variability in "instantaneous" EMG levels, which is likely to be affected by changes in such other factors as the number of firing motor units and their size, both increasing with an increase in movement speed. Variability of more general EMG indices (e.g., agonist and antagonist burst integrals) has been shown to decrease with an increase in movement speed during training (Corcos et al., 1990b) as is expected from the λ-model.

Let me emphasize that more "applied" aspects of motor control, like variability and motor learning, are also likely to benefit if their analysis is based on the Bernstein approach, that is, on a certain framework that explicitly states which variables are considered to be independently controlled by the central nervous system. Such a framework should not contradict known facts. The λ-model is just one example that seems helpful and effective for analysis of single-joint movements. Analysis of multijoint movements will probably require a somewhat different language. In the next chapter, we will see what kind of new problems emerge when one moves to the higher level of complexity associated with multijoint voluntary movements.

Chapter 7

Multijoint Movements

There is a body of experimental observations during multijoint movements but no hypothetical variables analogous to λ for single-joint movements. Kinematics of the most important point (working point) is likely to be primarily controlled while the reproducibility of individual joint trajectories and muscle forces is much lower. The Bernstein Problem—overcoming excessive degrees of freedom at a control level—is frequently substituted by a different problem of overcoming this redundancy at the level of performance. The Bernstein Problem can be illustrated at different levels, from control of a single muscle to control of multilimb movements and problems of postural stability. Analysis of multilimb coordination is an important source of experimental support for the dynamic pattern generation approach.

The Bernstein Problem

Until now, we have discussed problems of motor control for single-joint movements. As mentioned in chapter 4, single-joint movement (or isometric contraction) is an abstraction rather than a real-world phenomenon. This abstraction has proven to be helpful for formulating certain hypotheses and approaches and testing them in experiments. However, most of the actual voluntary movements are not confined to a single joint and do not take place in a standardized laboratory environment. They occur in a changing world and involve coordinated movements in a number of joints with corresponding finely tuned actions of postural muscles and, frequently, with an interaction between two or more limbs.

Is it possible to generalize the conclusions and even the approaches elaborated for single-joint movement studies for real-world multijoint and multilimb movements? The answer is both "No" and "Yes." Recently,

"No" has been receiving impressive support with the growing awareness of new problems arising during analysis of multijoint movements (Berkinblit et al., 1986a,b,c; Mussa Ivaldi et al., 1989; Flash, 1990; Hogan, 1990). On the other hand, "Yes" has been almost forgotten and abandoned as an apparent and misleading oversimplification. In chapter 4, I discussed certain reasons in favor of "Yes," and I still believe that single-joint movement is a good and helpful model. Nevertheless, the purpose of this chapter is to justify the "No" side of the problem and to discuss a variety of new factors and problems emerging with the level of complexity inherent in multijoint movements.

Two major sources of increased complexity of multijoint movement problems can be identified. The first one deals with a more complex physical description of the controlled peripheral system due, in particular, to important roles of new forces, for example, Coriolis and centrifugal, that were generally ignored during single-joint movement analysis. Biarticular muscles provide links between adjacent joints (Hogan, 1985, 1990; Zajac & Gordon, 1989; Gielen et al., 1990) thus introducing one more factor into joint interactions. Note that spring properties of these links can be controlled, presenting new options for control strategies (see "A Special Role for the Biarticular Muscles," later in this chapter).

The second source of complexity is the famous Bernstein Problem of overcoming excessive degrees of freedom (an elegant, comprehensive description of the Bernstein Problem can be found in a recent article by Turvey, 1990). During virtually all voluntary movements, the number of degrees of freedom (n) for the peripheral mechanical apparatus (e.g., limb) is higher than the number of degrees of freedom (n_0) necessary to execute a motor task or to unambiguously describe its execution. The number of available control variables (degrees of freedom) for the central controller (N) is likely to be less than n due to a number of factors including, in particular, anatomical constraints and biomechanical links between the joints (e.g., biarticular muscles). However, it is likely to be more than n_0 in most natural movements. The latter number is frequently three, corresponding to the three-dimensional space in which we happen to live. The Bernstein Problem is a problem of choice: How does the central nervous system choose a certain pattern for executing a motor task from an infinite number of possibilities?

In other words, the central nervous system is presented with n_0 equations that contain N unknowns ($N > n_0$). To unambiguously solve such a system, we need to add ($N - n_0$) equations or, equivalently, impose ($N - n_0$) constraints upon the system. Later in this chapter, I will discuss different approaches to the Bernstein Problem. However, it seems necessary now to stress a very important point: N is a number of degrees of freedom (a number of independently controlled variables) for a hypothetical central controlling system but not for the peripheral system. Therefore, the Bernstein Problem should be analyzed and solved in a central language but not in a language of motor performance.

Imagine, for example, that there is no λ-language, and muscle control is described in terms of changes in muscle force or in the number of recruited motor units and frequency of their firing. One would encounter a seeming Bernstein-type problem: How does the central controller choose which motor units to recruit and at what frequency in order to meet the task requirement? Different rules and optimization criteria can be suggested including, for example, the Henneman's size principle (Henneman et al., 1965). We are lucky, however, that the central nervous system doesn't need to deal directly with individual motor units but can use only one independently controlled variable (λ) for each muscle, thus avoiding the Bernstein Problem, at least at the single-muscle level.

A similar situation can take place for control of multijoint movements. In particular, let us assume that there exist hypothetical independently controlled variables (analogous to λ) that are used by the central nervous system to control three-dimensional movements. For a multijoint system, the number of these variables, N, can be less than the number of degrees of freedom of the peripheral mechanical system, n (cf. only one variable λ for a muscle containing a large number of motor units). Therefore, in certain situations, the Bernstein Problem can degenerate into a unique solution for a motor task.

Generally speaking, the Bernstein Problem is present for single-joint movements as well but is rarely explicitly recognized. Let us consider a static "pin" joint with only one degree of freedom. We need to specify only two parameters to define the joint's state (e.g., joint stiffness and position). If the joint is controlled by only two muscles, an agonist and an antagonist, the number of central variables (one per muscle) corresponds to the number of peripheral degrees of freedom, and the Bernstein Problem is avoided. However, most of the joints are controlled by more than two muscles, and there emerges a freedom of choice: Which muscles should be activated, and to what levels, to meet the task requirements? This single-joint Bernstein Problem is usually avoided although there are a number of exceptions (Buchanan et al., 1986, 1989; van Zuylen et al., 1988; Jongen et al., 1989; Flanders & Soechting, 1990; Hogan, 1990). The level of complexity increases considerably when one analyzes more realistic joint models (with several degrees of freedom) that may involve rotation. Classification of muscles into agonists and antagonists becomes rather artificial and even misleading, since virtually all the muscles appear to be multifunctional, that is, they take part in movements or isometric contractions in different directions (Buchanan et al., 1989; Zajac & Gordon, 1989).

The Problem of Inverse Kinematics

Now we are entering a very slippery way of reformulating the Bernstein problem using the language of motor performance. Unfortunately, one cannot suggest for multijoint movements even a hypothetical independently controlled central variable that would be analogous to λ for single-joint movements (see later in this chapter). So, we do not have a central language for multijoint movements similar to the λ-language. The Bernstein Problem is formulated as a problem of finding a solution for a motor task in a central language. Therefore, at the present stage of our understanding of motor control, this problem cannot be solved even theoretically.

If we want to continue studies of control of multijoint movements, there is no choice but to inadequately reformulate the Bernstein Problem using a language of performance. This is done in virtually all the studies dealing with multijoint movements. At least, let us put it up front: We are going to analyze a different problem—that is, how to overcome excessive mechanical degrees of freedom of a peripheral system to accomplish a motor task. Let me term this problem *pseudo-Bernstein*. This approach is likely to be misleading but, first, there seem to be no viable alternatives and, second, it may be productive in related fields (e.g., robotics).

Let us return to a notion of a working point as the most important point for executing a certain task. For example, for a grasping task, working point can be associated with the fingertips or with the palm; for kicking a ball in soccer, it can be on the front part of the foot; for hitting a nail with a hammer, it is likely to be located at the tip of the hammer, and so on. Presumably, this is the point about which the central nervous system is mostly concerned, because its trajectory is vital for executing the task. It has been shown that trajectory of the working point is better reproduced in successive trials than trajectories of individual joints (Bernstein, 1936; Soechting, 1984; Berkinblit et al., 1986a). However, even the trajectory of the working point is not a variable of the hypothetical central language. Imagine, for example, that an unexpected perturbation (change in external torque) occurs during a movement (e.g., you hit an unseen wall during a reaching movement). Trajectories of all the points, including the working point, will change immediately, while the central command remains presumably the same, at least for some time until motor corrections are introduced.

One of the pseudo-Bernstein problems deals with establishing a correspondence between working point trajectories in the external Cartesian space and movement trajectories in a space of joint angles. This is a fascinating problem by itself and is a good illustration of overcoming redundant degrees of freedom. Put the tip of your index finger on this page and move your arm without losing contact with the paper. There is an infinite number of combinations of joint angles corresponding to a fixed position of the working point (fingertip). The first question to be asked is: Does the system care about particular combinations of joint angles? The answer seems to be "No." Nevertheless, an inadequate formulation of the Bernstein Problem can be suggested in the form: How does the system choose a particular combination of joint angles for a position of the working point in the Cartesian space? Or, how does the system choose a particular time sequence of such combinations to match a desired working point trajectory?

This problem is known as the problem of inverse kinematics. It belongs to the class of "ill-posed" problems (cf. Mussa Ivaldi et al., 1989) in the sense that it cannot be solved without some additional information about the system or constraints imposed upon the system. Choice of such constraints fully resides at the researcher's discretion. I will return to this approach somewhat later.

The Problem of Inverse Dynamics

Robotics shares many common problems and approaches with human motor control. However, the goals are different. In motor control the goal is to understand how voluntary movements are controlled. In robotics the goal is to design a method of controlling a human like (or something else-like) mechanical system (android or robot). This difference makes approaches that are likely to have only a remote relation to motor control problems quite useful for robotics. One of these is inverse kinematics. Another is inverse dynamics.

If there is an ambiguity in relating movement trajectories in a joint space to working point trajectories in the Cartesian space, it is even more apparent in attempts to draw such relations with patterns of muscle forces. What time patterns of muscle forces are required to ensure a given working point trajectory? This question is even more ill-posed (if this is possible) because the number of excessive degrees of freedom is likely to be even higher (cf. Jongen et al., 1989; Zajac & Gordon, 1989). However, this question—known as the problem of inverse dynamics—is one of the most frequently addressed in robotics (Raibert, 1977, 1978; Hollerbach & Atkeson, 1987; An et al., 1988; Atkeson, 1989).

Engineers prefer to induce movements by applying forces with actuators. Therefore, a sequence of force-producing signals (commands) should be somehow calculated in order to bring about a desired movement of the working point. Different methods have been used to overcome the redundancy problem with considerable success (Jacobsen et al., 1986; Raibert, 1986; Kuperstein, 1987; Hannaford & Winters, 1990). Robots usually have means of movement corrections (i.e., feedback loops), and in general it appears that the principles underlying control of their movements are not that different from the principles of human motor control. However, the principles of multijoint human motor control are unknown whereas robots do exist and function. Recent progress in robotics has not stimulated a comparable progress in understanding human motor control. So, there should be a very basic difference. Independent of feedback corrections, movements of robots are based on direct control of force patterns applied by the actuators. Generally speaking, motor control in humans could be built on the same principle. But evolution has taken a different approach, which we call "dynamic." This difference is already obvious for single-joint movements (see chapter 1) and is likely to be no less profound for multijoint movements.

Kinematic Characteristics of Multijoint Movements

In this section, I will briefly review some kinematic findings during multijoint movements. Description of similar findings during single-joint movements (chapter 4) has been considerably facilitated by the availability of a general framework, the λ-model. Because there is no comparable framework for analysis of multijoint movements, the description is going to be segmented, jerky, and inconsistent, and it is likely to create an impression of total chaos (which is not far from reality).

Planar Movements

A number of general characteristics of multijoint planar movements have been described. Some of them are similar to characteristics of single-joint movements (cf. chapter 5); others are specific for the multijoint case.

If a task is to move a working point from a certain initial position to a certain final position on a plane, subjects prefer nearly straight trajectories of the working point independent of movement speed, load, and other factors (Morasso, 1981; Soechting & Lacquaniti, 1981; Abend et al., 1982; Atkeson & Hollerbach, 1985). Such movements are characterized by bell-shaped velocity profiles for the working point while speed profiles for individual joints can have a more complex, multiphasic form and depend on the area of external space where the movement is performed (Morasso, 1981). This finding suggests that the control system does not care much about movements in individual joints and corroborates the ill-posed nature of the problem of inverse kinematics (see previous discussion). A similar conclusion can be drawn from the study by Atkeson and Hollerbach (1985), who have shown that working point velocity profiles scale with movement speed; that is, they can be superimposed with a simple scaling procedure (see also Soechting & Lacquaniti, 1981). This obviously cannot result from a direct scaling of muscle forces. This finding of Atkeson and Hollerbach (1985) is somewhat similar to scaling movement trajectories described for single-joint movements by Adamovitch and Feldman (1984). It suggests that there may be a central variable whose changes result, in certain experimental conditions, in scaling working point movement speed (analogous to ω for single-joint movements; see chapter 5).

Deviations of working point trajectories from straight lines have been described in a number of studies (Soechting & Lacquaniti; 1981; Atkeson & Hollerbach, 1985; Kaminski & Gentile, 1986; Flash, 1987). These deviations are more pronounced for movements in certain areas of the working space. Flash (1987, 1990) has ascribed them to changes in compliance of the working point in different areas of the working space and modeled curved working point trajectories in the framework of the equilibrium trajectory hypothesis (see "The Equilibrium Trajectory Hypothesis" later in this chapter and Flash & Hogan, 1985; Flash & Mussa Ivaldi, 1990).

Trajectories in individual joints are, generally, less reproducible than the working point trajectory. However, they also demonstrate consistencies in behavior. In particular, peak speed of movements in individual joints scales with movement amplitude (Kaminski & Gentile, 1986). Soechting and Lacquaniti (1981) have shown that, during planar movements involving two joints, peak speeds in individual joints occur approximately at the same moments of time. Kaminski and Gentile (1986) have reported that the order of joint involvement depends on relative amplitude of displacement in individual joints for the future movement, so that the joint with bigger amplitude starts earlier. This finding does not correspond well to the mechanically more efficient proximal-to-distal sequencing of joint involvement. It may reflect the method for detecting the beginning of joint motion used by Kaminski and Gentile, a moment when joint speed exceeds 5% of the peak value observed during the movement. Their finding may reflect a difference in relative initial acceleration in the joints, so that they might reach the threshold at different times while their actual involvement could happen simultaneously or in a fixed proximal-to-distal order.

When the subjects are asked to perform curvilinear movements, the working point trajectories demonstrate clear segmentation (Abend et al., 1982). Movement time increases with a decrease in the radius of curved trajectories. Minimal movement time has been reported for straight trajectories. A quantitative relation between movement speed of the working point and radius of the trajectory has been proposed by Viviani and McCollum (1983) based on their study of drawing circles and ellipses (see also Viviani & Terzuolo, 1980; Morasso & Mussa Ivaldi, 1982). This "law of 2/3" can be expressed as

$$V = br^{2/3} \tag{7.1}$$

where V is speed of the working point, r is radius of the trajectory, and b is a constant. Later studies by Viviani (1985) have supported the law of 2/3 for movements along trajectories with different curvature, although the values of b could differ for different trajectory segments. This law has also been confirmed in studies of children (Fetters & Todd, 1987; Seiaky et al., 1987; Viviani & Schneider, 1991), who demonstrated consistent changes in b with age.

If the subjects are asked to draw simple planar figures like straight lines, triangles, and squares, movement time is relatively constant for drawing figures of different size but the same class (e.g., a big and a small triangle) (Accornero et al., 1984; Viviani & Schneider, 1991). This effect has been addressed as the principle of isochrony. An increase in the number of segments leads to a nearly standard increase in movement time by approximately 200 ms per segment. That is, drawing a triangle takes approximately 400 ms more than drawing a straight line, and drawing a square takes 200 ms more than drawing a triangle.

Writing.   Writing is an exciting illustration of multijoint highly coordinated movements. It may be considered a planar movement if we ignore changes in pressure exerted by the pen on the paper. The majority of subjects display low correlation between pen-point kinematics and axial pen force (Schomaker & Plamondon, 1990). On the other hand, the force exerted by the pen appears to be a discriminating parameter between individuals (Crane & Ostrem, 1983; Plamondon & Lorette, 1989). In previous centuries, fine control of pressure on the paper was considered an important component of the art of writing, and thickness of lines complied to certain laws of esthetics. In our computerized time, writing is becoming an obsolete method of transmitting information; it has become deprived of its artistic features. So, let us consider only contemporary writing, assuming that is consists only of the planar component.

Studies of relations between speed of the working point (tip of the pen) and radius of its trajectory (Viviani & Terzuolo, 1980; Morasso & Mussa Ivaldi, 1982; Lacquaniti et al., 1987) have corroborated major findings for drawing simple curved figures (see previous discussion). In particular, smaller speeds were observed at segments of trajectories with bigger curvature. Time for writing a letter has been shown to proportionally increase with the number of "elementary units" (strokes) (cf. Morasso & Mussa Ivaldi, 1982; Edelman & Flash, 1987) similar to the observations during the drawing of simple figures (cf. Accornero et al., 1984).

Hollerbach (1981; see also Denier van der Gon & Turing, 1965) has suggested that writing can be analyzed as an interaction of two sine oscillators moving the working point along Cartesian axes coupled with a constant shift of the arm along the paper. Frequency and amplitude of each oscillation can be modulated as well as their relative phase shift. Hollerbach modeled each oscillator as being controlled by a pair of nonlinear springs ("muscles") and simulated a number of features of actual writing, including writing at different speeds.

Edelman and Flash (1987) have advanced a mathematical model that considers writing as a combination of only a few elementary units similar to those practiced by children before they actually start to write. They consider that central structures perform "topological planning" of trajectory, that is, define basic features of its form (in kinematic terms). Motor hierarchy translates these commands into the language of muscle activation patterns using the "minimum jerk" principle (see Hogan, 1985 and chapter 8). This approach is not specific for writing but is an illustration of the general equilibrium trajectory hypothesis. Another approach to modeling handwriting has been suggested by Plamondon et al. (1990) based on interactions of curvilinear and angular velocity generators postulated to control velocity profiles of the working point.

Studies of writing and drawing have also demonstrated relatively independent functioning of proximal and distal joints (Soechting, 1984), which leads, however, to highly reproducible trajectories of the working point (tip of the pen). Even wrist trajectory is more variable than that of the working point. This finding can be interpreted as implying either very fast on-line corrections of motor commands to fingers (cf. Cordo, 1990) or, alternatively, supporting the idea that trajectory of the working point is the only directly controlled variable, and trajectories of other limb points are just by-products of the main program.

Let us try to formulate some conclusions from all these findings:

1. Movements are planned in some external working space, although we don't know in which variables.
2. Trajectory of the working point is more directly related to motor task than trajectories in individual joints. Examples of consistent behavior of individual joints are certainly outnumbered by inconsistencies, especially when compared to reproducibility of the working point trajectory.
3. There is likely to be a simple scaling procedure for performing the same task at different speeds.
4. More complex movements consist of elementary units glued together. Movement speed drops at the points of connection of the elementary units, which frequently correspond to points of increased curvature of the working point trajectory (cf. the law of 2/3).

Three-Dimensional Movements

Many of the findings for single-joint and planar movements have been confirmed for unrestrained multijoint movements in three-dimensional space (Morasso, 1983a,b). Trajectories of movements from one point to another are nearly straight with bell-shaped velocity profiles. Changing accuracy requirements can lead to a change in the symmetry ratio, resulting in a relative increase of deceleration time with an increase in the task difficulty (Milner & Ijaz, 1990). If a subject intentionally performs a movement along a curved trajectory, movement time increases. Curved trajectories have been shown to have an asymmetric segmented speed profile with minima of the speed corresponding to points with the minimal values of trajectory radius.

There are, however, specific features of three-dimensional movements that are not observed during planar movements. For example, subjects demonstrate apparent segmentation of movement trajectories when trying to change movement plane (Morasso, 1983b; Soechting & Terzuolo, 1987a,b). Without special instructions, they try to restrict movements to a plane. When being specifically trained to draw on a surface of a cylinder or sphere and then instructed to reproduce the same trajectory in free three-dimensional space, the subjects demonstrate slow movements consisting of many subcomponents (Soechting & Terzuolo, 1987b).

A series of studies by Terzuolo, Soechting, and Lacquaniti (Lacquaniti et al., 1986; Soechting & Terzuolo, 1986, 1987a; Soechting et al., 1986) considered multijoint movements in different planes consisting of sine components (cf. Hollerbach, 1981). They have, in particular, demonstrated that while working point trajectory can be modeled as a sum of two sine oscillations, movements in individual joints can deviate considerably from sine patterns.

A series of experiments by the same group (Soechting et al., 1986) studied arm movements in conditions when the number of degrees of freedom in the joint angle space was four (three in the shoulder and one in the elbow). The authors were searching for invariants that would suggest an additional constraint necessary to solve the pseudo-Bernstein problem, that is, corresponding movements in the joint space to working point movements in the external Cartesian space. Soechting et al. have managed to detect a phase shift between two of the angles, which was kept reasonably constant throughout a range of motor tasks that included drawing of circles, ellipses, stars, and figure eights.

Wiping Reflex in the Spinal Frog.   There are a variety of exciting studies in motor control using different animal preparations. However, I have decided to analyze only one group, namely studies of wiping reflex in the spinal frog. This choice is defined by both my own experience in such studies and the extremely intriguing performance of such a seemingly simple object (a spinal frog) during such a seemingly simple motor task (wipe a stimulus). Some of the findings in spinal frogs resemble the behavior of other preparations, for example, spinal turtles during scratch reflex (Stein, 1983; Mortin et al., 1985). In particular, different patterns of scratching depending on the location of the stimulus have been described (Robertson & Stein, 1988). Turtle studies have also shown that cutaneous stimulation can lead to long-lasting changes in excitability of spinal interneurons located in a single segment of the spinal cord (Currie & Stein, 1990), which can presumably lead to repetitive multisecond scratching movements in the turtle or wiping movements in the frog.

If an experimenter places a stimulus (a small piece of paper soaked in an acid solution) on the back of a sitting spinal frog, the frog, after a certain latent period, performs a series of finely coordinated movements, wiping the stimulus from the back and, sometimes, throwing it away from the body. This behavior was described by Pfluger in the middle of the 19th century. Recently, the wiping reflex has attracted the attention of the motor control scientists as a promising model of coordinated multijoint movement (Fukson et al., 1980; Berkinblit et al., 1986a; Giszter et al., 1989, 1990; Schotland et al., 1989; Bizzi et al., 1991). There are a number of unique attractive features in this type of reflex behavior: First, because the animal is spinal, there is no influence of a poorly controlled supraspinal neural inflow; second, the wiping movement is stimulus-directed, reproducible, and can easily be evoked; third, it is a multijoint movement (there are four major joints in the frog's hindlimb, even if one ignores joints of the toes and movements of the body)

Fukson et al. (1980) studied wiping movements of a spinal frog with different stimulus locations and found that the frog "took into account the scheme of its body." That is, if the stimulus was placed on the ipsilateral forelimb, accurate wiping movements were observed while changing position of the forelimb relative to the body. So, the spinal cord "knows" where the limbs are!

Later, the same group (Berkinblit et al., 1984, 1986a,c) performed more detailed studies of the trajectories of the working point (tip of the toes) and individual hindlimb joints. Depending on location of the stimulus, different patterns of wiping could be observed. When the stimulus was placed on the upper part of the back, the wiping movement consisted of several cycles each demonstrating a succession of phases (Figure 7.1): flexion (F), placing (P), aiming (A), wiping (W), and extension (E). Flexion and placing were relatively independent of position of the stimulus on the frog's back, while aiming moved the paw close to the stimulus and wiping removed the stimulus from the body surface. It is interesting that wiping of the same skin area in successive cycles could be performed in different directions, and orientation of the foot relative to the stimulus (attack angle) could also change. Trajectories of the working point were shown to be more reproducible than trajectories of individual joints in all the phases. The internal structure of the wiping cycle and synergies of its phases seemed to be quite stable. Changes in ambient temperature in a wide range (3 °C to 25 °C) led to considerable changes in movement speed (up to tenfold) without any visible effects on the architecture of wiping.

A series of experiments was performed investigating the effects of unexpected perturbations on the wiping patterns (Berkinblit, Feldman, Fukson, & Latash, unpublished). In the first series, a loose thread loop was placed on the hindlimb, preventing movements in the knee joint beyond a certain limit; maximal knee joint excursion was about 5°. In control experiments, changes in the knee joint angle during certain phases of wiping, including placing and aiming, were much bigger. The frog

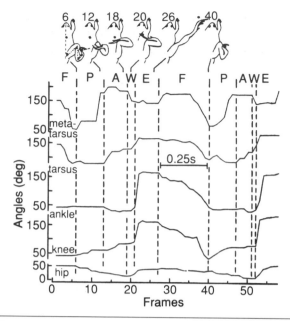

Figure 7.1 Phases of the wiping reflex in the spinal frog. Upper drawing shows typical hindlimb positions in the major phases: F—flexion, P—placing, A—aiming, W—wiping, and E—extension. Lower traces show corresponding changes in the major joint angles.

Note. From "Adaptability of Innate Motor Patterns and Motor Control Mechanisms" by M.B. Berkinblit et al., 1986, *Behavioral and Brain Sciences*, **9**, pp. 585-638. Copyright 1986 Cambridge University Press. Reprinted by permission.

was able to remove the stimulus from its back during the first attempt. Then, the knee was released and a cast was placed on the hindlimb, preventing movements in the next (more distal) joint. The frog once again wiped the stimulus at the first trial. Then, a lead bracelet was placed on the distal part of the hindlimb; weight of the bracelet was similar to the weight of the hindlimb itself. The frog still was able to wipe the stimulus accurately. In general, all the attempts to fool the spinal frog failed if it was physically able to lift the hindlimb and reach the target. The frog demonstrated reproducible errors in only one case, wiping below the stimulus, when its toes were forcefully clenched into a fist (a "glove" was put on it, preventing the toes from extending).

What do these findings mean? First, they conclusively show that motor programs, even on the spinal level, are not formulated in terms of contractions of individual muscles or even movements in individual joints. Second, they imply the existence of very fast on-line corrections of movement patterns, which are presumably built into the wiping program. Third, it seems that corrections are built-in only for conceivable perturbations (e.g., a small branch or leave on the hindlimb or the same branch on the trajectory of one of the hindlimb segments), but absolutely bizzare obstacles (toes clenched in a fist!) do not have correcting mechanisms.

More recently, other groups have turned to studying the wiping reflex in spinal frogs (Giszter et al., 1989, 1990; Schotland et al., 1989) and reported intriguing new findings. In particular, Giszter et al. (1990; see also Bizzi et al., 1991) used weak electrical stimulation through a microelectrode inserted in various frogs' spinal cord loci, containing numerous interneurons (upper and middle layers of the grey matter). They managed to find locations where the stimulation led to movements of the ipsilateral hindlimb. The same stimulation was used for different initial positions of the hindlimb. Independently of initial location, the stimulation induced a movement attempting to bring the toes of the hindlimb (working point) to the same final position (equilibrium point). Similar stimulation of ventral roots (α-motoneurons) could also bring about movements of the hindlimb, but the final locations of the working point were different for different initial hindlimb positions. Although the physiological relevance of electrical stimulation used by Giszter et al. is questionable, their data corroborate the idea that control of movements is based on the equilibrium-point principles. They also suggest that the equilibrium is not defined by certain fixed levels of α-motoneuron firing (cf. with the α-model, chapter 1) but by more complex mechanisms, probably involving polysynaptic reflexes (e.g., the λ-model).

Ostry et al. (1991) have studied withdrawal and crossed-extension reflexes in spinal frogs and withdrawal and swimming movements in intact frogs. They have found remarkably consistent straight paths in the space of hindlimb joint angles during various reflex movements and swimming. These straight paths suggest that the nervous system plans equilibrium trajectories for all the studied movements in a joint-level coordinate system, and that different kinds of motor behaviors may be built using a common set of relatively simple "bricks."

A Special Role for the Biarticular Muscles

Many distal-proximal pairs of adjacent joints have a common biarticular muscle. This kind of organization seemingly makes control more complicated. Let us try to think of a possible reason for it. Biarticular muscles provide adjustable spring-like links between adjacent joints, thus significantly influencing organization of limb synergies (cf. Nichols, 1989; Zajac & Gordon, 1989; Gielen et al., 1990; Hogan, 1990). The presence of biarticular muscles may simplify acquisition of some synergies, making them "more preferable" than the others. In a sense, biarticular muscles are a part of anatomical preprogramming of our motor apparatus.

Another important role of biarticular muscles is likely to be in providing a possibility to compensate for energy dissipation (negative work) in one joint by positive work in an adjacent joint. When a muscle is stretched by an external force and therefore performs negative work, the muscle cannot accumulate the delivered energy and dissipates it by spending some of its own energy sources (Aleshinsky, 1986; Zatsiorsky, 1986). If only uniarticular muscles are available, the organism has to spend energy for the mechanical work performed by each muscle, whether the work is positive or negative. A biarticular muscle can transfer energy and minimize the overall energy expenditure, as, for example, when two adjacent joints connected by a biarticular muscle are rotating in the same direction with a constant speed.

Biarticular muscles introduce a confusion into common physiological notions of agonist and antagonist. In particular, Zajac et al. (1986) have shown that a biarticular gastrocnemius may act either to accelerate the ankle into dorsiflexion and the knee into flexion, or the ankle into plantarflexion and the knee into extension, or the ankle into plantarflexion and the knee into flexion. The effects of gastrocnemius action also depend on dynamic interaction between the leg segments (foot and shank).

The link provided by a biarticular muscle leads to a dependence of position and/or torque in one joint on the changing patterns of activation of muscles controlling the other joint, even if the central command to the biarticular muscle is constant. Another source of such a dependence is dynamic effects of moving a segment of a limb on adjacent segments. Therefore, a command for a movement in only one joint would require a change in commands to seemingly noninvolved joints, which are connected with the involved joint by biarticular muscles. In the framework of the λ-model, motor control can be considered as regulation of reflex parameters for participating muscles. From this view, the adjustment of commands to "noninvolved" muscles can be realized at a relatively lower level of the motor control hierarchy (e.g., segmental level). Let us hypothesize that it involves built-in relations between the primary motor command, signals from peripheral receptors, and reflex parameters for the muscles not directly involved.

It seems reasonable to suppose that, for certain frequently occurring movement patterns, there exist built-in schemes providing for parallel changes in reflex parameters of muscles both involved and not directly involved in the movement. These changes would automatically take into account possible effects of biarticular muscles and adjust reflex parameters of all the muscles correspondingly. In terms of the λ-model, shifts in λ's for the muscles controlling one of the joints might lead to automatic shifts in λ for all the other muscles controlling both joints.

For example, in order to perform a single-joint movement, it is insufficient to change motor commands to uniarticular muscles controlling the joint. Motor commands to the muscles controlling the other joint (with a common biarticular muscle) have to be changed as well, in order to avoid an associated movement.

One could argue whether single-joint movements are common enough to merit a special synergy pattern, or whether they represent an artificial case common only for laboratory experiments (cf. Commentaries to Gottlieb et al., 1989a). However, one might expect that at least a nearly single-joint synergy might be a reasonably helpful block for constructing some of the everyday movements. Let us consider what kind of built-in relations between the reflex parameters (namely, parameters of the tonic stretch reflex) would be required to let the system perform single-joint movements.

Here, we are going to ignore dynamic effects of interaction between the segments (cf. Zajac & Gordon, 1989) and analyze only the contribution of the biarticular muscle, thus limiting the analysis to relatively slow movements. Generally speaking, however, activation of a uniarticular muscle can change torque in an adjacent more proximal joint. For a constant muscle force, this effect is more pronounced when a line connecting points of attachment of the muscle is far from a perpendicular to a limb segment of the adjacent joint (Figure 7.2). It is minimal, and can probably be ignored in comparison with the torques provided by the muscles controlling the more proximal joint, when this angle is 90°.

Let us consider a two-joint limb with five muscles: four uniarticular and one biarticular (Figure 7.3). Movements in joint α are controlled by muscles M1, M2, and M3, while movements in joint β are controlled by muscles M3, M4, and M5. The task is to avoid movements in joint β while moving joint α.

According to the equilibrium-point hypothesis, force produced by each muscle can be described as a function of muscle length and λ. Muscle length (L) for a uniarticular muscle is simply related to the joint angle and geometry of the muscle attachments to the bones (Figure 7.4a). Similarly, length of a biarticular muscle is a function of both joint angles (Figure 7.4b). We will use f functions for describing the dependencies of muscle length on joint angles, and F functions for describing the dependencies

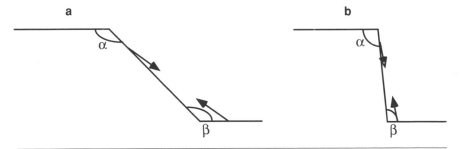

Figure 7.2   A single-joint muscle is acting around joint β. Its force vectors (shown by arrows) can create a nonzero muscle torque in a more proximal joint α that is not directly controlled by the muscle. This effect will be bigger in case a than in case b.

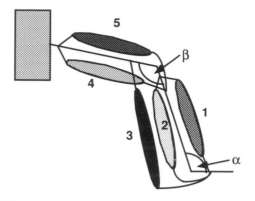

Figure 7.3   Two-joint limb is controlled by five muscles, four uniarticular (1, 2, 4, and 5) and one biarticular (3). A single-joint movement in such a system would require changes in muscle forces not only in the involved joint but also around the other joint. See equations in the text.

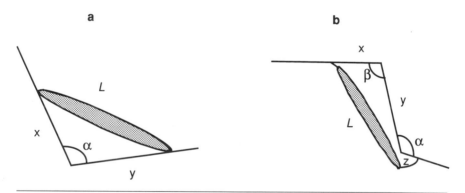

Figure 7.4   (a) Length of a uniarticular muscle (L) is unambiguously related to the joint angle (α). (b) Length of a biarticular muscle is defined by angles in both joints (α and β).

of muscle force (or torque) on muscle length and λ. Let us assume that, for small displacements, joint torques are linearly related to muscle forces (with k coefficients). Taking into account that an increase in a joint angle leads to an increase in length of some muscles and a decrease in length for others, we can write the following formulas for the muscle torques:

$$T_1 = k_1 F_1[f_1(\alpha°) - \lambda_1],$$
$$T_2 = k_2 F_2[f_2(180° - \alpha°) - \lambda_2],$$
$$T_3{}^\alpha = k_3 F_3[f_3(180° - \alpha°) + f_3{}'(\beta°) - \lambda_3],$$

$$(7.2)$$

$$T_3^\beta = k_3 F_3[f_3(180° - \alpha°) + f_3' \, (\beta°) - \lambda_3],$$

$$T_4 = k_4 F_4[f_4(\beta°) - \lambda_4],$$

$$T_5 = k_5 F_5[f_5(180° - \beta°) - \lambda_5],$$

where $\alpha°$ and $\beta°$ are initial joint angles, T_3^α and T_3^β are torques exerted by M3 in joints α and β correspondingly, and all the f and F functions are monotonically increasing.

The conditions for the equilibrium in both joints can be expressed as

$$T_1 = T_2 + T_3^\alpha,$$

$$T_5 = T_4 + T_3^\beta.$$

Let us imagine now that the control system of the mechanical structure shown in Figure 7.3 tries to perform a strictly single-joint movement in joint α. This movement can theoretically be performed with or without active participation of M3 (changing λ_3 or keeping it constant).

M3 is Passive. A flexion movement in joint α would lead to an increase in the length of M3 (due to the f_3 component). This would lead to an increase in T_3. Since M3 is passive, changes in T_4 and/or T_5 would be required to prevent movements in joint β. Namely, T_4 should decrease and/or T_5 should increase.

For an extension movement in the joint α, the opposite changes in T_4 and/or T_5 would be required to prevent movements in the joint β.

Prediction: When a movement in ankle joint is realized without active participation of gastrocnemius (M3), ankle dorsiflexion is likely to lead to a decrease in stretch reflexes of the knee flexors and an increase in stretch reflexes of the knee extensors; ankle plantar flexion would lead to opposite changes in the reflexes.

M3 is Solely Responsible for the Movement. An extension or flexion movement can be performed by changing λ_3.

For an extension movement, activation of M3 (a decrease in λ_3) would lead to a flexion movement in the joint β, which has to be prevented by either an increase in T_5 or decrease in T_4. For a flexion movement, a decrease in the level of activation of M3 (an increase in λ_3) would require the opposite changes in T_4 and/or T_5.

Prediction: When a movement in ankle joint is realized only due to λ changes for the gastrocnemius, ankle dorsiflexion is likely to lead to an increase in stretch reflexes of the knee flexors and a decrease in the reflexes of the knee extensors; ankle plantar flexion would lead to opposite changes in the reflexes.

Intermediate Case (All Muscles are Participating in α Movement). In this case, the effects on the β joint would depend on the "index of participation" of the uniarticular muscles, biarticular muscle, and external forces.

Theoretically, a matrix of the indices of participation can exist when activation patterns induce purely single-joint movements without changing central commands to the β-joint muscles (M4 and M5).

Prediction: Anything can happen.

Similar considerations can be used for analysis of a task to actively perform a flexion movement in joint β with the help of M3 without moving joint α leading to the following predictions: Activation or M3 (flexion in β) should lead to an increase in T_1 (activation of M1) and/or a decrease in T_2 (inhibition of M2). Inhibition of M3 (extension in β) should lead to opposite effects.

Some of these "predictions" correspond to the data of Nichols (1989), who studied interactions between autogenic and heterogenic reflexes in ankle and knee muscles of the cat's hindlimb. In particular, at certain levels of activation, Nichols observed paradoxical mutual inhibition between biarticular and uniarticular ankle extensors (medial and lateral heads of gastrocnemius and soleus). Suppression of soleus activity during activation of biarticular ankle extensors has also been reported for a number of experimental situations (Smith et al., 1980; Abraham & Loeb, 1986; Macpherson et al., 1986). Note that, in the framework of the present model (Figure 7.4), active contraction of gastrocnemius in static conditions corresponds to flexion in β, and for this case activation of M1 (tibialis) and inhibition of M2 (soleus) are predicted. Nichols concluded that heterogenic reflexes may be important for coordination of interjoint movements, in particular, for modulation of mechanical coupling of the ankle and knee joints. This conclusion is in good agreement with our original assumption for the role of biarticular muscles.

Attempts to Solve the Bernstein Problem

The Bernstein Problem is formulated as a quest for a sequence of motor commands to fit a motor task in the presence of "redundant degrees of freedom," that is, when the number of variables that can be specified by the control system is higher than the number of parameters defining the task. For multijoint movements, we do not have a central language similar to the λ-language for single-joint movements. Therefore, the Bernstein Problem should be reformulated in an inadequate, pseudo-Bernstein way, using a language of performance: To find a solution for a motor task when the number of mechanical degrees of freedom of the effectors is higher than the number of parameters defining the task.

There are two major approaches to the pseudo-Bernstein problem. One tries to get rid of excessive degrees of freedom by imposing certain mathematical constraints (equations) on changes in kinematic parameters. Usually, there are no clear a priori reasons why this and not another type of constraint is used, and the degree of correspondence between the model predictions and experimental findings is considered a justification for the choice of constraints. However, searching for a kinematic solution is close to the black box approach (i.e., unbiased attempts to find a complete and noncontradictory way of describing the system's behavior). It is done hoping that such a description will suggest the actual solution, that is, will let one introduce a hypothetical central language for multijoint motor control.

The other approach is based on introducing a "cost function" that is presumably optimized during voluntary movements. The choice of this function is rather arbitrary, but is frequently based on some reasonable-sounding engineering principles. This approach is based on an a priori assumption that the motor control system cares about a certain cost function more than others. Therefore, it looks more "biased" than the kinematic approach. As we will see, a striking feature of a number of criteria of optimization is similarity of their predictions: Many of them predict reasonable movement patterns qualitatively resembling those observed in the experiments. One of the possible conclusions from this finding is: Maybe the control system does not care about any of these criteria in particular, but natural movements do not considerably violate a variety of engineering principles, being reasonably smooth and safe (cf. Stein et al., 1988).

I will discuss several examples from both groups of solutions, with an emphasis on multijoint models, and continue this discussion in chapter 8.

Kinematic Solutions

The Berkinblit-Gelfand-Feldman Model. A kinematic solution for the pseudo-Bernstein problem has been suggested by Berkinblit et al. (1986b,c) based on the observations of movement patterns during the wiping reflex in spinal frogs (see earlier in this chapter). Expressed in lay terms, the major assumption of Berkinblit and his colleagues was: At any moment, movement in each joint is directed so that it brings the working point closer to the target with the speed proportional to efficacy of movement in this particular joint. Mathematically, this idea has been expressed using the product of two vectors, the first one (V_w) directed from the joint to the working point, and the second one (V_{kt}) directed from a joint number k to the target:

$$V_k = [V_w \times V_{kt}], \tag{7.3}$$

where V_k is speed vector in a joint number k. Similar equations are written for all the joints participating in a goal-directed movement (e.g., wiping). A system of Equations 7.3 defines speed in each joint at any moment of time, and therefore, yields a unique movement pattern, given initial limb configuration and target location.

This type of control can be termed *dynamic* because movements in each individual joint depend not only on position of this joint and target but also on the integrated influence of movements in all the other joints on the working point trajectory. Analysis of kinematic behavior of a multijoint limb controlled by such a system of equations has demonstrated reasonable-looking, smooth trajectories from an initial position to a target in a wide working space (Figure 7.5). Such a system demonstrates certain nontrivial properties of the experimentally observed multijoint movements, for example, successful performance of a task in conditions of fixation of one of the joints (Figure 7.5c) or during loading of one of the segments (cf. results on the spinal frog earlier in this chapter).

However, the model fails to address some of the findings in the spinal frog. For example, the placing phase of the wiping reflex is characterized by a movement in the knee joint that would bring the working point from the target if performed alone (Berkinblit et al., 1984, 1986a). Simultaneously, movements in more distal joints move the working point towards the target so that the resulting movement is to the target. Such organization may well serve a purpose, possibly dealing with stabilizing the working point trajectory and/or reducing the variability of the final working point position after the placing is complete. One of the pilot findings in the

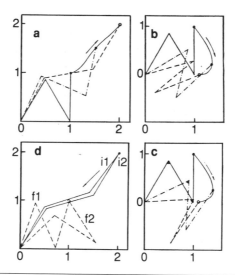

Figure 7.5 a, b: The model of Berkinblit et al. leads to smooth trajectories of the working point (the tip of the distal limb segment). c: Such a system can successfully reach a target when one of the joints (shown by filled triangles) is fixed. d: A target can be achieved starting from different initial limb configurations.

Note. From "A Model for the Aiming Phase of the Wiping Reflex" by M.B. Berkinblit et al. In *Neurobiology of Vertebrate Locomotion* (pp. 217-227) by S. Grillner et al. (Eds.), 1986, Wenner-Gren International Symposium Series, 45. Reprinted by permission.

experiments with joint fixation has suggested that fixation of the knee joint could lead to a movement in the next, more distal joint serving the same purpose (i.e., moving the working point from the target).

The model of Berkinblit et al. was also tested by Hasan and Karst (1989; see also Karst & Hasan, 1989) in experiments studying planar movements of the human arm. Their subjects were presented with visual targets at different locations and asked to move the hand to the target. Karst and Hasan studied directions in which the movements were initialized in the shoulder and elbow joints for various relative initial locations of the limb and the target. They found that, although in many cases the findings obeyed the rules of Equation 7.3, there was a considerable number of exceptions, when joint movements were initiated in a "wrong" direction.

Although the model of Berkinblit et al. does not explain all the available data, its simple and elegant mathematical formulation and also the ability to account for a number of nontrivial findings make this model attractive. It also provides an illustration that solutions for multijoint motor control problems should be sought in terms of interaction between control signals to individual joints and their combined effect on the working-point rather than in terms of combinations of such signals.

The Gutman Model. Another kinematic solution for the pseudo-Bernstein problem has been advanced by Gutman (Gutman & Gottlieb, 1989, 1990; Gutman et al., 1992) based on a quite different original consideration. Gutman introduced a notion of a nonlinear time used for planning movements in an external kinematic coordinate system, similar to nonlinearities in time perception observed during voluntary movements (Getty, 1975; Allan, 1979; Keele et al., 1987).

For single-joint movements, movement speed has been considered a linear function of distance between joint angle and target location with a factor expressed as a power function of time. Solving the resulting equations gave the following set of equations for movement trajectory, $X(t)$, velocity, $V(t)$, and acceleration, $A(t)$:

$$X(t) = D \exp\left(-\frac{t^\alpha}{\tau}\right),$$

$$V(t) = -\alpha \frac{D}{\tau} t^{\alpha - 1} \exp\left(-\frac{t^\alpha}{\tau}\right), \qquad (7.4)$$

$$A(t) = -\alpha \frac{D}{\tau} t^{\alpha-2} \left(\alpha - 1 - \alpha\frac{t^\alpha}{\tau}\right)\exp\left(-\frac{t^\alpha}{\tau}\right),$$

where $\alpha > 1$ is a constant reflecting the time nonlinearity, D is movement distance, and τ is a parameter reflecting movement speed (or time). The best correspondence with experimental data has been obtained with $\alpha = 3$.

Analysis of these equations has demonstrated a number of properties of natural single-joint movements, including bell-shaped velocity profile with a possibility of changing its acceleration-deceleration symmetry (cf. Nagasaki, 1989; Brown & Cooke, 1990; van der Meulen et al., 1990), emergence of Speed-Sensitive and Speed-Insensitive strategies (see chapter 4), and possibility of analysis of movement variability (see chapter 6).

The model has been generalized for multijoint movements, and trajectory of the working point has been expressed in a form:

$$\frac{dX_{wp}(t)}{dt} = A\,(X_t - X_{wp}(t))\,\frac{dt^\alpha}{dt}\,, \qquad (7.5)$$

where X_t is a vector corresponding to position of the target, X_{wp} is a vector corresponding to position of the working point, and A is a matrix. Equation 7.5 implies that velocity of working point shifts is proportional to the distance between the working point and the target. To transfer the working point trajectory into trajectories of individual joints, an inverse transformation must be used. Gutman used an inversion that had earlier been applied by Whitney (1969) to analysis of robot limb movements:

$$\frac{d\Phi(t)}{dt} = J^*(\Phi)\,(J(\Phi)J^*(\Phi))^{-1}\,\frac{dX(t)}{dt} \qquad (7.6)$$

where Φ is a vector of joint angles, J is a Jacobian matrix of mapping from the angular space into external physical space, $()^{-1}$ is a sign of matrix inverse, and $()^*$ is a sign of matrix transpose. Computer simulations of reaching movements using this approach have shown straight trajectories and bell-shaped velocity profiles.

Kinematic models of this kind are able to produce only some ideal drafts of planned movements (cf. virtual trajectories, chapter 2). The actual movements will depend on a number of dynamic parameters including inertial, elastic, and viscous properties of the effectors, and the field of external forces. Reconstructing actual trajectories from their virtual counterparts is no easy task, even for single-joint movements, because of a complicated nonlinear form of the equations (cf. Equations 2.9 and 2.11 in chapter 2). This task is likely to be much more complicated for multijoint movements. Nevertheless, having a procedure for calculating virtual trajectories would be a very important step.

Note, however, that in both models, virtual trajectories are calculated in terms of external physical space or combinations of movements in individual joints. This means that we still lack a central language for describing motor control of multijoint movements and, therefore, we still lack a framework to deal with the actual Bernstein Problem.

The Equilibrium Trajectory Hypothesis

Another way of introducing additional constraints on the motor system to solve the pseudo-Bernstein problem uses optimization criteria. These criteria are based on minimizing certain "cost functions" that are supposed to reflect different "principles of economy." For example, the system may want to perform a reaching movement in a minimal time, or using minimal inertial torques in the joints, or minimizing changes in some hypothetical central variables ("minimal effort"), or spending minimal energy. As we are going to see (chapter 8; for reviews see Nelson, 1983; Seif-Naraghi & Winters, 1990), some of these criteria lead to odd trajectories while the predictions of others are in good agreement with experimental observations. A striking feature of many criteria, including minimum energy, minimum jerk, minimum effort, constant stiffness, and combinations of some of these cost functions, is similar kinematic predictions (see Figure 8.2 in chapter 8).

One of the cost functions leading to a good correspondence of model predictions with experimental observations is a function of jerk, that is, time derivative of acceleration or third time derivative of movement trajectory (Hogan, 1984, 1985; Flash & Hogan, 1985; Flash, 1987). Minimizing an integral J for a given movement time T:

$$J = \frac{1}{2} \int_0^T \left(\frac{d^3\mathbf{r}}{dt^3}\right)^2 dt, \qquad (7.7)$$

where $\mathbf{r}$ is the vector of working point position, has been tested for single-joint movements, and shown to lead to nice-looking smooth trajectories.

Originally, minimization of the jerk function (Equation 7.7) was introduced for single-joint movements with a practical justification related to minimizing joint wear. For multijoint movements, this justification disappears, because minimization of the working point jerk does not necessarily mean minimization of this function for the individual joints, and the model becomes more close to the kinematic solutions described in the previous section. For single-joint movements, vector $\mathbf{r}$ may be equivalently related to either position of the working point or joint angle. Expansion of the minimal jerk model to the case of multijoint movements requires us to make a choice of coordinate system for measuring vector $\mathbf{r}$. Flash and Hogan (1985) have suggested that planning multijoint movements takes place in the Cartesian space rather than in the space of joint angles (cf. with classical observations of Bernstein of better reproducibility of working point trajectory than trajectories of individual joints). This approach has been termed the *equilibrium trajectory hypothesis*.

The equilibrium trajectory hypothesis is a natural expansion of the single-joint equilibrium-point hypothesis to multijoint movements. It implies that the central nervous system is able to shift an image of the working point along a desired trajectory expressed in external Cartesian coordinates. During a movement, this virtual trajectory is always ahead of the current actual position of the working point, and this disparity can be a measure of inertial forces during the movement. The equilibrium trajectory approach avoids the problem of inverse dynamics, because muscle forces are not calculated by the central nervous system but emerge as results of shifts of a central image of the working point.

According to the equilibrium trajectory hypothesis, planning a trajectory takes place in terms of kinematics in external space rather than in terms of hypothetical central variables. Ostry et al. (1991) have suggested that the equilibrium trajectory of a single-limb multijoint movement is planned in the space of the joint angles and, for a variety of motor tasks, represents a straight line in this space. The process of generation of equilibrium trajectory can probably be compared to the step of internal simulation according to the scheme suggested in chapter 6 (Figure 6.3). It is different from the notion of virtual trajectory that was introduced in the same scheme. For single-joint movements, the notion of virtual trajectory has been associated with hypothetical time functions $r(t)$ and $c(t)$, which are supposed to be based on the output of the internal simulation (i.e., on the equilibrium trajectory). Therefore, the equilibrium trajectory is a precursor of the virtual trajectory and requires some kind of translation to be interpretable for the lower levels of the hypothetical scheme of motor control. The difference between the single-joint and multijoint movements is that, for the former, there is a language for describing virtual trajectories, while for the latter, there is a gap between the equilibrium trajectory and actual peripheral patterns of movements.

Applying the minimum jerk principle to trajectory of the working point in multijoint movements has led to a number of theoretical predictions that include straight paths of the working point trajectories, smooth, unimodal velocity profiles (cf. bell-shaped profiles), and invariancy of the trajectory under translations, rotations, and scaling by speed or amplitude. A number of these predictions correspond to experimental observations. In particular, nearly straight paths and bell-shaped velocity profiles have been observed for various initial and final working point locations in external space and for movements at different speeds (Morasso, 1981; Flash & Hogan, 1985) corresponding to the predictions concerning invariancy of the working point trajectory. Application of the minimum jerk principle to analysis of curved movements (e.g., movement to a target around an obstacle) also yielded a number of nontrivial predictions corresponding to the published data (cf. Abend et al., 1982; Edelman & Flash, 1987; Lacquaniti, 1989).

A number of observations have been reported that seemingly contradict the minimum jerk principle, in particular deviations of working point trajectory from a straight path (Atkeson & Hollerbach, 1985; Flash, 1987; Flanagan & Ostry, 1990). Flash (1987; see also Flash & Mussa Ivaldi, 1990) has suggested that these deviations were due to the difference in mechanical properties of the effectors in different working point locations

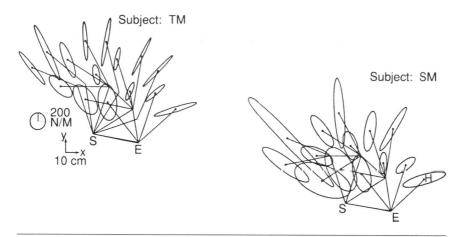

Figure 7.6   Hand "stiffness ellipses" for two subjects derived by regression of force and displacement vectors. Note the nonuniform distribution of stiffness depending on location of the hand in external space.
Note. From "The Control of Hand Equilibrium Trajectories in Multi-joint Arm Movements" by T. Flash, 1987, *Biological Cybernetics*, **57**, pp. 257-274. Reprinted by permission.

and arm configurations. Measurements of hand displacements induced by external forces in different directions in a horizontal plane let Flash reconstruct "ellipses of stiffness" (Figure 7.6), which show a considerable difference in stiffness values in different directions (see also Mussa Ivaldi et al., 1985). Therefore, if, during a movement, there is a lag between virtual trajectory and current working point location, the hand will not necessarily move exactly in the direction of the vector from the working point to the current position defined by the virtual trajectory but may exhibit some curvature due to the direction-dependent stiffness. Flash introduced corresponding corrections to the predicted minimum jerk straight paths and got somewhat curved trajectories very closely resembling those observed in the experiments.

A somewhat different approach has been developed by Mussa Ivaldi et al. (1989). It is also based on the idea that central planning of a multijoint movement can be expressed in working point shifts in the external space and, therefore, can be considered an expansion of the equilibrium trajectory hypothesis. However, the principle of control is very different from minimizing jerk. The authors have suggested that the central nervous system performs a "passive simulation" of an intended movement from a static position and defines a path that would be followed by the limb if it were moved to a new intended position passively. This path is unambiguously defined by the requirement of minimizing potential energy and corresponds to new combinations of joint elastic properties. Then, hypothetical control signals are tuning local elastic properties of the muscles (and, therefore, of the joints) correspondingly so that a movement takes place, similar to the simulated one. This method solves the problem of inverse kinematics, avoiding the pseudo-Bernstein problem by minimizing potential energy of the field of elastic forces in the joints. The model of Mussa Ivaldi et al. has a provision for movements performed in different external conditions. For example, isometric contractions are considered as results of a similar simulation process when a disparity between the virtual working point position and its actual position creates a resultant force in a desired direction.

Another model based on a passive simulation of an intended movement by calculating a "fictional force" acting on the working point has been introduced by Bock (1990). This analysis has also provided results compatible with a variety of experimental observations. However, it has required artificial-looking separate control of inertia- and gravity-related components of the controlled "fictional force."

Experiments With Jumping Targets

During the last 10 years, a relatively new paradigm for studying processes of motor control has proven to be very attractive to a number of research groups and produced a variety of intriguing findings. It includes an unexpected change in target location just before or soon after the movement initiation. In fact, this method was used in the 1970s in studies of both limb (Megaw, 1974) and eye (Lisberger et al., 1975; Becker & Jurgens, 1979) movements and had already led to important conclusions concerning the importance of continuous central processing of afferent information and the possibility of on-line corrections of a motor program.

The major idea of the experiments with "jumping targets" can be formulated in the framework of the equilibrium-point hypothesis for single-joint movements, or equilibrium trajectory hypothesis for multijoint movements. There is a central representation of a goal-directed motor command (virtual trajectory), which has a time structure. Therefore, if a stimulus for changing the motor program (due to a change in the target location) comes at a time T_0 before the virtual trajectory reaches its original destination, there are three major possibilities:

- The original program can proceed up to its planned termination and then be appended with a new program.
- The original virtual trajectory can be immediately aborted and replaced with a new virtual trajectory leading from a point corresponding to time T_0 to the new target.
- Something intermediate can happen. For example, the original virtual trajectory can be smoothly modified in order to bring the working point to the new target.

Note that these possibilities are formulated in "central" terms rather than in terms of external kinematics. Actual working point trajectories can demonstrate different patterns depending on the field of external forces and mechanical properties of the limb. For example, characteristic times for shifts of virtual trajectory during very fast single-joint movements are about 50-70 ms (Feldman et al., 1990a; Latash & Gottlieb, 1991a). Actual displacement of the working point during this time can be very small (less than 5% of the total movement amplitude), especially if the movement is inertially loaded (see chapters 4 and 5). However, taking into account the dynamic factors can help predict the hypothetical changes in virtual trajectory based on kinematic observations.

Two different patterns of kinematic changes in response to a rapid change in the target location have been described (Georgopoulos et al., 1981; Soechting & Lacquaniti, 1983; Gielen et al., 1984; van Sonderen et al., 1988, 1989; see also Alstermark et al., 1990). If a second target was presented shortly after presentation of an original target, a smooth movement to the second location was observed. If the delay before the target jump was more than a certain time interval (different critical values for the delay have been reported, generally being of the order of 50 ms), an apparent "notch" on the working point trajectory was observed. With even longer delays, a movement to the original target took place followed by a movement to the new location.

Let us illustrate these experiments for single-joint movements (Figure 7.7) when presentation of a first target is followed by presentation of a second target with a delay T_x. To simplify the description, let us assume that the movements take place in isotonic conditions, and that virtual trajectories can be modeled as monotonic fast shifts of the working point image (here it corresponds to the equilibrium point; see chapter 3) from an initial to a final position over a time $\tau = aD$, where D is movement amplitude and a is a constant corresponding to the rate of shift of the equilibrium point. Let us also assume that the delay between presentation of a target and resultant central changes in virtual trajectory is constant and independent of movement speed and direction.

If the second target is located farther from the initial working point position than the first one $(D_2 > D_1)$, the simplest way to react to the second target presented with a delay $T_x < \tau_1$ is to continue shifting the equilibrium point at the same rate until it hits the second target (Figure 7.7a). This is likely to lead to a smooth trajectory to the second target, as if it were planned in advance. If the second target is presented with a delay $T_x > \tau_1$, the first motor program has already been terminated and the subject needs to initiate a new program, shifting the equilibrium point from the location of the original target to the new location (Figure 7.7b). This is likely to lead to a visible notch on the working point trajectory.

If the second target is located closer to the initial position than the first one $(D_2 < D_1)$, the critical time for a smooth change in the motor program is τ_2. If $T_x < \tau_2$, a smooth movement to the second target will be observed

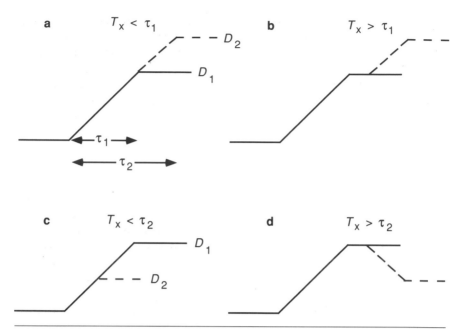

Figure 7.7 A hypothetical central variable controlling an aimed voluntary movement is shifting at a constant rate from an initial to a final value. Movement to a first target is controlled by its shift to D_1 over τ_1 while movement to the second target is controlled by its shift to D_2 over τ_2. A second target is presented with a delay T_x. Depending on the relation between T_x, τ_1, and τ_2, a smooth movement to the second target can be observed (a and c), or a visible "notch" on the trajectory (b and d).

(Figure 7.7c). If $T_x > \tau_2$, a modification of the central command is likely to be reflected in the kinematic trajectory (Figure 7.7d).

These simple considerations are in good agreement with the kinematics of single-joint movements in the experiments with jumping targets. Similar considerations can be used for describing multijoint findings invoking the equilibrium trajectory hypothesis, namely that there is a central representation of the working point gradually shifting to the target, and that this process of generating virtual trajectory can be modified at any time (cf. Georgopoulos et al., 1981; Soechting & Lacquaniti, 1983; van Sonderen et al., 1988, 1989).

Georgopoulos and his colleagues (1986, 1989) studied arm movements in monkeys in the jumping target paradigm. Generally, kinematic performance of the monkeys was in good agreement with the suggested interpretation. In some of these experiments, activity of individual motor cortex neurons was also recorded, and the authors have formulated a hypothesis of mental rotation of an imaginary movement vector that has been directly associated with the recorded vector rotation for a population of active cortical neurons. On the one hand, the hypothesis that motor cortex neurons directly encode movement direction looks very appealing as suggesting direct links between well-defined neurophysiological structures and movement planning. On the other hand, however, it creates an impression of drawing too close a parallel between morphology and function. An alternative explanation of the same data has recently been suggested by Mussa Ivaldi (1988), who has demonstrated that similar findings could result from cortex cells encoding variables characterizing individual muscles rather than movement direction in the external space.

Analysis of activity of motor cortex cells during preparation and execution of visually guided arm movements in monkeys has shown direction-dependent activity in some of the cells while activity of the others was more closely related to target location and/or moving limb (Alexander & Crutcher, 1990a,b; Crutcher & Alexander, 1990). The authors have come to a conclusion resembling that of Georgopoulos et al. (1989) that direction-dependent activity of motor cortex cells can be related to movement parameters at a relatively abstract level (kinematics, trajectory) rather than to muscle forces necessary to produce the movement.

A recent study of arm planar movement to jumping targets by Flash (1990) has supplied both expected and unexpected results. If a second target was presented after initiation of the movement, velocity profiles of movements at moderate speeds showed bell-shaped monophasic or biphasic patterns depending on the delay before presentation of the second target. Flash tested a hypothesis that the resultant movement was a superposition of two movements shifted along the time axis: The first one from the initial position to the first target and the second one from the first target to the second target (Figure 7.8). Velocity profiles for movements from the initial position to the first target were measured experimentally, and the profiles for movements from the first target to the second target were modeled using the minimum jerk principle. Algebraic summation of these two functions with a time shift demonstrated a close resemblance to the profiles observed during movements in conditions of target jumping. These observations indirectly corroborate the adequacy of the minimum jerk approach for modeling natural movements. However, movements performed at moderate speeds presumably do not have a large time lag between the virtual and actual trajectories. Therefore, when target location changes, an image of the working point is supposedly not very far from the actual working point position. If this happens during an initial phase of the movement, it seems more natural to immediately modify the virtual trajectory than to continue execution of the first motor program up to the original target and only then to start planning a movement to the second target. In particular, the model of Flash predicts that if a distant target changes its location and jumps back close to the point of movement initiation, movement time will be very long. This is not obvious and should be tested.

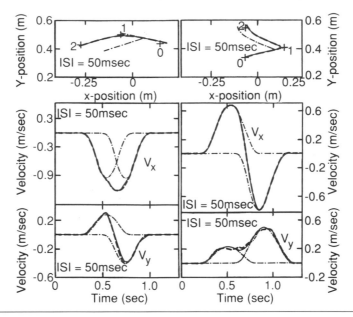

Figure 7.8   A subject performed a movement to a target 1 from an initial position 0. After 50 ms, the target was switched to 2. Measured hand trajectories are shown by solid traces. According to the model, these trajectories represented superposition of a trajectory from point 0 to point 1 and a trajectory from point 1 to point 2 shifted by 50 ms. The modeled trajectories are shown by dashed lines.

Note. From "The Organization of Human Arm Trajectory Control" by T. Flash. In *Multiple Muscle Systems. Biomechanics and Movement Organization* (pp. 282-301) by J.M. Winters and S.L.-Y. Woo (Eds.), 1990, New York: Springer-Verlag. Reprinted by permission of Springer-Verlag.

As you can see, there are a lot of intriguing findings and promising models—but we still lack a multijoint λ.

The Problem of Postural Stability

The fact that human beings are able to maintain vertical posture is, by itself, a miracle. One would have a hard time trying to imagine a mechanical system that is less stable in the field of gravity. During analysis of control of the vertical posture the human body is sometimes modeled as an inverted pendulum (Gurfinkel & Osovets, 1972; Hayes, 1982) that is not easy to equilibrate, especially in the presence of external perturbations and changes in its orientation with respect to the field of gravity. However, the problem is much more complicated due to the presence of a number of joints along the axis of the pendulum. In physics, stability of a mechanical system in the field of gravity requires that projection of its center of mass falls within the area of support (cf. Horstmann & Dietz, 1990). The area of support for a human being is relatively small (of the order of one square foot), requiring fine tuning of interaction between movements in different joints along the human body in order to maintain the equilibrium. Another miraculous phenomenon is that we can perform limb movements without falling down. Assessments of dynamic forces during natural limb movements suggest that they are more than sufficient to destroy the fragile postural equilibrium (Bouisset & Zattara, 1990; Ramos & Stark, 1990a,b). And, finally, there are no adequate words to express the profound awe that is experienced by motor control physiologists when they see a walking person. A multilink inverted pendulum can walk and run and even maintain vertical posture during stumbling!

These fantastic features of the hypothetical system controlling vertical posture did not escape the attention of Bernstein (1947, 1967), who formulated a number of problems and approaches that are still discussed. Bernstein suggested that programming of a voluntary movement must include two distinct components: The first is related to the movement itself, and the second is related to maintenance of the vertical posture (see also Gelfand et al., 1971a). Bernstein also considered maintenance of the vertical posture an illustration of the concept of synergies, that is, built-in coordinated sequences of motor commands to a number of joints leading to a desired common goal (e.g., not falling down). Presence of synergies is assumed to simplify the control of vertical posture, thus solving (at least partially) the problem of mechanical redundancy.

We are going to discuss two aspects of postural stability: First, how the vertical posture is maintained; second, how it is related to voluntary limb movements. In fact, both problems have no solutions.

Postural Synergies

There seem to be two polar views on how postural synergies to perturbations of the vertical posture are organized. On the one hand, perturbation leads to movements in the joints of the body and, consequently, to changes in the length of the muscles. Muscles are known to exhibit a variety of reflex reactions to length changes (different components of the stretch-reflex, see chapter 1), generally resisting the external perturbation, which, by themselves, lead to reactive movements at a relatively short latency. So, postural corrections could be solely due to these reflexes. On the other hand, there can be preprogrammed combinations of motor commands to different joints designed to maintain the vertical posture in cases of conceivable external perturbations (Nashner, 1976; Horak & Nashner, 1986; Nashner et al., 1989; cf. "Preprogramming in Motor Control" in chapter 1). These preprogrammed synergies can be triggered by specific peripheral signals.

These two views are certainly extremes. For example, demonstration of activation of shortening muscles during certain perturbations of the vertical posture (Gurfinkel et al., 1974) has suggested that there should be something aside from the muscle stretch-reflexes contributing to postural reactions. However, the reflexes are always there and play their role in changing muscle activity, so the hypothetical preprogrammed synergies must be tunable to take into account the reflex contribution. There is a temptation to say that the equilibrium-point hypothesis has a fair chance to solve all these problems because, in its framework, commands to individual muscles are formulated in terms of changing reflex parameters.

However, as I already mentioned, the equilibrium-point hypothesis is an essentially single-joint model and cannot be directly generalized to multijoint movements (e.g., postural synergies). For example, seemingly identical perturbations of vertical posture can lead to different individual joint kinematic trajectories successfully solving the task (Allum et al., 1989; Keshner & Allum, 1990). By the way, we do not even know exactly what the task is. One of its components is, obviously, not to fall down. There may, however, be other components including, for example, minimizing energy spending (cf. Riccio & Stoffregen, 1988), or minimizing head displacement or acceleration (cf. Gurfinkel et al., 1988; Keshner & Allum, 1990). The situation is similar to what we have just seen for control of goal-directed multijoint movements: We do not have even a hypothetical adequate language for describing the processes of motor control during maintenance of the vertical posture. There is only one solution: use an inadequate language. It seems that the language of the λ-model (combinations of r and c functions) at least has an advantage of providing a common framework for the reflex and preprogrammed components in postural reactions, and I will use it here. So, let us assume that postural synergies consist of combinations of functions $r(t)$ and $c(t)$ (see chapter 4) for individual joints. Thus, reflex reactions are not in competition with preprogrammed sequences of motor commands (synergies) but are an inherent part of the latter.

Postural synergies are frequently described as combinations of muscle activation patterns (EMGs) specific for a given perturbation (Nashner & McCollum, 1985; Diener et al., 1988; Keshner et al., 1988) and modulated by local sensory information (Dietz et al., 1989; Di Fabio et al., 1990; Horstmann & Dietz, 1990). Earliest of these reactions appear at a latency of less than 100 ms, suggesting their preprogrammed rather than voluntary nature. They are supposed to be triggered by multimodal sensory inputs with an important contribution from proprioceptive, visual, and vestibular receptors (Nashner, 1977; Allum & Keshner, 1986; Diener et al., 1988). Some of these reactions seem to be rather general, for example, coactivation of antagonist muscle pair stabilizing the joint irrespective of direction of the perturbation (Allum et al., 1989; Keshner & Allum, 1990). Other reactions are specific to the type and direction of perturbation (Keshner et al., 1988; Woollacott et al., 1988).

Yang et al. (1990) have suggested that postural synergies may involve or be based on certain fixed relationships between torques in the involved joints. They used a three-segment model (hip, knee, and ankle) and showed that such a relationship is preserved irrespective of the location, direction, and magnitude of the postural perturbation. Such an invariance may be an important feature of the maintenance of vertical posture. However, it may not be directly generalizable to the central mechanisms of postural control because, as we have already noted, control signals are likely to be formulated in more abstract terms than joint torques. Description of postural synergies with the language of muscle activation patterns (or joint torques) can, in some cases, be misleading. For example, if the same preprogrammed synergy, that is, a combination of $r(t)$ and $c(t)$ functions to the involved joints, is released in response to two different perturbations (e.g., translation and rotation of the platform), EMG pictures can be quite different due to different changes in length of the postural muscles induced by the perturbations. Therefore, differences in the EMG patterns to different perturbations do not, by themselves, prove that different synergies have been used (cf. chapter 5).

Using the concept of preprogrammed synergies for postural corrections to perturbation requires a variety of preplanned combinations of motor commands by some hypothetical function generators for all conceivable perturbations that may occur during maintenance of the vertical posture or various associated activities. This seems to place a heavy burden on the memory. An alternative approach has been formulated in the framework of dynamic organization of movements (Saltzman & Kelso, 1985, 1987; see the next section). According to this view, maintenance of the vertical posture (as well as other coordinated motor activities) is realized by introducing a certain dynamic order that ultimately defines output control signals to individual joints [e.g., $r(t)$ and $c(t)$ functions]. The system fluctuates around an equilibrium defined by the task and inherent dynamic properties of the system. Synergies emerging in response to a perturbation may be considered emergent properties of the postural system defined by its dynamic organization (cf. also Litvintsev, 1972) which do not need any preprogramming and/or triggering. In other words, a dynamic system in a state of equilibrium is already preprogrammed to return to equilibrium in cases of not very large perturbations.

Anticipatory Postural Reactions

Voluntary limb movements are virtually always associated with changes in activity of postural muscles (Lee, 1980; Cordo & Nashner, 1982; Bouisset & Zattara, 1983, 1990; Brown & Frank, 1987; Crenna et al., 1987). Some of these changes occur prior to the movement and can be addressed as anticipatory. Their assumed role is to minimize perturbations of the vertical posture that would otherwise be induced by the movement. Another group of reactions in postural muscles occur later. This group can be considered compensatory reactions to balance perturbations induced by the intended movement. At least some of these reactions are likely to be due to purely mechanical coupling between the upper and lower body segments (Ramos & Stark, 1990b).

There are clear differences in the mode of control and function between these two groups of associated changes in activity of postural muscles (cf. Hugon et al., 1982; Burbaud et al., 1988; Bouisset & Zattara, 1990; Ramos & Stark, 1990a; Rogers & Pai, 1990). Anticipatory reactions are initiated by the subject, while later, compensatory reactions are initiated by sensory feedback triggering signals as other postural corrections to perturbations (see the previous section). Anticipatory reactions try to predict postural perturbations associated with a planned movement and minimize them, while compensatory reactions deal with actual perturbations of balance that occur due to suboptimal efficiency of the anticipatory components. In particular, in reproducible experimental conditions, repetition of a motor task would be expected to lead to better preprogramming, less variable anticipatory reactions, and smaller compensatory reactions. However, dissociating these two components may not be easy because of the overlapping EMG patterns.

In accordance with the presented assumption of the function of anticipatory postural reactions, Bouisset and Zattara (1987, 1990) have suggested that these reactions initiate an acceleration of the body that would oppose the expected perturbation due to an intended limb movement. In particular, they studied postural reactions to unilateral and bilateral arm movements and have concluded that these reactions are preprogrammed and specific to the forthcoming movement, in particular to its relation to the body's axes of symmetry. Similar conclusions have been drawn from studies of anticipatory activity of neck muscles preceding a voluntary movement (Gurfinkel et al., 1988). So, both anticipatory and compensatory reactions are assumed to be preprogrammed and differ in their relative timing with respect to the limb movement and method of triggering; feedforward or feedback.

Getting back to the view on dynamic organization of postural corrections (Saltzmann & Kelso, 1985, 1987), one may expect compensatory components to reflect dynamic properties of the perturbed system. However, anticipatory reactions cannot represent a system's reaction to a future perturbation and seem to require preprogramming. This preprogramming can, however, be based on some kind of internal simulation of a future perturbation (cf. with the passive motion hypothesis by Mussa Ivaldi et al., 1989) leading to elaboration of an appropriate pattern of postural corrections. This pattern can be realized before the perturbation in the form of an anticipatory reaction.

In general, postural reactions present a useful tool for studying hypothetical mechanisms of preprogramming in general and clinical populations. Deficits in preprogramming reflected in impaired mechanisms of postural correction have been reported for various groups of patients including those with Parkinson's disease, Down syndrome, cerebral palsy, and others (see chapter 9). However, there have been very few attempts to dissociate changes in the abilities to adequately use two major groups of preprogrammed reactions, anticipatory and compensatory.

Problems of Multilimb Coordination

There are two components in any motor task involving movements of two or more limbs. First, each limb must demonstrate an appropriate movement pattern. Second, movements of different limbs should be appropriately timed to reach a common goal. Let us address the first component as "motor" and the second as "clock" (cf. Kugler & Turvey, 1987; Turvey et al., 1989). While describing the motor component during repetitive movements, a notion of "central pattern generator" is frequently invoked, implying a hypothetical interneuronal structure producing a certain pattern of muscle activation sequences within a limb (for a review see Stein, 1984). Movement patterns induced by a central pattern generator are assumed to have a relatively stable structure and only a few parameters that can be tuned by the "higher" nervous structures and/or peripheral feedback signals leading to certain modifications of the pattern (although the number of mechanical degrees of freedom for the peripheral apparatus can be much higher). One of the adjustable parameters is time scale or frequency; for example, walking can be performed at a variety of speeds while preserving a general pattern of both kinematic and EMG events. Locomotion is probably the favorite type of motor activity for studying central pattern generators. However, extensive reviews on locomotion and its patterns can be found elsewhere (Grillner, 1975, 1979), and here I would like to focus on examples that look more appropriate for studying the clock component.

Von Holst (1954), in his classical studies, considered relative timing of limb movements a consequence of interaction of two mechanisms, "maintenance tendency" and "magnet effect." The maintenance tendency reflects the inclination of a limb to perform repetitive movements at its own preferred frequency, while the magnet effect describes the inclination of a limb to be driven at a frequency of another limb. The magnet effect has been studied for a variety of motor tasks including, in particular, bimanual movements (Klapp, 1979; Yamanishi et al., 1980; Kelso et al., 1981; Turvey et al., 1986) and simultaneous vocal and manual tasks (Peters,

1977; Klapp, 1981). It has been shown that there exist only two stable rhythmic patterns for two limbs, when they are in-phase and when the phase shift is 180°. These two stable patterns persist even if the two limb movements are performed by two different persons looking at each other (Schmidt et al., 1990). This last series of experiments illustrates an important role of "informational observables" in defining patterns of voluntary movements (cf. Kugler & Turvey, 1987; Turvey, 1990b).

During discrete simultaneous movements performed by two limbs, they tend to exhibit virtually simultaneous initiation and termination of movements leading to nearly identical movement times (isochrony principle: Kelso et al., 1979a; Marteniuk et al., 1984). When the subjects are asked to perform considerably different movements in two limbs, for example unidirectional flexion in one joint and a complex sequence of flexions-extensions in the contralateral joint (Swinnen et al., 1988, 1991; Walter & Swinnen, 1990), the isochrony principle breaks down, and the subjects exhibit sequential rather than simultaneous movements in the two limbs despite the instruction explicitly requiring simultaneous movements. In the next section, we will return to the more general question of how many variables the nervous system can control simultaneously.

Lennard (1985) has suggested that timing and central pattern generator functions during multilimb movements are separate. This assumption has recently been corroborated by Turvey et al. (1989), who studied bimanual wrist rhythmic movements in a regime of "absolute coordination" (identical frequency of movement in both limbs, cf. von Holst, 1954). They analyzed "clock" and "motor" variance using a method of Wing and Kristofferson (1973) in which the subjects swung pendulums with variable mass and length. Only the motor variance depended on the deviation of the period of absolute coordination from the system's characteristic period (defined by the parameters of the pendulums), and the right and left clock variances were related. These findings let Turvey and his colleagues (see also Kugler & Turvey, 1987) suggest that absolute coordination does not simply introduce an order between two local timing mechanisms for both limbs but rather provides a common clock at a higher level.

Dynamic Pattern Generation

Analysis of the timing of rhythmical multilimb movements has stimulated development of a new controversial and fashionable approach to motor control that can be termed *dynamic pattern generation*. This approach has a rather sophisticated mathematical and physical foundation, but its experimental verifications have until now been limited to only a few types of motor behavior, in particular, patterns of relative phase during a number of bimanual tasks at different frequencies (Kelso et al., 1979a,b, 1980, 1981, 1983; Kelso & Tuller, 1984; Kay et al., 1987, 1991; Kelso & Schoner,

1988; Schmidt et al., 1990; Scholz & Kelso, 1990; Schoner, 1990; Turvey, 1990b; deGuzman & Kelso, 1991) and during single limb multijoint movements (Kelso et al., 1991). In particular, one of the major findings is the existence of only two stable patterns of bimanual rhythmic movements (in-phase and out-of-phase) with possible transitions induced by a small change in movement frequency. In chapter 2, I described a series of findings during single-joint oscillatory movements also suggesting the existence of similar abrupt phase transitions between two observables (virtual and actual trajectory) induced by a slow increase in movement frequency.

Most of the relevant findings create an impression of being rather simple and analyzable in more conventional frameworks. This dissociation between a sophisticated theory and a limited "down to earth" experimental confirmation does not mean that the theory is wrong but inadvertently leads to a suspicious attitude from more classically oriented researchers. We have already discussed certain specific features of dynamic pattern generation that make it different from more conventional approaches, in particular the seeming inapplicability of the notions of motor program (chapter 6) and preprogrammed postural corrections ("The Problem of Postural Stability" in this chapter). Let us now try to describe some of the specific features of the dynamic pattern generation avoiding complicated mathematics and excessive use of obscure terminology.

Let us define a *complex system* as a system whose behavior cannot be predicted from the properties of its components. For example, a car is a complex system because it is impossible to predict its behavior based on the properties of the elementary particles that are inherent parts of the car. City traffic is a complex system whose behavior cannot be predicted based on performance characteristics of its components (individual cars). However, behavior at each level can be described more or less adequately using a special language. Such languages are based on certain specific variables that describe general characteristics of behavior at a chosen level. For example, percentage of time pressing the gas pedal versus time pressing the brake pedal during rush hour may be a good variable for describing city traffic, although it is not directly related to individual characteristics of the cars.

In a human body, a neuron is a complex system as compared with elementary particles. The system of motor control of one muscle is complex as compared with individual neural and muscle cells. The system of motor control of multijoint movements is complex as compared with control of individual muscles. For control of a single muscle, the equilibrium-point model suggests a variable describing the muscle behavior, the threshold of a hypothetical TSR, λ. Control of movements in a joint, in the absence of biarticular muscles, can be described with time changes in a few variables [$\lambda_i(t)$, where i is the number of muscles crossing the joint]. We do not have even hypothetical realistic candidates for the control variables for multijoint movements (see earlier in this chapter). However, this does not mean that there are no control variables. Therefore, one can analyze multijoint motor control as if these variables were known.

One of the very important notions of the dynamic pattern generation approach is that of the *order parameter* (cf. also *collective variable*, Kelso & Schoner, 1988; Schoner & Kelso, 1988; Schoner, 1990; Turvey, 1990b). Generally, the number of order parameters is relatively small so that using them makes description of a complex system look simple. In other words, they reduce relatively high dimensional space states to low dimensional control states. The most frequently cited example of order parameter in the motor control literature is relative phase shift between two periodic processes. The processes can represent movements of two contralateral limbs of one person (Kay et al., 1987, 1991; Kelso & Schoner, 1988; Schoner & Kelso, 1988; Turvey, 1990b; Walter & Swinnen, 1990; Swinnen et al., 1991), two limbs of two persons looking at each other (Schmidt et al., 1990), two joints of one limb (Kelso et al., 1991), and, as we showed in chapter 2, virtual and actual trajectories during oscillatory single-joint movements (Latash, 1992). Stable patterns are commonly characterized by relatively small fluctuations of the order parameter around a certain value (180° or 360°, although other values can be observed during specialized training, as shown by Zanone & Kelso, 1992). A change in a global control parameter (movement frequency) in a certain range does not affect the order parameter. Further change in the control parameter can induce an abrupt shift in the order parameter and its stabilization at a new value. Order parameter, as suggested by the term itself, describes order within a system leading to certain stable external movement patterns that may differ considerably in time patterns of both control and performance variables but still have something very important in common, that is, the same general picture of dynamic interactions within the system.

An important step toward elaborating an appropriate language for analysis of multijoint motor control (identifying collective variables) is to single out significant units of movement (Greene, 1971; Szentagothai & Arbib, 1974), which can contain a number of muscles, or a number of joints, or even a number of limbs depending on a particular task. Behavior of these so-called "action units" (Bernstein, 1967; Whiting, 1984) or "coordinative structures" (Turvey et al., 1988; Kelso et al., 1979a,b) can be described by a few order parameters. Their presence is revealed in certain invariant relationships among kinematic, kinetic, and/or EMG events that have been described for a variety of motor activities including locomotion, writing, speech, and others (Grillner, 1979; Viviani & Terzuolo, 1980; Schmidt, 1982; Kelso et al., 1983; Nashner & McCollum, 1985; cf. with the notion of muscle invariant characteristic forming the basis of the λ-model, chapter 1). Action units are organized in a task-specific manner. They introduce an order between the components of a complex system imposing certain laws of interaction that force these components (e.g., individual muscles and joints) to act cooperatively. Such laws of interaction are likely to be defined in terms of internal dynamics of the complex system rather than in terms of movement kinematics.

Another important notion for describing dynamic pattern generation is equilibrium. A system is in a stable equilibrium if, after a small transient perturbation, it returns to the same state. The equilibrium is unstable if a small perturbation leads to major changes in the system's state and, eventually, to a new equilibrium. The perturbations can be either external or generated inside the system [e.g., by small changes in the control variable(s)]. A system can be in an equilibrium without being stationary (e.g., it can move along an equilibrium trajectory). This trajectory can be stable or unstable as defined by the system's reactions to small transient perturbations.

Let us imagine a system containing zillions of neurons (e.g., the central nervous system of a human being) that are all connected through a different number of synapses, in the absence of any explicit laws of interaction between the elements. Such systems are analyzed in physics, and there are certain laws of their chaotic behavior that may lead to spontaneous formation of stable patterns (Garfinkel, 1983; Haken, 1983; Haken et al., 1985). Imposing a set of laws of interaction on the elements (formation of action units) can introduce order into the behavior of such a system so that it comes to an equilibrium (or oscillates about an equilibrium). An equilibrium point—or an equilibrium trajectory—to which the system is attracted as a result of the imposed laws of interaction, is termed an *attractor*. Depending on the imposed laws of interaction, the attractor can represent a point, be periodic, or be "strange" (Abraham & Shaw, 1982; Garfinkel, 1983). In the last case, there will be certain concealed regularities in an otherwise seemingly random behavior of the system (cf. human electroencephalogram, Osovets et al., 1983).

In this framework, voluntary control of movements can be considered as formation of action units by imposing certain laws of interaction leading to a stable equilibrium behavior defined by a motor task. There are a number of ways to experimentally analyze such systems. For example, if one limb is tapping a rhythm, this can be considered a result of action of a certain set of laws of interaction. A transient external perturbation can distort the peripheral pattern, but, if the perturbation is small, the rhythm will be restored after a transitory process. This example demonstrates a self-regulatory, autonomous ability of the action unit. If another limb starts to tap at a different frequency or at the same frequency with a phase lag, controlling two independent tapping processes may be too much for the control system (cf. Baldissera et al., 1982), and a new action unit may emerge, embracing both limbs. The new action unit will correspond to a new pattern common for both limbs, thus decreasing the number of centrally controlled variables. Patterns of such transitions can be recorded and compared with similar transitory patterns observed in physical systems that obey similar dynamic laws or be predicted by an appropriate mathematical apparatus. This kind of comparison is virtually the only source of experimental data supporting the dynamic pattern generation approach (Kelso et al., 1979a,b, 1980, 1981, 1983; Kelso & Tuller, 1984; Kay et al., 1987; Kelso & Schoner, 1988; Scholz & Kelso, 1990; deGuzman & Kelso, 1991).

It is rather obvious that controlling movements with a smaller number of variables is more economical ("simple") than with a larger number. However, we used to think about the central nervous system as something virtually omnipotent. Then, why cannot it control two oscillatory processes with different frequencies or with a phase shift different from 180°? There are two answers. First, the central nervous system may just be lazy and prefer to simplify the task when possible. Then the observed pattern changes reflect a deliberate strategy that may be even stronger than the free will of a subject who genuinely tries to follow the instruction. Second, the central nervous system may be limited in its ability to simultaneously control a number of variables. In the second case, introducing new, larger action units looks not like a strategy but rather a necessary step to keep control. How many variables can our central nervous system control simultaneously? The answer may be surprisingly small (maybe, a magic number 7? For example, I can eat, drink, read, watch TV, talk to my wife, play with my daughter, and think about motor control simultaneously, but this is likely to be the limit). Then, controlling a movement that involves more muscles, or more joints, or more limbs, enforces creating new action units and introducing new sets of laws of interaction that would decrease the number of independently controlled variables.

Let us illustrate the contents of this section with an example. We cannot formulate the laws of control of multijoint or multilimb movements. However, it is not difficult to control general characteristics of motor behavior of large groups of people, for example, in the army. An officer can shout "Stop!" or "Forward!" and, with this simple variable, control multilimb leg and arm movements of the soldiers. An army looks like the simplest example of an action unit where the laws of interaction are extremely explicit and unambiguous. Note that behavior of this action unit is dependent on the laws of interaction between the subunits (soldiers and officers). To control an army may be much simpler than to control your own arm when you try to pick up a cup of tea.

Chapter 8

Optimization

One of the ways to try to deal with the Bernstein Problem is to suggest how to overcome excessive degrees of freedom based on some "reasonable" considerations, that is, to introduce a cost function and optimize it. There are a number of cost functions based on engineering and/or psychophysiological considerations that lead to reasonable kinematic trajectories. They, however, suggest how to build a robot that would move like a human being rather than how human beings control voluntary movements. Adding one more axiom to the equilibrium-point hypothesis (limiting rate of change of the control variables) leads to the possibility of relativistic transformations at the control level that can be observed experimentally.

One of the characteristic features of living systems is the availability of a number of solutions for a given task. Free will can be asserted even when choosing a trajectory for picking up a cup of tea. Each time a task is presented, only one solution is chosen. So, the question emerges: How does one choose a solution from an available "menu?" There should be some concealed "secondary" considerations that make this choice possible. Or, maybe, these concealed considerations are primary, and the "main" task is perceived by the subject's decision-making apparatus as something relatively unimportant. For example, avoiding spilling hot tea on one's clothes may be more important than getting the teacup to the mouth. A number of relevant examples dealing with the rehabilitation of patients suffering from certain motor pathologies can be found in chapter 9.

This problem is similar, although not identical, to the Bernstein Problem of overcoming redundancy in motor control (see chapter 7). One of the differences has been discussed in chapter 7, where I introduced the notion of pseudo-Bernstein problems that are related to peripheral rather than central motor redundancy. Pseudo-Bernstein problems can be formulated as finding a combination of joint angles corresponding to a given position of the working point in the external space when the number of mechanical degrees of freedom for the effectors exceeds the number of parameters necessary to define position of the working point. If one considers movement dynamics, another problem emerges. Even if there is no "spatial redundancy" (i.e., there is a unique correspondence between joint angles and working point positions), there is an infinite number of trajectories leading from an initial to a final position. For example, during single-joint, single degree of freedom movements, joint angle and position of the working point are unambiguously related. However, movements between two positions can be performed with different velocity profiles that probably correspond to different time patterns of hypothetical central control variables. How does the system choose the velocity profile observed in the experiments?

There are two ways to deal with this problem. First, we could observe the system's behavior and try to describe it in a noncontradictory way, hoping that such a description will eventually lead to understanding the laws of the system's behavior (cf. kinematic solutions in chapter 7). Second, we could use our knowledge of physical and engineering principles and offer the system our solution for its problems based on some "reasonable" considerations. These considerations usually involve a process of optimization of some function. There is no clear border between optimization and kinematic approaches. Some of the models use such vague and arbitrarily defined notions as "central effort" or "comfort," rather than better defined engineering or physical variables, and optimize either a hypothetical central function or an arbitrary function of performance. Although my heart belongs to the first approach, this chapter will mostly discuss the methods and results of the optimization techniques.

System Dynamics and Cost Functions

Let us analyze movements of an idealized mechanical nonredundant system with only one degree of freedom, for example, rotations in a pin joint. The pseudo-Bernstein problem does not apply in this case. Note, however, that the original Bernstein Problem can emerge if the joint is controlled by several muscles, and each muscle is controlled by the central nervous system independently (Buchanan et al., 1986, 1989; Jongen et al., 1989; Hogan, 1990; see also chapter 7). If one analyzes the purely mechanical behavior of the system, there is only one dimension and only one degree of freedom. A second order differential equation (a parallel mass-dashpot-spring model) can be used for description of movements in the joint:

$$T = m\frac{d^2x}{dt^2} + b\frac{dx}{dt} + k(x - x_0) + T_e, \tag{8.1}$$

where T is total muscle torque, T_e is external torque, x is position, t is time, m, b, and k are coefficients of inertia, viscosity, and stiffness, respectively, and x_0 is zero length of the elastic component (spring). We have already discussed some of the problems that may emerge when using Equation 8.1 for analysis of control of voluntary movements (chapter 2). However, most of the optimization models do not deal with problems of motor control and possible role of feedback mechanisms, but rather analyze a peripheral mechanical system whose movements are induced by some ideal actuators, such as the movements of an ideal robot (cf. Atkeson, 1989). The actuators can theoretically bring about any time patterns of T giving rise to different kinematic patterns leading from an initial position to a target.

Let us consider the task of moving from an initial to a final position as "primary." To discriminate between different trajectories leading to a desired change in position, it is helpful to introduce the notion of a "cost function" (F_c) that reflects "secondary" considerations affecting movement kinematics. Mathematically, these secondary considerations are expressed as requirements of minimization of a cost function. The spectrum of cost functions, analyzed in the literature, is rather wide. They can be tentatively classified into four groups. The first group contains those based on some "reasonable" engineering principles expressed in terms of performance (i.e., using measurable movement parameters like movement time, muscle torques, impulse, energy, etc.). Cost functions of the second group are based on some hypothetical central measure of "effort" and reflect the views of the authors on mechanisms of motor control. The third group combines functions of both central and peripheral parameters into a cost function. The fourth group is in fact closer to the kinematic models discussed in chapter 7, although it introduces an explicit cost function that is, however, based on some psychological rather than engineering or physiological principles.

Minimizing Indices of Performance

In this section, we will discuss models of optimization based on engineering criteria (for reviews see Nelson, 1983; Seif-Naraghi & Winters, 1990). Cost functions in these models have been based on measurable parameters of performance. However, cost function, by itself, does not define unambiguously the behavior of the system. As pointed out by Seif-Naraghi and Winters (1990), different optimal solutions can emerge for the same cost functions depending on the relation between the coefficients m, b, and k in Equation 8.1, in particular if the system is overdamped or underdamped ($b^2 > 4mk$ or $b^2 < 4mk$, correspondingly).

Movement Time

For simplified versions of the model (Equation 8.1), the optimal solution for making a movement of a certain amplitude within a minimal time is "bang-bang," known also as the teenager driving principle; that is, using the maximal level of control signals that are generally associated with maximal muscle forces or levels of acceleration and deceleration (Nelson, 1983). Velocity profile for this mode of control does not look like a typical bell-shaped pattern observed in the experiments (see Figure 8.1a). For purely inertial systems, the optimal switching time between the accelerating and decelerating "bangs" is exactly in the middle of the movement. Introducing viscosity leads to a shift of the optimal switching time depending on movement parameters and relations between m and b (Athans & Falb, 1966). A similar solution occurs for a task requiring movement over the maximal distance in a given time. Consideration of the nonsimplified version of the model (Equation 8.1) suggests different optimal solutions depending on the relation between the coefficients of viscosity, inertia, and stiffness. For example, an overdamped second-order system ($b^2 > 4mk$) requires a "bang-bang-hold" pattern of control (Seif-Naraghi & Winters, 1990).

Impulse

Impulse of the system is defined as time integral of force. For systems with negligible viscosity, minimizing the system's impulse is equivalent to minimizing the peak velocity. For a given movement time, the optimal strategy will be to quickly accelerate the system to an optimal speed, keep the speed constant, and quickly decelerate the system. This type of control is addressed as "bang-zero-bang" (Figure 8.1b). A change in movement time leads to a change in the "zero" period and to a corresponding adjustment in the optimal velocity level without changing the general trapezoid shape of the velocity pattern. If the movement must be performed "as fast as possible" (in minimal time), the solutions for minimal impulse coincide with the solution for minimal movement time, and the bang-zero-bang control becomes bang-bang. Consideration of a more full model leads to an asymmetry of accelerating and decelerating bangs, as pointed out by Seif-Naraghi and Winters (1990), who considered the requirement of minimal impulse equivalent to minimizing neural "effort."

These two optimization approaches lead to cartoon kinematic patterns and are certainly oversimplifications. However, the next several criteria lead to plausible kinematic predictions that cannot easily be distinguished from the actually observed kinematics and from one another.

Energy

A cost function, tentatively associated with energy, has been suggested in the form

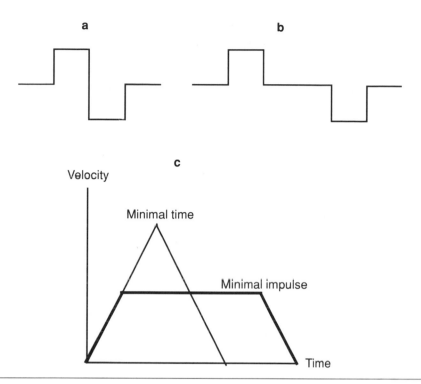

Figure 8.1   Examples of inputs associated with muscle forces, levels of acceler-ation, or "neural drives" for the minimal movement time (a, "bang-bang") and minimal impulse (b, "bang-zero-bang"). Corresponding velocity profiles for a fixed distance are shown in c.

$$E = \frac{1}{2A} \int_{0}^{T} a^2(t) \, dt, \qquad (8.2)$$

where $a(t)$ is acceleration, A is the maximal level of acceleration, and T is movement time (Nelson, 1983). This cost function is based on the assumption that the input power requirement for muscles is proportional to the square of the muscle force output (Hatze & Buys, 1977) and that muscle force is proportional to acceleration (purely inertial system). Velocity profiles for this kind of optimization are bell-shaped, and dem-onstrate an increase in peak velocity with a decrease in movement time. It is important to realize that if the hypothetical central controller calcu-lates and minimizes E, it needs to know movement time (T) in advance to use this criterion. The same restriction applies to the next two criteria based on cost functions related to jerk or torque changes in individual joints.

Jerk

Jerk is the rate of change of acceleration or the third time derivative of displacement. Hogan (1984) has suggested minimization of a jerk-related function as a criterion for optimization. Originally, this assumption was introduced based on the practical idea that jerk can be related to stress in the joints during movements, and that minimizing it reduces joint wear. Later, the minimum jerk principle was generalized to multijoint movements in a more kinematic form, minimizing the jerk of the working point. The following cost function has been introduced:

$$J = \frac{1}{2} \int\limits_{0}^{T} \left(\frac{d^3x(t)}{dt^3} \right)^2 dt, \text{ or } J = \frac{1}{2} \int\limits_{0}^{T} \left(\frac{da(t)}{dt} \right)^2 dt, \tag{8.3}$$

where $x(t)$ is vector of position of the working point in external Cartesian space, $a(t)$ is acceleration, and T is movement time. This criterion implies the existence of smooth rather than pulse-like neural inputs and leads to a possibility of nonmonotonic N-shaped virtual trajectories (i.e., time functions of a hypothetical central variable) resembling those observed in the experiments with reconstruction of virtual trajectories during fast single-joint movements (chapter 2). The minimum jerk model has been used as a basis for the equilibrium trajectory hypothesis for multijoint movements (Flash & Hogan, 1985; see chapter 7). In the framework of this hypothesis, the hypothetical central variable has been associated with an ideal planned trajectory of the working point in the external Cartesian space. The minimal jerk criterion leads to straight movement trajectories with bell-shaped velocity profiles. Later, in order to account for the experimentally observed curved trajectories, Flash (1987) introduced corrections based on nonuniform spatial distribution of the working point stiffness depending on its location in the space and limb configuration (see chapter 7). This approach has also been shown to be compatible with a number of experimental observations, including movements performed against different loads and asymmetry of the acceleration and deceleration phases (Nagasaki, 1989; Stein et al., 1988). Although the cost function for the original version of the minimal jerk model was based on an engineering principle (reducing joint wear), its multijoint version is closer to kinematic models, being expressed in terms of kinematics of the working point.

An approach resembling the minimal jerk model has been suggested by Uno et al. (1989; see also Kawato et al., 1990) in the form of a minimum torque-change model. They introduced a cost function similar to Equation 8.3:

$$J_T = \frac{1}{2} \int_0^T \sum_{i=1}^n \left(\frac{dZ_i(t)}{dt}\right)^2 dt, \qquad (8.4)$$

where $Z_i(t)$ is torque in joint i and n is the total number of joints. Since the muscular-skeletal system has complex nonlinear dynamical properties, Equation 8.4 is not equivalent to Equation 8.3. However, Equation 8.4 cannot be considered a more general case. There is a profound difference between these two approaches. The minimal jerk approach for multijoint movements is basically a kinematic model that imposes certain restrictions on the working point kinematics and does not specify which parameters are supplied by the control system to the effectors. The minimum torque-change model tries to take into account movement dynamics (in particular, effects of loading, perturbations, etc.) and is based on motor commands directly associated with muscle torques, which, as we have seen many times, is not very appealing for a motor control physiologist, although it is quite acceptable in robotics. Taking into account dynamic factors improves the predictive abilities of the minimal torque-change model, but it requires very accurate assessments of dynamic parameters of the system (in particular, coefficients of viscosity and inertia), which may not be readily available.

The authors of the model have applied an iterative scheme for calculating optimal trajectory and motor commands (muscle torques). They have demonstrated slightly curved movement trajectories resembling those reported by Morasso (1981) and Abend et al. (1982) with a bell-shaped velocity profile. When compared with the minimal jerk model, predictions of the minimum torque-change model correspond better to actual data for certain motor tasks; for example, for some initial limb configurations or when the task is to move to a target through an intermediate point located laterally with respect to a straight line connecting the initial and final positions. The minimum torque-change model has also been able to account for the data during constrained arm movements when the working point (hand) was connected to a spring. A different approach for minimizing J_T has been applied by Flash (1987) and has led to somewhat different predictions, possibly reflecting the difference in the estimates of inertial parameters for the upper limb. Since J_T depends on arm dynamics, the results of minimization are sensitive to the assessments of parameters characterizing dynamic properties of the system.

Figure 8.2 presents velocity patterns for different optimization criteria and for an undamped linear-spring model (Nelson, 1983). Note a very close resemblance of the linear-spring (K), minimal jerk (J), and minimal energy (E) patterns. All are in good agreement with experimental observations. A natural conclusion from this finding is that values of all the cost functions are kept close to optimal, that is, human movements do not

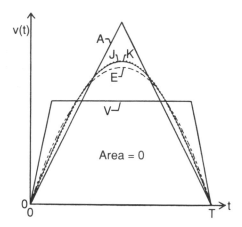

Figure 8.2   Velocity profiles corresponding to different optimization criteria. A—minimum peak acceleration, E—minimum energy, J—minimum jerk, K—constant stiffness, V—minimum peak velocity or impulse. Note very similar profiles in the J, K, and E cases.

Note. From "Physical Principles for Economies of Skilled Movements" by W. Nelson, 1983, *Biological Cybernetics*, **46**, pp. 135-147. Reprinted by permission.

violate any of the suggested engineering and/or kinematic principles (cf. Stein et al., 1988; Oguztoreli & Stein, 1990). If the aim of optimization modeling is to understand how motor control in humans is organized (and not to design a reliable robot), other factors should be taken into account, aside from the degree of correspondence of model predictions to the data. In particular, a number of questions should be asked:

- Does the optimization procedure specify parameters that should be controlled by the central nervous system?
- If the answer to the first question is "Yes," which parameters should be independently controlled?
- Can these parameters be theoretically controlled based on what we know about the human system of motor control?

Minimizing "Effort" and Maximizing "Comfort"

Hasan (1986a) has suggested optimization of a hypothetical central function as a mechanism of solving the redundancy problem and defining movement kinematics. The function was associated with the "effort" applied during the movement. This approach has been based on the λ-model of Feldman (chapter 1), and accordingly, effort has been identified with the central neural drive (i.e., centrally defined changes in the joint equilibrium position). This definition of *effort* is in good agreement with the model of position and torque perception suggested by Feldman and Latash (1982a,b) which has also been based on the equilibrium-point hypothesis. The following cost function has been introduced:

$$F_c = \int_0^T k\left(\frac{dr}{dt}\right)^2 dt, \tag{8.5}$$

where k is joint stiffness (which is assumed to be constant), r is equilibrium position, and T is movement time. Once again note that to calculate and minimize F_c, the hypothetical central controller needs to know movement time (T) in advance.

Simulations with different constant values of joint stiffness k have demonstrated virtual trajectories [functions $r(t)$] with a double-peaked first time derivative. At certain values of the stiffness, the virtual trajectories have shown a clear N-shaped pattern similar to the experimentally observed N-curves (Latash & Gottlieb, 1991c; see chapter 2). If joint stiffness is allowed to change independently, criterion 8.5 suggests the existence of the "best" stiffness when the function F_c has its absolute minimum. Hasan has assessed the best joint stiffness and has found good agreement with experimental results (Bizzi et al., 1984; Hogan, 1984). Assessments of a number of kinematic parameters, based on Equation 8.5, have given values experimentally hardly distinguishable from those reported by Hogan (1984) based on the minimal jerk model.

The optimization criterion of Hasan has recently been further developed by Lan and Crago (personal communications) who introduced the possibility of changing joint stiffness during the movement. Lan and Crago have shown that the "optimal" patterns during relatively fast single-joint movements involve a transient increase in joint stiffness in the middle of the movement and a similar to N-shaped virtual trajectory (cf. "Shifting Joint Compliant Characteristics During Fast Movements" in chapter 2). This model predicts nice-looking bell-shaped velocity profiles and even phasic burst-like activation patterns of agonist and antagonist muscles.

The notion of *comfort* has been invoked by Cruse and Bruwer (1987; Bruwer & Cruse, 1990) for creating a cost function. Their optimization approach is ideologically very close to kinematic solutions (see previous discussion) in that it represents an attempt at describing the system's behavior rather than prescribing how the system should behave. Cruse and Bruwer assumed that, for every joint, there exists a position of "maximal comfort." They performed a series of psychophysiological experiments (Cruse et al., 1990), and have identified these positions for the wrist, elbow, and shoulder joints. Not unexpectedly, these positions happen to be somewhere in the middle of the physiological range. They also asked the subjects to assess the level of discomfort at other positions in the same joint and have reconstructed comfort functions. These functions have a close to parabolic form, sometimes being nonsymmetrical, and are rather stable, independent of changes in other joints. The authors have suggested that comfort functions represent a "psychological cost" for the subject to keep a not very comfortable position in a joint. A sum of the comfort functions for all the joints was considered the cost function for the multijoint limb, and voluntary movements have been assumed to be based on minimizing the cost function (i.e., minimizing "discomfort"). This approach certainly does not belong to engineering optimizations, and all the psychological reasonings look like no more than justifications for introducing a rather arbitrary-seeming "rule" for calculating optimal combinations of joint angles that happen to be in good agreement with experimental observations.

Cruse et al. (1990) have analyzed a variety of natural multijoint pointing arm movements performed with and without additional loadings and changes in the limb geometry (movements with a pointer). The basic ideas underlying some of these experiments are close to those described previously for studies of the wiping reflex in spinal frogs. However, basing the cost functions on values of individual joint angles has led to a discrepancy of the predictions with the data, which is not very unexpected since the reproducibility of individual joint positions for multijoint movements performed in different external conditions is far from ideal (see chapter 7). In fact, a model based on only kinematic parameters (joint angles) should not be expected to be able to account for the data in conditions of changing dynamic factors (e.g., loading). Therefore the attempts by the authors to explain these discrepancies by referring to vague notions of dependence of the cost functions on some physiological costs associated with holding a loaded joint at a certain position do not look persuasive.

This model has been realized with the help of both an algorithmic procedure and network modeling (Bruwer & Cruse, 1990). The authors have used their cost function for building two network models similar to those of Kohonen (1982; see also Ritter et al., 1989) and of Josin (1988; see also McClelland & Rumelhart, 1988). Both modeling approaches have been shown to solve the "static" problem (i.e., finding an optimal combination of joint angles for a given target position). The second model has also been able to account for some of the features of movement dynamics, including slightly curved trajectories (cf. Atkeson & Hollerbach, 1985; Cruse & Bruwer, 1987; Flash, 1987). However, the network models experience considerable difficulties that could be overcome by the algorithmic approach, for example when the initial position corresponded to a rather uncomfortable combination of joint angles.

Complex Cost Functions

Reviewing different optimization approaches creates a general impression of their equivalency in the sense that most of the models demonstrate a good correspondence to the data, and predictions of a number of them seem to be experimentally indistinguishable. This is not too surprising because models whose predictions apparently contradict the data are likely to be either not submitted for publication or get negative reviews and be rejected. So, there are a number of "good" cost functions and, correspondingly, a number of good optimizing strategies (cf. Pedotti & Crenna, 1990). Apparently, if one combines several good cost functions and tries to optimize their linear combination with variable coefficients (weights), the fit to the data can be further improved. In other words, if the system, depending on the motor task, can decide to pay $X\%$ attention to jerk, $Y\%$ to energy spending, and $Z\%$ to the internal effort, choosing appropriate (X,Y,Z) combinations can account for virtually any peculiarities of the data. There is only one question left: Does it make any additional sense aside from stating that the system can theoretically be omnipotent?

Such an approach has been used by Seif-Naraghi and Winters (1990; see also Martin et al., 1989). Their model is based on optimizing a complex function with a number of major components related to movement time, error in the final position, "input effort," "input energy," and some "metabolic/mechanical energy considerations." This approach has proven to be quite useful for describing various optimization models by making all the coefficients zero, except one. Therefore, the model of Seif-Naraghi and Winters is, by itself, an excellent review of different optimization techniques suggested before. Its predictive ability looks quite impressive and includes kinematic, dynamic, and electromyographic patterns during voluntary movements; emergence of the two strategies, Speed-Sensitive and Speed-Insensitive (see chapter 4); changes in motor patterns in different loading conditions and in cases of perturbations; and so on. However, this improvement in the predictive ability, as compared with earlier "unimodal" optimizations, is quite expected and close to trivial. It seems that, for a breakthrough in the area of optimization, we need something qualitatively different than just an accumulation of optimized functions.

Relativistic Motor Control

As we have already seen, the number of models attempting to describe how human voluntary movements are controlled is monotonically increasing with time. However, this proliferation does not necessarily imply a similar increase in understanding of how the motor control system functions, because a sizable proportion of the models is related to a confined body of experimentally observed phenomena or to some general principles of multilink multiple degrees of freedom mechanical systems without, first, asking the questions: Can such a scheme theoretically function in a living body?, and: Is the model specific to motor control in humans (or animals)?

In an attempt to put limitations to such models, Adamovitch and Feldman (1989) have recently formulated a number of prerequisites for the theories of motor control. These prerequisites are not model-specific; they are specific to what we know about the central nervous system and the control systems in general. In particular, they require that such theories consider movements as shifts in the equilibrium state of the system and indicate parameters whose regulation allows the brain to execute the shift. Incorporating such limitations seems a very timely idea. We are going to add one more axiom to these limitations, which is applicable to any model irrespective of which variable is regulated.

The rate of change of a controlled variable in the motor control system is limited.

This seemingly obvious and innocent statement leads to rather dramatic consequences, including the possibility of relativistic transformations in cases of superimposed motor programs (see Latash et al., 1991).

According to the λ-model, central regulation of single-joint movements can be described with functions $r(t)$ and $c(t)$. I am not going to analyze changes in $c(t)$ in this section. In the simplest version of the λ-model, smooth single-joint movements are considered to be results of unidirectional shifts of r at a constant rate (see Figure 2.12a in chapter 2). More complex patterns of r changes have been suggested and experimentally demonstrated (Latash & Gottlieb, 1990b, 1991c). However, even in this case, one may assume that the rate of the r changes (ω) can be regulated by the central nervous system to perform movements with different speeds.

The introduced axiom requires existence of a maximal rate of r shift, ω_{max}. Let us now imagine that a person tries to perform a movement in a joint on a background of another movement in the same joint (e.g., two smooth flexions). If performed separately, the central programs for these movements could be described with functions $r_1(t)$ and $r_2(t)$. An attempt to simultaneously realize both programs may theoretically lead to a simple algebraic summation of the r functions:

$$R(t) = r_1(t) + r_2(t). \tag{8.6}$$

However, the axiom puts limitations on the resultant function $R(t)$. These limitations can be realized in different ways. For example, algebraic summation 8.6 can take place at all times when

$$\frac{d[r_1(t) + r_2(t)]}{dt} = \omega_1 + \omega_2 < \omega_{max}.$$

If $\omega_1 + \omega_2 \geq \omega_{max}$, saturation occurs, so that $dR(t)/dt = \omega_{max}$. This mechanism (illustrated in Figure 8.3) would lead to a constant rate of $R(t)$, when $\omega_1 + \omega_2 \geq \omega_{max}$ (shown by bold solid lines). This would eliminate specific

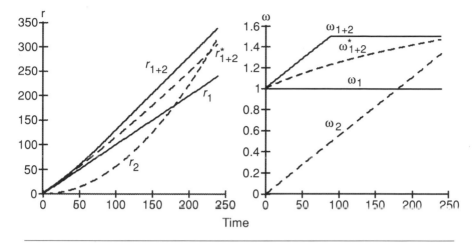

Time

Figure 8.3   Examples of superposition of two motor programs. The first one (r_1) represents an increase in r at a constant rate ω. The second one (r_2) increases ω linearly with time r_{1+2} and ω_{1+2} show the results of superposition with saturation, while r^*_{1+2} and ω^*_{1+2} correspond to superposition with a "relativistic" transformation.

Note. From "Relativistic Effects in Single-Joint Voluntary Movements" by M.L. Latash et al., 1991, *Biological Cybernetics*, **65**, pp. 401-406. Reprinted by permission of the publisher.

features of both programs, which are due to specific functions $r_1(t)$ and $r_2(t)$, replacing them with a constant rate increase in $R(t)$. It seems preferable to have a mechanism limiting ω_{max} and simultaneously preserving specific features of the motor programs.

One possible alternative was proposed by Lorenz in the 19th century and has formed the basis of Einstein's theory of relativity.

Suppose that a hypothetical central control system issues commands $r_i(t)$ in some internal system of coordinates. There are in fact only two coordinates, time (t) and r because we are not going to discuss changes in c. This internal coordinate system is "motionless" if no command is running. If there is already a command $r_1(t)$, it leads to a corresponding shift of the coordinate system. Generating another command $r_2(t)$ in the moving coordinate system in the presence of ω_{max} leads to a distortion of algebraic summation (Equation 8.6). This distortion can be described in the same terms and equations as the nonlinear summation of velocities in our universe in the presence of an upper limit, the speed of light c. Simple considerations, which can be found in any college textbook on physics, lead to the following formula for summation of ω:

$$\omega_{1+2} = \frac{\omega_1 + \omega_2}{1 + \dfrac{\omega_1 \omega_2}{\omega_{max}^2}}. \tag{8.7}$$

Equation 8.7 for summation of ω forces similar transformations for time, that is, transformations of internal time of the program realization as compared with the external ("body") time. If a program is taking place on the background of another program, $r_1(t)$, "shifting" the system of coordinates [$\omega_1 = dr_1(t)/dt$]:

$$t_1 = \frac{t}{\sqrt{1 - \dfrac{\omega_1^2}{\omega_{max}^2}}} . \tag{8.8}$$

In this equation, both t and t_1 represent internal variables used for generating the control functions $r(t)$ and $c(t)$ [or $r(t_1)$ and $c(t_1)$]. Time t is assumed to be used for controlling a single motor act in the absence of superposition of motor programs and to have a direct relation to the external time, t_{ext}. For example, there may be a constant coefficient k transforming t into t_{ext} ($t = kt_{ext}$). Here we will assume $k = 1$. However, since we will be dealing only with relative changes in time, the conclusions are independent of k. Another variable, t_1, is used when a movement is executed on the background of another motor program. Its relation to t_{ext} will depend on the characteristics of the background program (ω_1) as suggested by Equation 8.8. An example of the "relativistic" superposition of two motor programs $r_1(t)$ and $r_2(t)$ is shown in Figure 8.3 by the bold dashed lines.

Assessments of the maximal rate of r changes for nearly isotonic elbow flexion movements of relatively large amplitude (50°-70°), performed by two groups (Feldman et al., 1990a; Latash & Gottlieb, 1991a), have led to similar values: about 700°/s - 800°/s. Although these values are relatively high, they allow experimental demonstration of the relativistic effects, described by Equations 8.7 and 8.8, during common and even not very fast movements.

A relatively slow flexion movement in the elbow joint can be described with a shift in the central parameter r at a constant rate ω. Using a method for reconstructing shifting joint compliant characteristics during single-joint movements described in chapter 2, one can reconstruct virtual trajectory of such a movement and assess ω. Figure 8.4a shows an unperturbed trajectory of a slow movement (over 1000 ms) and a virtual trajectory. Note that the virtual trajectory demonstrates a relatively rapid change in the beginning of the movement, which is probably necessary to accelerate the limb. Figure 8.4b shows superimposed velocities of the movement and virtual trajectory. Figure 8.4c shows relative changes in time calculated from Equation 8.8 assuming $\omega_{max} = 800°/s$. Note an approximately 5% change at the beginning of the movement, and much smaller changes during the rest of the movement.

In order to measure the predicted relativistic time changes, the subjects were instructed to perform slow elbow flexion movements on the background of relatively fast oscillations in the same joint. If a nonlinear

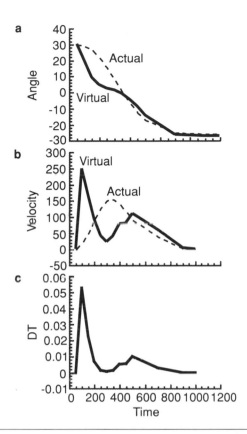

Figure 8.4 Actual and virtual trajectories (a, in degrees) and their velocities (b, in °/s) during a relatively slow elbow flexion. The lower graph (c) shows predicted time distortion (DT, in %).

Note. From "Relativistic Effects in Single-Joint Voluntary Movements" by M.L. Latash et al., 1991, *Biological Cybernetics*, **65**, pp. 401-406. Reprinted by permission of the publisher.

superposition of two motor programs corresponding to Equations 8.7 and 8.8 takes place, the period of oscillations on the background of a smooth flexion should increase (an exaggerated example is shown in Figure 8.5).

Examples of the data for each individual subject (after certain normalization procedures) and the averaged data are shown in Figure 8.6. Note an increase in the period of the oscillations during the first phase of the movements. This increase was statistically significant for each subject at $p < 0.05$ (Student's two-tailed t criterion). Note a consistent increase of the period during the first half of the movement. The peak increase in the period, averaged across the subjects, is about 5%-7% and is in good agreement with the predictions.

One might suggest several interpretations of the data. For example, different psychophysiological reasons may be invoked for interpreting

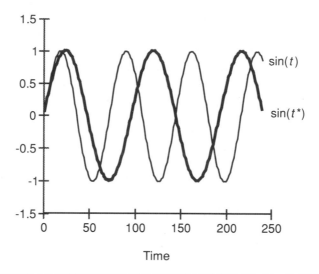

Figure 8.5   An oscillatory movement is modeled by a sine function sin(*t*). When this movement is performed on the background of a unidirectional flexion in the same joint at a constant speed, relativistic transformation leads to a function sin(*t**) (bold line) whose period is bigger than that of the original function sin(*t*).

Note. From "Relativistic Effects in Single-Joint Voluntary Movements" by M.L. Latash et al., 1991, *Biological Cybernetics*, **65**, pp. 401-406. Reprinted by permission of the publisher.

the observed shifts in the oscillation periods, including vague notions such as "distraction of attention."

However, because the present approach directly predicts shifts in the periods in the same direction of approximately the same magnitude, there is no reason to search for more complicated interpretations. In fact, if the speed of light were, say, 100 meters per second, this interpretation would seem obvious, because it would be corroborated by observations of relativistic effects in everyday life. Our body is its own universe, and its "speed limit" (the highest speed of information transmission, which is limited by the speed of nerve impulse conduction) is comparable with characteristic speeds of fastest movements. So, relativistic effects can and are likely to play a role in the human body.

Another example of relativistic effects has been presented by V. Smolyaninov and A. Karpovich in their studies of locomotion in different species (personal communications). In these studies, relativistic effects were predicted when frequency of limb movements reaches a limit posed by the speed of signal transmission along the body. The authors also attributed the effects to the limited speed of propagation of neural signals along the body axis.

One should not be scared of the term "relativistic," as something from a different world, because the world of our body is different.

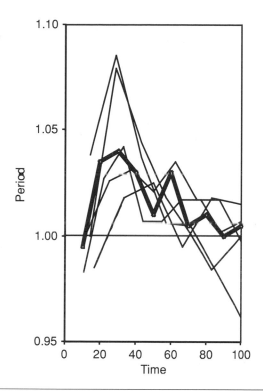

Figure 8.6   Values of the oscillatory movement period averaged across individual trials are shown by thin lines for individual subjects. Data averaged across the subjects are shown by a bold line. Note an increase in the period during the first half of the movement.

Note. From "Relativistic Effects in Single-Joint Voluntary Movements" by M.L. Latash et al., 1991, *Biological Cybernetics*, **65**, pp. 401-406. Reprinted by permission of the publisher.

Chapter 9

Examples of Motor Disorders

Analysis of motor deficits following certain disorders within the central nervous system suggests that a sizable part of these deficits are consequences of adaptation to the primary disorder rather than of the disorder itself. Spasticity, a predominantly spinal disorder, has relatively more primary motor deficits, some of which are reversible with intrathecal baclofen. In Parkinson's disease, most of the motor deficits may be considered adaptive, while in Down syndrome, all the problems in motor control are likely to be consequences of adaptation of the central nervous system to the primary impairment in cognition.

In this chapter, we are going to attempt to reconcile, at least partially, two seemingly irreconcilable aspects of motor control: Modeling and Clinics. Clinical observations present a serious challenge to any model of motor behavior. First, the language of clinical reports differs substantially from the language of theoretical or even experimental motor control studies and requires some kind of translation to make the data interpretable. Second, movement pathologies are usually defined in terms of signs and symptoms rather than in terms of mechanisms; this is due, in part, to the virtual lack of knowledge about the underlying mechanisms. At best, some of the commonly studied motor pathologies are associated with dysfunction of a relatively well-localized anatomical formation of the central nervous system, e.g. with pathological changes in basal ganglia in the cases of Parkinson's disease and Huntington's chorea. However, we should remember the warning of Bernstein (1967, see chapter 1) against making too direct a parallel between anatomical formations and motor functions.

The cybernetic (black box) approach, used for describing models of motor control in the previous chapters, introduces some kind of order into data presentation and attempts to elaborate a universal language to make results of different studies comparable and interpretable in the same framework. However, the same approach alienates the models from the "real" world by introducing concepts that frequently do not have straightforward analogues in the realm of clinical studies or even a defined neurophysiological substrate. As a result, clinical data become very hard to analyze from theoretical standpoints, and the lack of understanding between motor control physiologists and clinical neurophysiologists gets deeper.

However, if motor control physiologists hope to relate their models to "real life," they should take into account the available (although limited) information about the hypothetical controller: the central nervous system itself. Even if this information is limited in quantity and suboptimal in quality, it nevertheless imposes certain restrictions upon a priori unlimited imagination of the theoreticians. Generally, there are two major ways of getting relevant information.

One approach to this problem is animal studies, which allow the researcher to introduce more or less well-defined pathological changes or to electrically stimulate more or less well-defined anatomical formations. There are two problems with animal studies, however. The first problem is that you cannot ask an animal to perform a certain task, although training in standardized experimental conditions sometimes compensates for this disadvantage. The second problem is that observations in animal experiments quite frequently prove to be not directly generalizable for motor control in humans. This refers not only to well-known specific features of human (or primate) motor control, for example, a developed system of monosynaptic cortico-motoneuronal projections to certain muscle groups of the upper extremities controlling fine finger movements (Asanuma, 1973; Merton & Morton, 1980; Muir & Lemon, 1983), but also to seemingly identical systems. For example, stable spasticity associated with a profound loss of voluntary motor control is, unfortunately, a frequently observed phenomenon in humans. In animals, however, it is much harder to get a good model of spasticity: In many cases, the animal either does not survive or restores its motor function to a close to normal level.

Another approach is related to clinical motor control studies that have two important aspects. The first is practical: Testing the efficacy of new methods of treatment and improving our knowledge of the mechanisms of motor disorders. This approach is based on applying our knowledge about intact motor control systems for understanding a pathology. The second aspect is more selfish: To study a pathology in order to improve our knowledge about the intact motor control system. I am going to present several examples of both approaches for various motor disorders.

After describing the findings, I will attempt, when possible, to translate them into the language of the equilibrium-point hypothesis. Particular attention will be paid to three questions:

- What are the changes in voluntary motor commands in certain motor disorders?
- What are the interactions between changes in muscle reflexes and voluntary motor commands?
- Which pathological changes are the results of the "primary neurological disorder" and which are more likely to reflect secondary, compensatory changes in the motor control system?

Spasticity

Spasticity is a common component of a variety of motor disorders resulting from brain trauma (including cerebral palsy), spinal cord injury, and certain systemic degenerative processes (including multiple sclerosis). It is defined as a disorder of spinal proprioceptive reflexes manifested as profound changes in reflexes to muscle stretch with a strong velocity-dependent component, emergence of pathological reflexes and uncontrolled spasms, an increase in muscle tone, and impairment of voluntary motor function (Hughlings Jackson, 1889; Lance, 1980a,b). As one can see, spasticity is defined in terms of signs and symptoms rather than in terms of underlying mechanisms. Although it has been commonly assumed that spasticity is associated with a deficit in spinal inhibitory mechanisms including both postsynaptic and presynaptic inhibition (Pedersen, 1974; Pierrot-Deseilligny, 1983; Davidoff, 1985; although cf. Burke, 1988), there is no consensus about what causes these deficiencies in inhibition. To say that they are due to disruption of normal functioning of certain descending systems does not help much because these systems are not well-defined and their role in voluntary motor control is vague.

Relations between spasticity and muscle reflexes are not as unambiguous as implied by the presented definition. For example, spasticity can be associated with exaggerated, unchanged, and even absent monosynaptic reflexes including the H-reflex (see "Monosynaptic Reflexes" in chapter 1). The Babinski reflex, a common correlate of spasticity, can be absent in certain patients. The same is true for clonus in ankle and other joints. A variety of changes in different components of muscle reactions to stretch have been reported (Herman, 1970; Burke et al., 1971; Lance, 1980a; Berardelli et al., 1983). No reproducible differences have been found between gains of the long-latency reflexes to muscle stretch (including the hypothetical tonic stretch reflex) in spastic and control subjects (Berger et al., 1984; Rack et al., 1984) and the increased resistance to muscle stretch has been partially attributed to purely peripheral changes in muscle stiffness

(Dietz et al., 1981; Dietz & Berger, 1983; Hufschmidt & Mauritz, 1985; Thilmann et al., 1991b). A recent study by Lee et al. (1987) has suggested that threshold of the tonic stretch reflex (λ) may be changed in spasticity. In these experiments, muscle reactions to slow passive joint movements were studied in control and spastic subjects, and earlier emergence of reflex α-motoneuron activation was demonstrated in spasticity. In the framework of the modern version of the λ-model ("λ"-model, see "Muscle Reactions to Length Changes" in chapter 3), this is an expected result since changes in muscle reflexes are, by definition, mediated by "λ" changes. This interpretation is not that obvious, however, in the original λ-model, which permits recruitment of α-motoneurons by other, "non-λ" reflex pathways that might bring about earlier recruitment of α-motoneurons in spastic patients without any changes in the λ-loop. Note that a more recent study of patients with spastic hemiparesis by Thilmann et al. (1991a) has demonstrated an increase in the gain of the stretch-reflex in spastic muscles as compared with contralateral "normal" muscles and no changes in the reflex threshold.

The lack of uniformity in reflex changes in spasticity does not permit using changes in a certain reflex (or in a number of reflexes) for reliable quantitative assessment of this motor disorder. Suppression of monosynaptic reflexes by muscle vibration, which is presumably mediated by presynaptic inhibitory mechanisms (Hagbarth & Eklund, 1968; Gillies et al., 1969; Delwaide, 1973), has been suggested as a quantitative index for spasticity (Ashby & Verrier, 1976; Ongeboer de Visser et al., 1989). This method, however, does not reflect the state of postsynaptic inhibition and, in some cases, poorly correlates with clinical status. Also, monosynaptic reflexes are sometimes absent in spasticity, making this method inapplicable. Two clinical scales have been successfully used for quantitative assessment of spasticity. The most frequently used is the Ashworth scale, which reflects degree of muscle resistance to passive limb movements (Table 9.1; cf. Hattab, 1980). The other is the so-called "spasm scale," which reflects frequency of spasms, their duration, and their general or local character (Table 9.2, Penn et al., 1989). Both scales are subjective and reflect general impression of the patient's state on the physician, which is likely to be important from the clinical view but is not very helpful for understanding the mechanisms of spasticity.

Even more uncertain is the relation between spasticity and voluntary motor control. Hughlings Jackson (1889) considered the positive signs of spasticity to be an increase in muscle reflexes (a release phenomenon) and the motor disability to be a negative symptom reflecting damage of α-motoneurons. In his view, eliminating the signs of spasticity would not be expected to help motor function. More recently Landau (1974, see also Sahrmann & Norton, 1977) has restated this position and argued that motor function cannot be improved with any procedures that might reduce spasticity. He has warned that such expectations will not be met

Table 9.1   The Ashworth Scale

Score	Degree of muscle tone
1	No increase in tone
2	Slight increase in tone, giving a "catch" when affected segment is moved in flexion or extension
3	More marked increase in tone, but affected segment is easily flexed and extended
4	Considerable increase in tone; passive movement is difficult
5	Affected part is rigid in flexion or extension

Note. From Penn et al., 1989; reproduced with permission of the *New England Journal of Medicine.*

Table 9.2   The Spasm Scale

Score	Frequency of spasms
0	No spasms
1	Mild spasms induced by stimulation
2	Infrequent full spasms occurring less than once per hour
3	Spasms occurring more frequently than once per hour
4	Spasms occurring more frequently than ten times per hour

Note. From Penn et al., 1989; reproduced with permission of the *New England Journal of Medicine.*

because they do not deal with the primary problem of diminished neural input to segmental control of the final common path.

Landau's view has been challenged by a number of studies in which hyperactive reflexes and co-contraction of antagonist muscles appear to interfere with proper motor function (Mizrahi & Angel, 1979; Knutsson, 1983; Berger et al., 1984). In particular, Corcos et al. (1986) demonstrated that the hyperactive soleus reflex in spastic patients could interrupt voluntary dorsiflexion. In this case, it seems that more normal movement might occur if the hyperactive reflex were suppressed. However, to conclusively support this view, it was necessary to demonstrate actual improvement in voluntary motor control after eliminating the spastic signs and symptoms.

Until recently, however, most of the therapeutic approaches to spasticity, including oral medications, were of limited success (Hattab, 1980; Merritt, 1981), forcing some of the patients to turn to destructive neurosurgical procedures (see reviews in Merritt, 1981; Dralle et al., 1988), which, by themselves, could be severe enough to cause further deterioration of residual voluntary motor function if it had been present before.

In more recent studies, intrathecal infusions of baclofen, an agonist of gamma-aminobutyric acid (GABA), have been shown to effectively reduce muscle spasms and exaggerated reflexes (Muller et al., 1987; Latash et al., 1989b; MacDonell et al., 1989). Clinical significance of this new, intrathecal way of drug delivery with implantable pumps has been widely recognized (Penn & Kroin, 1985, 1987; Muller et al., 1988a,b; Penn et al., 1989), especially after a carefully controlled double-blind study with a long-term follow-up (Penn et al., 1989), which has unambiguously demonstrated the effectiveness of this method. Another exciting possibility that emerged as a result of using intrathecal baclofen has not attracted comparable attention: Effectiveness of baclofen and its fast action (Muller et al., 1987; Latash et al., 1989b, 1990) provide a rare opportunity in clinical motor control studies to use a spastic patient as his or her own control in two states, with and without spastic signs.

Intrathecal Baclofen: A Wonder Drug and a Wonder Tool

The site of action of baclofen in the spinal cord is not clear. It has been shown to be a selective ligand for bicuculline-insensitive GABA receptor sites that occur widely in the central nervous system (Davies, 1981; Bowery et al., 1984). Baclofen-sensitive GABA receptors have also been found on primary afferent terminals (Price et al., 1984). These findings are consistent with observations of pronounced suppression of both monosynaptic and polysynaptic reflexes with intrathecal baclofen in animal models (Henry, 1980; Curtis & Malik, 1985) and in man (Muller et al., 1987; Latash et al., 1989b). For example, studies of short-term effects after a single intrathecal injection of baclofen have demonstrated an effective suppression of monosynaptic reflexes (H-reflex) up to its total elimination during the first 30-45 minutes after the injection (Latash et al., 1989b; MacDonnell et al., 1989). This was followed by a more gradual suppression of polysynaptic reflexes to passive joint movements and Babinski's reflex.

A double-blind study of intrathecal baclofen was carried out in patients with multiple sclerosis and spinal cord injury and spasticity resistant to all available nondestructive therapies including oral baclofen (Penn et al., 1989). This study has demonstrated a pronounced effect in reducing scores for both Ashworth and spasm scales accompanied by clear clinical gains. The effect of intrathecal baclofen can be maintained with a programmable drug delivery through an implanted pump as demonstrated by a long-term follow-up (up to 5 years in some of the patients; Figure 9.1 shows

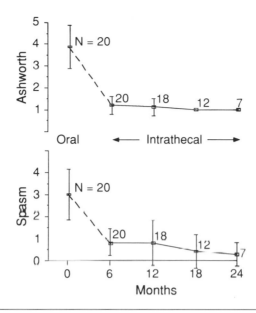

Figure 9.1   Ashworth and Spasm scores averaged across the subjects during the long-term follow-up. Zero on the time scale corresponds to the scores on the maximal dosage of oral baclofen before the pump implantation. The number of patients decreases because the drug pumps were implanted at different times.

Note. From "Intrathecal Baclofen for Severe Spinal Spasticity" by R.D. Penn et al., 1989, *New England Journal of Medicine, 320*, pp. 1517-1521. Reprinted by permission of the *New England Journal of Medicine.*

earlier data, when the longest follow-up time was 2 years). Effectiveness of baclofen in predominantly spinal cases of severe spasticity has now been well documented although its therapeutic value for spasticity of supraspinal origin is still waiting to be conclusively proven.

Besides being a "wonder drug," baclofen also appears to be a "wonder tool" for motor control studies, presenting the investigators with a unique chance to compare a patient with himself in two dramatically different states: Two identical motor testing sessions are performed, one before and one 1.5 to 3 hours after a bolus intrathecal baclofen injection. This time interval is long enough to ensure profound suppression of pathological reflexes (Latash et al., 1989b) and short enough to perform the testings within one day (in the hope that all the changes in the patient's motor performance are due to the baclofen injection).

Let me illustrate the effects of intrathecal baclofen on voluntary movements in a patient with a long-standing spastic hemiparesis following head trauma. The patient was tested prior to and 2 hours after a bolus injection of baclofen. All the pathological reflexes in the muscles of the

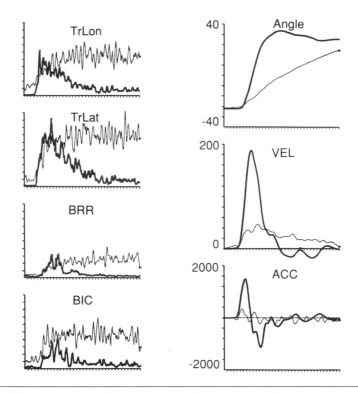

Figure 9.2 Attempts at elbow extension "as fast as possible" averaged across six trials before (thin traces) and after (thick traces) an intrathecal bolus injection of baclofen. Note a dramatic increase in peak speed and a considerable drop in the co-contraction of the antagonist muscles. EMG scales correspond to the maximal level observed during attempts at isometric maximal voluntary contractions, kinematic scales are in degrees, °/s, and °/s², time bin is 50 ms.

affected side, including ankle clonus and Babinski's reflex, were virtually abolished. Attempts at fast voluntary movements in the elbow joint before baclofen were accompanied by profound coactivation of antagonist muscle groups (cf. Knutsson, 1983; Dimitrijevic, 1988) and slow, clumsy trajectories (Figure 9.2, thin traces). After the injection, the antagonist coactivation was considerably reduced, the EMG patterns showed a normal-looking triphasic pattern (cf. chapter 4), and the peak speed increased nearly four times (Figure 9.2, thick traces). This was accompanied by a marked improvement in usage by the patient of his affected arm. Elbow joint compliant characteristics were reconstructed in the patient before and after the injection, using the classical "do not intervene voluntarily" paradigm (see chapter 2). The results are illustrated in Figure 9.3. There is a marked decrease in joint stiffness after baclofen, despite identical initial joint torques (bias) and magnitudes of torque perturbations, suggesting a decrease in the gain of the underlying tonic stretch reflex. This

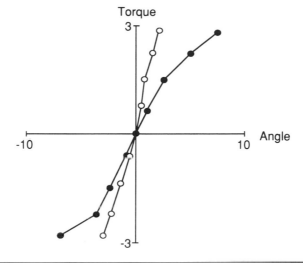

Figure 9.3   Elbow joint compliant characteristics reconstructed in a spastic patient with a hemi-syndrome before (open circles) and after (filled circles) intrathecal baclofen. During the experiment, the subject was instructed: "Do not change the command to your muscles; let the motor move your arm." Loadings and unloadings were mixed pseudorandomly. Each point was obtained by averaging torque and angle values in three trials. Angle scale is in degrees, torque scale is in Nm.

finding contradicts a number of earlier reports of the lack of changes in the tonic stretch reflex gain in spasticity (Berger et al., 1984; Rack et al., 1984; Lee et al., 1987) but is in good agreement with the study by Thilmann et al. (1991a). A number of other patients have demonstrated an improvement in the selectivity of muscle activation on baclofen (Latash et al., 1990). In patients with sufficient residual voluntary motor control, this was accompanied by an improvement in EMG and kinematic patterns of voluntary movements.

A potentially important and intriguing finding in patients with hemi-syndromes is that they did not notice any weakness in the unaffected limbs despite a high intrathecal baclofen dose. This apparent difference in the effects of baclofen on pathological muscle reflexes and voluntary motor control suggests that the effects of intrathecal baclofen cannot be regarded as a nonspecific widespread inhibition throughout all spinal structures. Since the site of baclofen action is mainly considered to be presynaptic (Bowery et al., 1980; Johnston et al., 1980; Zieglgansberger et al., 1988), the observed data can be explained by either a different representation of baclofen-sensitive receptors on terminals of different descending systems or their different susceptibility to intrathecal baclofen due to anatomical and pharmacokinetic factors (including diffusion of baclofen throughout spinal cord structures). The lack of apparent changes in intact muscles in the hemi-syndrome case suggests that one of the long-term reactions of spinal structures to a spasticity-inducing pathology can consist of an increase in the number of GABA-sensitive receptors or sensitization of the existent receptors on pathologically active reflex inputs (cf. Hultborn & Malmsten, 1977; Noguchi et al., 1979). This would account for both the strong effects of baclofen on reflex inputs to α-motoneurons and its lack of effect in intact muscles.

Presently, there is a limited number of motor control studies in spasticity with intrathecal baclofen. Hopefully, the situation will change, and in-trathecal baclofen will be acknowledged as a "wonder tool" by the clinical motor control researchers as it is now being acknowledged as a "wonder drug."

Spasticity—a Disorder or an Adaptation?

We have already discussed briefly the possibility that some of the reactions within the central nervous system to a traumatic denervation may be considered compensatory (cf. Burke, 1988). Let us go into more detail in this direction and try to understand which of the central nervous system reactions to a neurological trauma "make sense" and which of them are likely to be nothing more than a nuisance complicating life and not adding anything helpful.

Figure 9.4 shows schematically some of the possible consequences of a neural trauma. Any trauma is likely to lead to the lack of both descending inhibition and descending excitation to the segmental levels. The lack of excitatory inputs may be expected to lead to a decrease in the centrally induced levels of α-motoneuron activity, and consequently, to a decrease in voluntary muscle force (weakness or paresis). We cannot expect, at least at the present level of knowledge, to adequately correct this deficiency. Functional electrical stimulation is a way to induce stronger muscle contractions, but it represents a substitution of the function rather than its correction.

The lack of descending inhibition has multiple consequences, including the characteristic features of spasticity: spasms, exaggerated reflexes, and increased muscle tone. According to our assumption, it may also lead to hypersensitivity to the lacking inhibitory mediators, including GABA, below the level of trauma due to an increase in the number of corresponding receptors and/or their affinity to the mediators. The last reaction may probably be considered compensatory because it increases the effectiveness of remaining supplies of GABA and its agonists, including baclofen. Increased muscle tone may also be considered part of the adaptive reaction because it prevents muscle atrophy and blood circulatory problems in the extremities, and also can be used by some of the patients for ambulation.

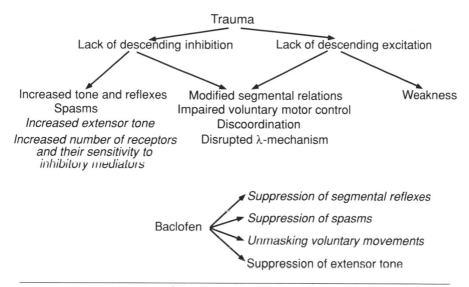

Figure 9.4   Possible consequences of a trauma of the central nervous system. Effects of baclofen are schematically shown below. Potentially "useful" consequences are italicized.

The lack of both descending inhibition and excitation is likely to change the segmental relations and introduce a new order (which is commonly addressed as "disorder") into the motor control system. This may, in particular, include destruction of the hypothetical tonic stretch reflex (λ-mechanism) so that the residual voluntary motor control, if it is present, needs to be organized differently and relearned (cf. Latash, 1986). Intermuscular, interjoint, and interlimb coordination will also suffer, but since we do not have an idea how these mechanisms work in healthy humans, we are very unlikely to even suggest a framework for studying these groups of abnormalities.

Let us create a "wish-list" for an idealized optimal method of treatment of spasticity without asking for an unrealistic, exact restoration of all the damaged neural structures and pathways. First, this method should be able to suppress the hyperexcitable reflexes and eliminate spontaneous spasms and excessive tone that prevent many of the patients from occupying certain positions (e.g., sitting in a wheelchair), disrupt their sleep, and frequently cause pain. Second, this method should not induce additional muscle weakness and, if possible, reverse the originally present paresis. Third, it should be adjustable, so that part of the muscle tone can be preserved during the daytime and used for ambulation, and suppressed during the night to ensure adequate sleep. Fourth, this method should not affect the healthy, nonspastic muscles and the way they are controlled. Fifth, we would like this method to restore normal segmental relations used for multimuscle, multijoint, and multilimb coordination. And, sixth, this method should not have central or other side-effects.

Table 9.3 compares the wish-list with the effects of several antispastic procedures, including oral baclofen, destructive surgery, and intrathecal baclofen, using a very crude and subjective scale: 0—none, 1—some, 2—a lot. The table is self-explanatory and illustrates how close intrathecal baclofen is to, and how far other methods of treatment are from, the ideal drug. However, even intrathecal baclofen fails to do everything desirable, which is quite understandable because consequences of a profound central nervous system trauma cannot be reduced to the lack of GABA.

Motor Disorders in Parkinson's Disease

There are four basic clinical features of Parkinson's disease: tremor, brady-kinesia, rigidity, and deficit in postural reflexes (cf. Fahn, 1990). Tremor is characterized by 5 to 6 Hz alternating activity of antagonist muscles controlling a joint leading to alternating joint movements that can be seen both at rest and during voluntary movements in the joint. Bradykinesia usually refers to slowness of voluntary movement and difficulty in its initiation, although deficits in spontaneous and/or automated movements are also sometimes addressed as bradykinesia. It can affect any part of the body and be more or less generalized. Rigidity is a sustained increase in the resistance to externally imposed joint movements. Deficits in postural reflexes reveal themselves as poor modulation of anticipatory and/or preprogrammed changes in activity of postural muscles associated with voluntary movements or in response to external perturbation (cf. "Prepro-gramming in Motor Control" in chapter 1 and "The Problem of Postural Stability" in chapter 7).

This classification of motor disorders associated with Parkinson's disease may be helpful for clinical description of the disease but apparently needs some clarification to be related to the hypothetical mechanisms of voluntary motor control. Let us tentatively identify three functional levels related to generation of voluntary motor command and peripheral charac-teristics of muscle activity (Figure 9.5). The first level deals with generation of a hypothetical control signal that, for single-joint movements, may be associated with time functions $r(t)$ and $c(t)$ (see chapters 1 and 2). The second level is responsible for preprogramming and executing corrections to possible perturbations (cf. "How Do the Electromyograms Emerge?" in chapter 3 and "The Problem of Postural Stability" in chapter 7). The third level includes segmental mechanisms (reflexes and intraspinal con-nections) that normally act in a coordinated and predictable fashion and are used by the first two levels as an alphabet for specifying motor com-mands. Basic clinical features of Parkinson's disease are likely to reflect malfunctioning at more than one of the three levels. For example, some

Table 9.3 Comparison of the Effects of an "Ideal Drug" and Some of the Common Methods of Treating Spasticity

Method	Restoration of normal segmental relations	Suppression of spasms and excessive tone	Additional weakness	Effects on healthy muscles	Adjustability	Side effects
Ideal drug	2	2	0	0	2	0
Intrathecal baclofen	0	2	1	0	2	0
Oral baclofen	0	1	1	?	1	2
Destructive neurosurgery	0	2	2	2	0	1
Local cooling	0	1	0	0	1	1
Phenol block	0	2	2	0	0	1

Note. 0—none; 1—same; 2—a lot.

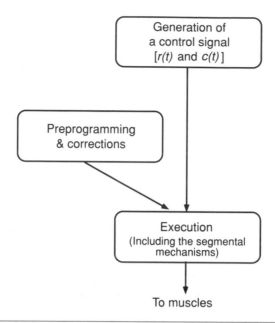

Figure 9.5   The scheme shows three levels of generation of signals to muscles. Different clinical features characteristic for motor disorders in Parkinson's disease are likely to get contributions from defective functioning at different levels.

of them are likely to be related to problems with preprogramming (deficits in postural reactions) and generation of hypothetical motor commands (bradykinesia). Rigidity and tremor can be observed at rest, when the patient is not trying to perform a voluntary movement and, therefore, may be attributed to a dysfunction at the third level. Let me try to describe differences and similarities in motor phenomena in Parkinson's disease and control populations relating them to one of the three functional levels.

Voluntary Movements

Motor control studies in Parkinson's disease have revealed a number of differences from the control population in performing movements of different complexity. While performing a simple movement, when all the parameters of movement are known in advance, patients with Parkinson's disease initiate and perform the movement slower. In particular, they demonstrate an increase in reaction time (Heilman et al., 1976; Evarts et al., 1981; Stelmach et al., 1986; Brown & Marsden, 1991), which increases with an increase in movement complexity (Sanes, 1985). Availability of advance task information does not seem to help the patients reduce the reaction time (Flowers, 1978a,b; Bloxham et al., 1984) although they can employ a predictive strategy in tracking tasks (Day et al., 1984a; Beuter et al., 1990) and in discrete aiming tasks (Worringham & Stelmach, 1990). During target-directed limb movements, movement time in Parkinson's disease demonstrates a linear increase with an increase of the "index of difficulty" (Halsband et al., 1990) conforming to Fitts' law in control subjects (cf. "Speed-Accuracy Trade-Offs" in chapter 6). However, the slope of this relation is considerably higher in Parkinson's disease than in the control population, leading to longer movement times for the whole range of movement amplitudes and target sizes. Contrary to these findings, Teasdale et al. (1990; see also Flowers, 1976) have reported that patients with Parkinson's disease are unable to modify their movement velocity in response to changing environmental demands; in particular they observed nearly constant peak velocity when movement amplitude was changed. An increase in movement time in Parkinson's disease is accompanied by a considerable asymmetry of the acceleration and deceleration phases (Inzelberg et al., 1990) and is mostly due to a prolongation of the deceleration phase. Note, however, that Connor et al. (1989) have reported normal duration of oral speech movements in Parkinson's disease, suggesting a difference between orofacial and limb movement impairments in Parkinson's disease.

Both temporal and spatial motor variability of targeted limb movements have been shown to be higher in Parkinson's disease (Sheridan et al., 1987), which the authors attribute to an increased "inherent" variability in muscle force production (see also Wierzbicka et al., 1991). Sheridan and Flowers (1990) have suggested that Parkinson's bradykinesia may in part result from the increased variability to preserve an acceptable level of accuracy, that is, be a result of a compensation rather than a primary deficit (see later in this chapter).

Fast single-joint voluntary limb movements of patients with Parkinson's disease are characterized by relations between load, kinematics, and EMGs similar to those observed in the control population (cf. chapter 4). In particular, movement peak velocity is scaled with movement amplitude (Halsband et al., 1990; although see Teasdale et al., 1990). An increase in movement amplitude or load leads to an increase in the first agonist EMG burst (Berardelli et al., 1986). On the other hand, fast movements of patients with Parkinson's disease are typically hypometric, especially for movements of large amplitude (Flowers, 1976; although see Sheridan & Flowers, 1990). The entire targeted movement is frequently constructed of several discernible segments (Hallett & Khoshbin, 1980; Berardelli et al., 1984, 1986; Inzelberg et al., 1990). Correspondingly, the EMG patterns demonstrate a number of repeated cycles of agonist-antagonist bursts. A considerable amount of co-contraction of antagonist muscles (Ohye et al., 1965; Hayashi et al., 1988) can also be a factor disrupting both EMG and kinematic patterns during voluntary movements.

These findings, together with observations of slow buildup of EMG activity during voluntary movements of patients with Parkinson's disease (Evarts et al., 1979) have suggested a number of formulations for the hypothetical deficit in the mechanism of voluntary motor control underlying these differences. Note, however, that all these formulations imply that the EMGs and/or muscle forces are adequate measures of the voluntary motor command (cf. the description of "pattern-imposing models" in chapter 4) thus ignoring possible contribution of feedback signals from the proprioceptors and segmental changes, which could have affected the EMG patterns and muscle forces independently of the hypothetical voluntary motor command. In particular, it has been suggested that the mechanism controlling EMG magnitude during rapid movements is impaired (Pullman et al., 1990). The failure to generate sufficient muscle forces has also been attributed to a basic failure to "sufficiently energize the muscles" (Berardelli et al., 1986). Marsden (1984) has concluded that, although the overall form of motor programs is preserved in Parkinson's disease, the details of the number and frequency of activated motor neurons can be inaccurate (cf. Dengler et al., 1990). The last formulation suggests that the differences in voluntary motor patterns in Parkinson's disease are due to changes at the third ("segmental") level rather than at the level of generation of motor command.

However, despite the slow buildup of EMGs during isometric voluntary contractions, patients with Parkinson's disease ultimately achieve the correct final level (Stelmach & Worringham, 1988b). Berardelli et al. (1986) report that, during wrist flexions, the EMGs do not saturate but can be modulated, within limits, to the amplitude of the movement. Patients with Parkinson's disease are able to produce accurate force levels (Stelmach & Worringham, 1988b), although the control of the rate of force increase and decrease seems to be more affected (Wing, 1988). These observations suggest that the problem is not in achieving absolute levels of EMGs but rather in the dynamics of muscle activation, which is likely to considerably depend on the action of reflex feedback loops and changes within the segmental apparatus.

The differences in motor performance become especially pronounced for sequential multijoint movements (Hallett & Khoshbin, 1980; Evarts et al., 1981; Marsden, 1982, 1984; Accornero et al., 1985; Sanes, 1985; Benecke et al., 1986, 1987; Harrington & Haaland, 1991). In particular, the intervals between components of sequential movements are prolonged (Berardelli et al., 1986; Benecke et al., 1987). Impaired performance during attempts at integrating two or more motor programs has been reported in a number of studies (Schwab et al., 1954; Horne, 1973; Benecke et al., 1986). It is interesting, however, that this impairment becomes less obvious when two tasks are performed simultaneously by two different limbs (Stelmach & Worringham, 1988a; Horstink et al., 1990) especially when the patients don't have to rely on "internal control" because of the availability of external cues. Both control subjects and subjects with Parkinson's disease demonstrate a tendency toward performing two movements nearly simultaneously (Kelso et al., 1979a,b; Marteniuk et al., 1984). This can be considered as a shift to a new level of specifying parameters of a hypothetical motor program (new order parameters; cf. "Dynamic Pattern Generation" in chapter 7) when both limbs are controlled as a single unit. Then, the problem of superposition of motor programs disappears, because there remains only one motor program that has two distinct peripheral patterns.

Differences in Preprogrammed Reactions

Preprogrammed reactions (long-latency reflexes) were described in chapter 1. Remember that they presumably represent short sequences of motor commands associated with a current general motor task and are released in response to a specific signal (triggering signal) about an external perturbation. For example, if a subject is holding a position in a joint against an external load and is instructed to return to the original position "as fast as possible" in cases of torque perturbations, unexpected changes in load give rise to a sequence of EMG reactions that include a burst in lengthening muscles with a latency of about 70 ms. Reactions to external perturbations with similar latencies are observed during a variety of motor tasks, including maintenance of the vertical posture and locomotion. Generally speaking, such reactions can be triggered internally in the absence of an apparent external signal. We have already discussed the possibility of both internal and external triggering during maintenance of the vertical posture (chapter 7).

Patients with Parkinson's disease demonstrate profoundly different preprogrammed reactions, suggesting an impairment in the hypothetical mechanism of preprogramming. Stretching of a quiescent or voluntarily activated muscle of a patient with Parkinson's disease leads to the long-latency muscle responses, whose amplitude has been shown to be considerably higher than in the control population (Berardelli et al., 1983; Rothwell et al., 1983; Cody et al., 1986; Hunter et al., 1988). This increase has been attributed to an overcompensation in transmission in a hypothetical receptor-motor cortex-muscle loop (transcortical loop, Lee & Tatton, 1975) and has been considered a possible mechanism of parkinsonian rigidity (Tatton & Lee, 1975; Mortimer & Webster, 1979; Berardelli et al., 1983; Rothwell et al., 1983). The capability to voluntarily suppress the preprogrammed reactions induced by postural perturbations has been shown to be impaired in Parkinson's disease (Schieppati & Nardone, 1991). On the other hand, anticipatory postural corrections *before* a voluntary movement have been observed in only a small fraction of persons with Parkinson's disease (about 5%) as compared with 100% among control subjects (Bazalgette et al., 1986; cf. Traub et al., 1980a). Postural reactions occurring in the course of a voluntary movement are not specific to the movement and occur bilaterally in cases of both unilateral and bilateral voluntary movements (Bouisset & Zattara, 1990).

The first group of observations suggests a poorly controlled increase in a group of preprogrammed reactions that are readily generated in conditions when control subjects do not usually demonstrate them, while the second group indicates an impaired ability to generate anticipatory postural reactions, which can be considered preprogrammed reactions to an internal stimulus (cf. chapter 7). This seeming contradiction can be resolved by trying to dissociate between motor abnormalities that are likely to be primary consequences of the disease and those that are likely to reflect a process of compensation. It is generally assumed that Parkinson's disease leads to an impaired ability to program and initiate movements (Martin, 1967; Flowers, 1975, 1976, 1978b; Wing & Miller, 1984). Then, the lack of anticipatory postural reactions can be considered an example of this impairment.

If the ability to preprogram motor corrections is impaired and there is no compensation, the most commonly used motor programs may become useless, because any external perturbation would lead to their global disruption. Let us suppose that the central nervous system still "wants" to use some of the programs that require continuous corrections, such as walking and maintenance of the vertical posture. The necessary preprogrammed reactions are stored in memory but the mechanism of their adequate triggering is defective. Then, the central nervous system may try to compensate for the impaired ability to adequately preprogram by decreasing the triggering threshold for the preprogrammed corrections and/or increasing their gain. Overcompensation is likely to happen. One of its consequences can be a new perturbation giving rise to a triggering signal leading to a preprogrammed reaction in the opposite direction. Several results can be expected from such a compensatory mechanism. First, stiffness of the system will increase (cf. rigidity). Second, oscillations can occur with a period corresponding to slightly more than doubled latency of the preprogrammed reactions due to the time necessary for the peripheral receptors to react to a perturbation induced by a preceding preprogrammed reaction. This assessment corresponds to oscillations at about 6 Hz (cf. parkinsonian tremor). Third, walking and standing will be possible although they are likely to look awkwardly rigid.

Changes in Segmental Reflexes

Since the primary cause of Parkinson's disease is undoubtedly supraspinal (see the next section), it has been suggested that motor disorders in Parkinson's disease are due to changes in the descending motor commands while the segmental apparatus is generally intact. This view is corroborated, in particular, by observations of unchanged tendon jerk reflexes in patients with Parkinson's disease (Rothwell et al., 1983) and generally normal short-latency action of Ia muscle afferents (Matthews et al., 1990). However, a number of changes in presumably segmental mechanisms have been reported. They include a deficit in reciprocal inhibition that could, in particular, lead to a considerable co-contraction of antagonist muscles during voluntary movements (Ohye et al., 1965; Hayashi et al., 1988), increased reflex activity during tracking phases in which the muscle is lengthening (Johnson et al., 1991), and a "paradoxical" Westphal phenomenon (Andrews et al., 1972; Lee et al., 1983; Berardelli & Hallett, 1984; Matthews et al., 1990). The Westphal phenomenon represents an abrupt reflex excitation of a muscle in response to an externally imposed movement leading to a decrease in the muscle length. In a sense, it is an inverse of the stretch-reflex.

Let us try to consider these changes from the view that they are results of a process of compensation rather than "genuine" disorders. I have already suggested that the hypothesized deficit of preprogramming in Parkinson's disease may benefit from an overcompensation that makes the system generally more rigid. One of the possible mechanisms is an increase in the gain and/or decrease in the threshold of the preprogrammed reactions (see earlier in this chapter). A decrease in reciprocal inhibition may be another way to promote simultaneous coactivation of antagonist muscles, thus increasing peripheral joint stiffness (although decreasing the central, reflex-mediated stiffness; cf. "Reciprocal and Renshaw Inhibition" in chapter 3) and making the peripheral receptors less responsive to external perturbations. The Westphal phenomenon may represent a preprogrammed reaction in a shortening muscle, which can sometimes be seen in control subjects (Rademaker, 1947). A general increase in the preprogrammed reactions in Parkinson's disease can make the Westphal reaction more commonly observed.

Possible Mechanisms

Dysfunction of the basal ganglia leads to Parkinson's disease (for recent reviews see Jellinger, 1986; Graybiel et al., 1990). The function of basal ganglia in control of voluntary movements, as well as in other brain activities, is virtually unknown. Many of the hypotheses related to the role of basal ganglia are based on observations in Parkinson's disease (for reviews see Marsden, 1982; Sheridan et al., 1987). The general scheme of drawing such hypotheses is as follows: If something is different in the motor performance of patients with Parkinson's disease from the control population, this difference is assumed to be a reflection of a dysfunction of basal ganglia; therefore, basal ganglia take part in this aspect of motor behavior in healthy people. Following this line of reasoning, it has been hypothesized that basal ganglia are involved in assembling sequences of movements (Stein, 1978), in integration of several simultaneous motor programs (Stelmach & Worringham, 1988a), and in the transfer of information across a period of time before a response is initiated (Buchwald et al., 1975). Studies of primates have suggested that activity of basal ganglia neurons is related to direction of intended or executed movement (Crutcher & DeLong, 1984; Alexander, 1987; Mitchell et al., 1987) and/or to the "amount" of voluntary muscle activity (Alexander & DeLong, 1985; DeLong et al., 1987). Correspondingly, central theories of the motor abnormalities in Parkinson's disease include a faulty transmission of motor commands from the "decision-making" level (cf. Figure 1.1) (Angel et al., 1970) and a loss of the ability to generate preprogrammed (Flowers, 1975, 1976, 1978a,b) and ballistic movements (Kornhuber, 1971; Marsden, 1982). Peripheral theories complement description of motor disorders in Parkinson's disease with a delay in proprioceptive feedback (Dinnerstein et al., 1962) and overcompensating transcortical long-latency loop (Lee & Tatton, 1975).

We have already discussed the possibility that some of the motor disorders in Parkinson's disease may reflect a process of compensation rather than be direct consequences of the "primary disorder." From this view, disorders of motor performance in patients with Parkinson's disease must be first classified (at least, tentatively) into "primary" and "compensatory." The latter group cannot be considered "motor disorders" but is, rather, a reflection of a new order introduced in an attempt to minimize the consequences of the primary dysfunction.

Assume that the primary dysfunction in Parkinson's disease includes problems in preprogramming and in integration of movement components. Then three out of four major clinical signs associated with Parkinson's disease (bradykinesia, tremor, rigidity, and deficit in postural reflexes) are likely to be reflections of the processes of compensation. If the central nervous system cannot adequately use motor corrections in cases of unexpected perturbations, it is likely to prefer to move at lower speeds (bradykinesia) and to increase resistance of the body to external perturbations (rigidity). Bouisset and Zattara (1990) have formulated a fundamental question: Is bradykinesia a consequence of the pathological postural reactions or are the postural reactions reduced secondary to the decrease in movement speed? They favor the first hypothesis, which agrees with our general scheme. The long-latency loop (equivalent to the loop leading to preprogrammed reactions) is likely to be in a state of overcompensation leading to tremor. In this framework, only the loss of postural reflexes seems to be a reflection of the primary dysfunction in preprogramming. Generally speaking, the system may prefer to function suboptimally (but relatively reliably) rather than risk total failure. Another observation supporting this view is a much higher safety margin demonstrated by Parkinson's disease patients in precision grip tasks (Muller & Abbs, 1990).

Let me explicitly state that the primary goal of this controversial and "lopsided" discussion is to formulate the problems rather than to solve them unambiguously. For example, there are observations suggesting that tardive dyskinesia can be a reflection of the primary cause of Parkinson's disease; in particular, injection of GABA antagonists in the pars reticulata of the substantia nigra in rats could lead to a facial dyskinesia (Arnt & Scheel-Kruger, 1980), which can hardly be attributed to a "compensation." However, are results of these (and other) animal experiments generalizable to the Parkinson's disease patients?

General Characteristics of Movements in Down Syndrome

Down syndrome individuals have motor deficits apparent in their everyday activities that might be described as "clumsy." They also deviate from the control population in performing laboratory motor tests. In particular, persons suffering from Down syndrome demonstrate longer simple motor reaction times (Berkson, 1960; Cowie, 1970; Anson & Davis, 1988; Anson, 1989), lower muscle tone and correlated lower voluntary muscle contraction force (Rarick et al., 1976; Morris et al., 1982), inverted order of joint involvement during multijoint movements (Anson, 1989), excessive forces in tests like finger tapping or gripping (Frith & Frith, 1974; Henderson et al., 1981; Cole et al., 1988), increased variability in motor performance (Henderson, 1985), and a lack of adaptation to changes in sensory information (Shumway-Cook & Woollacott, 1985; Cole et al., 1988; Nativ & Abbs, 1989).

Most of the studies of motor control in persons with Down syndrome involved relatively complicated motor tasks requiring coordinated movements (or muscle contractions) in several joints. However, organization of multijoint movements remains a very poorly understood issue even for the general population (see chapter 7). Single-joint movements are better understood and more extensively studied, and there are general theoretical approaches to their organization and control (see chapters 1-5). A study of simple single-joint motor tasks in persons with Down syndrome was performed (Latash et al., 1989a; Latash & Corcos, 1991) in an attempt to answer the question: Is there anything wrong with their basic mechanisms of control of single-joint movements?

In general, subjects with Down syndrome performed single-joint voluntary flexion movements at much lower speeds than the control subjects (Figure 9.6a) demonstrating a close to linear scaling with distance typical for the control population (see chapter 4). Average peak speeds for all the distances differed threefold. Coefficients of variability (C_v) of the peak speed were calculated for each subject and each distance separately. Averaged values across the subjects demonstrate higher values of C_v for the

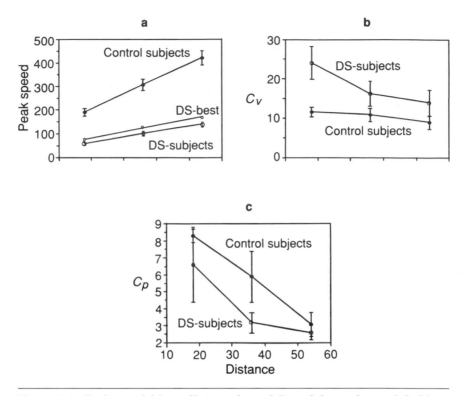

Figure 9.6 Peak speed (a), coefficient of variability of the peak speed C_v (b), and coefficient of variability of the final position C_p (c) averaged across Down syndrome (DS) and control subjects for three different elbow flexion amplitudes (18°, 36°, 54°). DS-best in (a) shows data averaged across the DS-subjects for the highest speed trials. Distance scale is in degrees, peak speed scale is in °/s, C_v and C_p scales are in %.

Note. From "Kinematic and Electromyographic Characteristics of Single-Joint Movements in Down Syndrome" by M.L. Latash and D.M. Corcos, 1991, *American Journal of Mental Retardation*, **96**, pp. 189-201. Reprinted by permission of the publisher.

subjects with Down syndrome (Figure 9.6b). They demonstrated slightly lower values of coefficients of variability of the final position (C_p, Figure 9.6c).

Subjects with Down syndrome could demonstrate in consecutive trials very different patterns of muscle activation, ranging from "normal-looking" biphasic or triphasic patterns (cf. chapter 4) to multiphasic irregular EMG bursts leading to wobbly kinematic trajectories. Figure 9.7 shows an example of two consecutive movements of a subject with Down syndrome. One has a smooth trajectory and is accompanied by a well-defined phasic agonist burst. The other movement's trajectory has obvious "bumps"; the EMGs have multiple additional bursts in both agonist and antagonist muscles. However, despite this general lack of consistency, some subjects

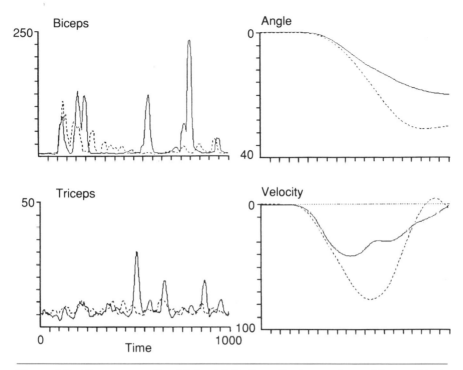

Figure 9.7 Two consecutive trials at 36° elbow flexion performed by a subject with Down syndrome. Note a high variability of both EMG and kinematic patterns. Time scale is in ms, EMG scales are in μV, angle and velocity scales are in degrees and °/s respectively.

Note. From "Kinematic and Electromyographic Characteristics of Single-Joint Movements in Down Syndrome" by M.L. Latash and D.M. Corcos, 1991, *American Journal of Mental Retardation,* **96**, pp. 189-201. Reprinted by permission of the publisher.

were able to demonstrate agonist EMG and kinematic patterns characteristic for the Speed-Insensitive strategy according to a classification introduced by Gottlieb et al. (1989a; see chapter 4).

In experiments with unexpected loadings and unloadings, subjects with Down syndrome demonstrated all the typical EMG components seen in the control subjects. These included a short-latency (under 50 ms) response, presumably monosynaptic in part, and a longer-latency (about 70-80 ms) reaction (preprogrammed reaction). This reaction started as an increase in the agonist activity in cases of loading and a decrease in cases of unloading. When the instruction to the subjects with Down syndrome was changed from "react" to "do not react," the preprogrammed EMG components did not demonstrate any visible modulation in nine out of ten subjects in contrast to a profound modulation observed in control subjects (Latash & Corcos, 1991; see also "Preprogramming in Motor Control" in chapter 1). In a later study (Latash et al., 1992), we spent considerably more time explaining the task and training the subjects and were able to detect instruction dependent modulation of preprogrammed reactions in all the subjects with Down syndrome. The pattern of this modulation, however, could be different from the one observed in a control population. In particular, subjects with Down syndrome apparently preferred to use a coactivation strategy—that is, coactivated agonist and antagonist muscles irrespective of the direction of the perturbation.

Possible Mechanisms

The interpretations of the apparent motor deficits in individuals with Down syndrome range from proposing basic abnormalities in motor control mechanisms to attributing the deficits exclusively to problems in cognition and training. Dysfunction of spinal motor control mechanisms was proposed as a primary factor for the deficits by Parker and Bronks (1980) and Gilman et al. (1981). Davis and Sinning (1987) have hypothesized that subjects with Down syndrome lacked the ability to use all the range of the centrally regulated parameter for the muscles. They identified this parameter with the threshold of the muscle tonic stretch reflex (λ) according to the equilibrium-point hypothesis (see chapters 1, 2, and 3). The inability of subjects with Down syndrome to develop and use motor programs for rapid movements has tentatively been attributed by Frith and Frith (1974) and Kerr and Blais (1985, 1987) to cerebellar dysfunctions that might be causally related to the reduced weight of the cerebellum reported in these individuals (Woollacott & Shumway-Cook, 1986).

On the other hand, several groups of authors have suggested that individuals with Down syndrome have generally intact control mechanisms but have problems with proper modulation of voluntary motor commands and preprogrammed reactions (Shumway-Cook & Woollacott, 1985; Cole et al., 1988; Latash et al., 1989a; Latash & Corcos, 1991).

The Problem Is in ω Rather Than in λ

The subjects with Down syndrome in our experiments did not demonstrate problems in regulating final position (reflected by relatively low coefficients of variation), which can be interpreted as an adequate control of the final λ_{ag} or r (cf. with findings by Kerr & Blais, 1985). However, they were very different from the control subjects in absolute values of peak speed and its variability. This finding suggests that the problem is in regulating the second basic variable ω, the rate of the λ_{ag} (or r) shift (see chapter 5).

This conclusion is different from the one drawn by Davis and Sinning (1987), who proposed that the low voluntary muscle contraction force and the lack of effects of training in subjects with Down syndrome in their studies were due to a narrow range of λ_{ag} changes. However, the testing procedure in their study involved exerting a force in an unusual position (supine, pressing on a handle over one's head), and the training (biceps and triceps curl) took place in a different, more conventional setup. Their subjects did show improvement during the training (which is obvious from their data) but failed to demonstrate it during the control testing. I think that the accessible range of λ changes should not dramatically depend on body position, and the observations were mainly due to the fact that the subjects with Down syndrome, who were more impaired compared to our group, failed to understand what action was appropriate to press with the maximal force during the control testings. Some other factors might also have played a role, including exaggerated caution; this was apparent in our subjects, who were generally reluctant to perform large (72°) movements, although they were able to do so.

Possible Effects of Practice

Note that subjects from the general population demonstrate improvement in performing single-joint, fast elbow flexion movements in the course of extensive training (Gottlieb et al., 1988; see also "Effects of Practice on Kinematic and EMG Patterns" in chapter 6). These observations suggest that during the first series of trials, control subjects did not use the ideal command (such as the one shown in Figure 9.8a), even for these relatively simple tasks. Therefore, the nonideal performance of the subjects with Down syndrome does not necessarily imply a pathology in the control mechanism. In fact, the possibility of observing kinematic and EMG patterns qualitatively indistinguishable from the control subjects' suggests that, at least in this respect, the subjects with Down syndrome are simply at the lower border of a continuum that includes the general population.

The observations of Kerr and Blais (1987) suggest that there is ample room for improvement of motor performance in subjects with Down syndrome. A recent study of the effects of prolonged training of fast single-joint elbow movements in a standardized experimental situation suggests that even considerably impaired individuals with Down syndrome can reach a level of performance similar to that of untrained control subjects (Almeida et al., 1991). A dramatic increase in peak velocity and a corresponding decrease in movement time was accompanied by virtually unchanged variability of the final position, which remained very low (cf. Figure 9.6). Such effects were observed in both teenagers and individuals with Down syndrome aged over 30, suggesting that there is room for improvement independent of age.

Possible Causes of Low Speeds

The relatively low peak speeds observed in subjects with Down syndrome can be due to three factors: multistep shifts in λ_{ag} (Figure 9.8b,c), low

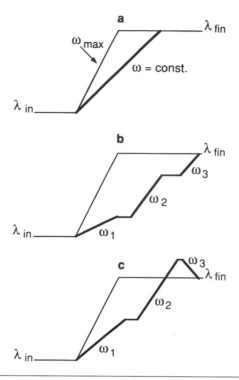

Figure 9.8   Hypothetical patterns of changes in the centrally controlled variable λ corresponding to smooth "normal" movements (a) and to "clumsy" movements with multiphasic EMG patterns (b, c).
Note. From "Kinematic and Electromyographic Characteristics of Single-Joint Movements in Down Syndrome" by M.L. Latash and D.M. Corcos, 1991, *American Journal of Mental Retardation*, **96**, pp. 189-201. Reprinted by permission of the publisher.

values of ω, and lack of muscle force due to peripheral factors (Rarick et al., 1976; Morris et al., 1982). The last factor might lead to slow movements even in cases of very rapid shifts in λ_{ag}. The first factor was obvious in the trials with multiphasic agonist EMG bursts (Figure 9.7). The second factor (value of ω) is the least controlled one; the experimenter can never be sure if a subject is using the highest available value of ω or a submaximal value. However, the up to threefold increase in peak velocity after prolonged training (Almeida et al., 1991) suggests that, before the training, naive individuals with Down syndrome did show low values of ω and were able later to increase them.

Preprogrammed Reactions

The subjects with Down syndrome demonstrated all the basic EMG components in the reactions to perturbations, including the preprogrammed reactions (see chapter 1). However, nine out of ten subjects with Down syndrome failed to demonstrate any visible modulation of these reactions when the instruction was changed. This finding corroborates the general idea that individuals with Down syndrome are deficient in their ability to adapt their motor commands to changes in sensory information (Shumway-Cook & Woollacott, 1985; Cole, 1988). However, all the subjects from another study (Latash et al., 1992) were able to modulate their preprogrammed reactions, although differently from the control subjects. Preprogramming makes sense only if it is done correctly. In particular, preprogramming an increase in activity of a "wrong" muscle group can lead to aggravating the effects of the perturbation. The more universal, coactivation strategy (i.e., coactivation of the muscles leading to an increase in joint stiffness) seems to be preferred by individuals with Down syndrome as compared with the general population. This strategy leads to attenuating the effects of perturbations independently of their direction. On the other hand, the strategy is always suboptimal, because it cannot in principle lead to total compensation. This may be the reason why this strategy has not been reported for control subjects who prefer to use the more effective, although more challenging, reciprocal activation pattern.

Let us discuss the following three findings in relation to everyday activity of individuals with Down syndrome: slowness of movement, higher variability of movement speed, and impaired modulation of preprogrammed reactions.

The slowness of movement, which is a correlate (or even component) of general "clumsiness," can in fact be considered an adaptive reaction similar to the previous discussion of bradykinesia in Parkinson's patients. Based on their everyday experience, individuals with Down syndrome may be reluctant to move at higher speeds. One of the reasons for this may be the lack (or impairment) of modulation of the preprogrammed reactions used by healthy individuals for corrections of unexpectedly perturbed movements. Individuals with Down syndrome may have problems in dissociating predictable conditions of laboratory motor tasks from the everyday environment when the need for the preprogrammed reactions can occur at virtually any time, for example stumbling during walking or running, postural corrections during lifting or throwing objects, balance perturbations in a moving bus or train, and so on.

It seems that individuals with Down syndrome prefer to perform at submaximal speeds rather than risk total failure if something unexpected takes place. Naive control subjects also do not move at their highest speeds when first taking part in a motor study in a new laboratory environment (Gottlieb et al., 1988). However, both Down syndrome and control subjects improve their performance rather quickly. Part of this improvement is likely to be based on the subject's confidence in the safety of the setup and the lack of unexpected events.

Relatively high variability of movement speed may be another reflection of the same sense of uncertainty in subjects with Down syndrome leading to their performance at submaximal speeds. If a subject is performing close to his or her optimal level, the variability of the peak speed is likely to be low because the range of peak speed variation is limited. Another source of variability in the peak speed may be suboptimal control patterns (Figure 9.8) that are likely to be used by untrained subjects with Down syndrome.

In general, subjects with Down syndrome do not show any signs of major qualitative deficits in their motor control mechanisms. They could even be trained to perform like the controls. However, some of the findings suggest that they might have problems in proper modulation of central motor commands. Extensive training is possibly a way of compensation for some of these motor deficits.

Some General Implications for Rehabilitation

The major goal of the process of rehabilitation is frequently considered to be bringing motor patterns of patients as close to normal as possible. As a result, the patients with various motor disorders, both central and peripheral, are trained to move "more correctly." The three types of motor disorders previously discussed represent cases of basically different primary causes for abnormal patterns of voluntary movements from disorders of the segmental apparatus (spasticity), to those in functioning of a well-defined brain structure (Parkinson's disease), and to impairment of cognition (Down syndrome). However, in each case, there seems to be a clear difference between primary and compensatory components in the disordered behavior of the motor system. The share of the compensatory components in the general clinical picture seems to increase for the diseases involving functionally higher levels of the central nervous system. For example, a sizable proportion of the abnormal features characteristic for spasticity are likely to reflect the primary cause; in particular, spinal cord injury commonly leads to a paresis or plegia below the level of injury that apparently cannot be considered compensatory. In Parkinson's disease, a larger proportion of the clinical signs are already likely to reflect compensation. And in Down syndrome, virtually all the motor "abnormalities" seem to be compensatory. Attempts at correcting the compensatory adjustments that take place in response to a primary impairment without, first, dealing with the primary cause are likely to lead to a further deterioration of movements rather than to their improvement.

For example, patients with motor disorders may demonstrate muscle activation and kinematic patterns considerably different from those seen in the control population. However, the patterns that are optimal for healthy people may not be optimal for a patient with significantly changed motor and/or neural apparatus. Let us consider an individual with Down syndrome whose capacity to make quick decisions is impaired. His or her everyday experience suggests that unexpected things can happen in the external world requiring quick corrections of movements (e.g., during stumbling). If the movement is too fast, the impaired decision-making mechanism may not be able to supply the correction in time, and a major disruption in the motor pattern can occur (e.g., falling down). So, the central nervous system of an individual with Down syndrome may prefer to deliberately slow down the movements, especially in an unusual laboratory environment, not because it cannot make them faster but because of a seemingly "secondary" factor, which may in fact be more important than the primary one. Attempts to encourage this individual to move faster in the unpredictable environment may be, first, unsuccessful and, second, even damaging.

Another example is presented by patients with relatively mild vestibular disorders who demonstrate a peculiar pattern of walking that resembles cross-country skiing or Groucho Marx's gait, during which the head displacements in the vertical direction are minimized. Since head stabilization is a major component of the postural control system during walking (Pozzo et al., 1990), observations of such peculiar walking patterns might have been the consequence of an impaired ability of these patients to stabilize the head. Attempts at bringing the walking pattern of these patients closer to normal may have negative influences; in particular, the head may start to exhibit high-amplitude movements resembling those seen in patients with severe vestibular disorders. A similar conclusion can be drawn from the observations of patients with vestibular disorders who use an "ankle strategy" for postural corrections but lack a "hip strategy" (Shumway-Cook & Horak, 1989; Horak et al., 1990). A reason for this may be that a shift in the body center of gravity performed by changing position in the ankle joints leads to smaller head displacements than a similar shift performed by a movement in the hip joints.

A recent paper by Winter et al. (1990a) presents a number of cases of abnormal walking patterns in patients with various primary causes, including amputation, joint replacement, and spastic hemi-syndromes following brain injury. Analysis of gait biomechanics led the authors to conclude that many of the atypical features represent results of adaptation and should not be considered pathological. Winter et al. write: "In cases of major surgery . . . or in long-term therapy it is important not to treat the adaptations (secondary problems) but to treat the primary problems" (p. 692).

If a patient with a long-lasting motor disorder demonstrates peculiar patterns of voluntary movements, a therapist should realize that these patterns might well have been developed by the patient's central nervous system taking into account its deeply concealed goals and the state of the disordered motor system. There is always a chance that the patient's central nervous system has failed to find the optimal solution and perfectly adjust to the changed properties of the motor system. Then, the physical therapist can try to improve the situation. However, in doing so he or she must be aware of a basic underlying assumption: That he or she is smarter than the central nervous system. This seems to be a very strong assumption, especially in relation to control of voluntary movements, because the central nervous system "knows" how to control them, while human beings are far from even formulating viable hypotheses in this field. This certainly does not mean that the patients should be left to themselves. For example, sometimes pain prevents the patients from performing exercises that are necessary to keep the motor system in an active state and whose positive effects become apparent only in the long run. In certain pathological cases, pain can be strong enough to prevent the central nervous system from accepting the optimal strategy for developing compensations and rehabilitation in general. In such situations, the physical therapist can in fact be smarter than the central nervous system. However, in most of the cases, let us first think and only then try to correct.

Chapter 10

Language and Movement

The system of production of voluntary movements is likely to have many similar features with the system of production of phrases including the presence of deep and surface structures. The laws of coordination may be considered similar to the laws of grammar; both generate a number of possible solutions for a problem that are grammatical or coordinated. A number of not very scientific, everyday observations suggest certain features of the deep and surface structures and the process of transformation. The tonic stretch reflex mechanism, which forms the basis of the equilibrium-point hypothesis, may be considered the last universal transformation common for all voluntary movements.

Language and movement are very similar. We learn their basic laws at approximately the same age. If this age passes without the skill being acquired, it is a great problem to compensate for it (compare learning a foreign language or a complex gymnastic exercise). Our inborn cortical asymmetry is reflected both in motor and lingual functions; a right-handed person feels equally clumsy while playing tennis with his or her left hand and trying to verbally express his or her interpretation of Mozart's symphonies or Chagall's paintings.

Movement and language similarities are not confined to these superficial points. More deep analogues include ambiguity of response to seemingly identical problems (stimuli), presence of deep and surface structure, and the important role of perception (cf. Bellman & Goldberg, 1984; Kelso & Tuller, 1984). The following quote (my translation of the Russian original) from a paper by Gelfand et al. (1971a) did not appear by chance: "One may say that forces of individual muscles represent letters of the language of movements, and synergies unite these letters into words the number of which is much less than the number of letter combinations.

Richness of the vocabulary subserves all the variety of possible movements."

One should probably accept the fact that contemporary linguistics is ahead of contemporary motor physiology in creating a general theory. At least, a rational approach in linguistics dominates an empirical one, while in motor control studies this is not yet the case. This leap in linguistics development was to a great extent due to works by Chomsky (1971, 1975). Chomsky introduced the concept of generative grammar as a set of rules whose application to a lexicon creates all possible grammatical phrases, avoiding ungrammatical ones. Two structures were specified in language: *deep* and *surface*. The process of generating a surface structure from a deep structure has been termed *transformation*. Our theoretical abilities (*competence*) have been separated from our actual activity (*performance*).

A similar approach in motor control theory was originally introduced by Bernstein (1935, 1947, 1967), who formulated a number of basic characteristics of a system able to carry out regulation of everyday voluntary movements (see the first section in chapter 1). In particular, Bernstein stressed the necessity of several control levels and the importance of sensory corrections during execution of a movement. Bernstein also introduced notions of competence and performance although not as explicitly as Chomsky (notions of movement engrams and of movements themselves; cf. the first section in chapter 6). These ideas are becoming widely recognized (e.g., Brookhart, 1979; Saltzman, 1979; Talbott, 1979; Kugler et al., 1980; Eccles, 1981; Schmidt, 1988; Turvey, 1990a). Let me quote an article by Turvey (1990a): "movement system's 'deep structure'—the states that guide and shape the 'surface' behavior of the observables" (p. 66).

The following speculations on certain problems of motor control are based on the concepts of deep and surface structures and transformation borrowed from linguistics, and on the Bernstein approach to control of voluntary movements. Some of the ideas have undergone certain changes during the analysis; in particular, the role of sensory corrections during the execution of voluntary movements. The central place is going to be assigned to memory: It is proposed that virtually all voluntary movements in all the variety of their external conditions of execution are based on memory rather than on continuous corrections in the course of their execution (cf. Keele, 1968; Szentagothai & Arbib, 1974; Schmidt, 1975, 1988).

On the Laws of Coordination

When looking at a person who performs a voluntary movement, one may ask why we observe this particular pattern of muscle activation during the execution of this certain task in the current external conditions. This question is ill-posed (cf. Mussa Ivaldi et al., 1989 and chapter 7). The task and conditions of its execution (to the extent that we are able to control them) do not unambiguously define the movement. If one accepts the idea of the essentially nonequilibrium nature of the motor control system, external movement patterns are going to be unique even if the task and conditions are reproduced ideally (cf. "Dynamic Pattern Generation" in chapter 6). As a result, repetition of a motor task leads in different trials to different movement trajectories. This means that from trial to trial an unambiguous correspondence between motor task (including external conditions) and peripheral motor patterns (including kinematic, kinetic, and EMG variables) cannot be carried out even for the simplest movements.

However, all normally realized movements associated with a given motor task have one common feature, which can be termed coordination. Formulating basic differences between coordinated and noncoordinated sequences of motor commands would allow one to define all possible sequences of motor commands leading to fulfillment of a given motor task.

Similarly, in the field of linguistics, it is virtually never possible to define why a particular person uses one set of particular words and constructions and not some others equivalent for expressing the same meaning. Linguistic problems are quite different: How can one separate grammatical and nongrammatical constructions? By solving this problem one would get a set of rules which, being applied to a lexicon, generate all possible grammatical constructions, avoiding nongrammatical ones.

Probably, in the case of voluntary motor control, the basic problem should also primarily deal with formulating the laws of coordination, that is, rules that, when applied to a given motor task and external conditions, would generate all possible combinations of motor commands appropriate for the task (Figure 10.1). Note that these laws are supposed to guide or constrain the movements rather than cause them (cf. Turvey, 1990b). The actual choice of a sequence of motor commands is performed by the central nervous system based on a lot of uncontrolled (and, as a rule, poorly defined) factors, including instantaneous state of interneurons, "will," "mood," "attention," and so on. According to the dynamic pattern generation approach (Kelso, 1981, 1984; Kay et al., 1987; Saltzman & Kelso, 1987; Kelso & Schoner, 1988; Schoner & Kelso, 1988; Scholz & Kelso, 1990; see "Dynamic Pattern Generation" in chapter 6), this choice is always ambiguous due to the nature of the motor control processes, which are assumed to be based on stochastic dynamic interactions inside a nonequilibrium system.

Imagine a person who wants to take a cup of tea from a table. Let us assume that her visual perception is direct and adequate (cf. Gibson, 1966). Optical information from the cup and surrounding furniture is an observable that is going to constrain her future movement (cf. Warren et

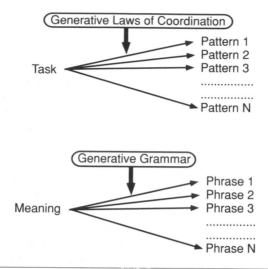

Figure 10.1   Laws of coordination lead to generation of peripheral movement patterns based on the desired task similar to how the generative grammar leads to different phrases based on the desired meaning.

al., 1986). We assume, however, that there are other factors that are not directly observable that are also going to constrain the virtually unlimited menu of joint trajectories and muscle activation patterns for fulfilling the task. These factors are the laws of coordination.

Investigations of the wiping reflex in a spinal frog (Fukson et al., 1980; Berkinblit et al., 1984, 1986a; see chapter 7) have demonstrated that notions of coordination and movement generation can be used at the spinal level as well. Movements of a spinal frog hindlimb were studied in response to acid irritation of the skin of the back or of the forelimb. Several stages have been defined in the wiping reflex: At first, flexion of the hindlimb brings it to a standard position; then a complex sequence of movements in different joints brings the hindlimb distal portions to the stimulus; and finally, the hindlimb performs a series of rapid extension movements resembling those seen during intact frog swimming or jumping. Different extension movements in the series performed wiping of the stimulus in different directions. One may say that the extensor wiping movements in different directions were generated by coordination laws applied to the given task (perhaps the same laws ruling swimming or jumping were effective).

The spinal frog successfully performed wiping of a stimulus during the first attempt in conditions of fixation of one of the joints (e.g., knee joint) of the working hindlimb. This example demonstrates that the effect of the interaction of coordination laws with the task cannot be unambiguously expressed as a fixed sequence of motor commands to individual muscles or even to muscle groups regulating the movements in individual joints.

The concept of coordination laws can probably be applied even at the level of single muscle contraction. In this case, one may consider the so-called "size principle" of motor unit recruitment (Henneman et al., 1965) as a coordination rule. This rule is coordinative in the sense that it permits only definite courses of recruitment, forbidding nongrammatical sequences of motor unit recruitment. Note that the size principle works if the muscle is a "prime mover" for a given task and can be violated if the muscle has a secondary role (Desmedt & Godaux 1981; Garnett & Stephens, 1981; Haar ter Romeny et al., 1982).

Deep Structure

Assume that any complex movement can be described, at a peripheral level, by changes of a parameter R whose values correspond to definite combinations of joint torques and angles. For the k-joint movement let us denote this parameter as R_k. R_k is a k-component vector; its every component corresponds to the state of a single joint. Setting dependence of R_k on time (t) unambiguously defines changes in time of all the joint torques and angles (i.e., defines a k-joint movement).

However, virtually any movement can be executed in different external conditions when the same motor task can lead to considerably different values of joint torques and angles. This can be due to different initial conditions, different force fields, different previous movements, and so on. Usually, it is absolutely impossible to control all these variables. Let us, therefore, introduce an idealization of a real movement U_k corresponding to R_k. The new vector cannot, generally speaking, be expressed in terms of joint torque and angle changes. However, one can presume that in certain cases it is possible to extract from U_k components corresponding to movement planning in single joints:

$$U_k = \{V_1, V_2, \ldots, V_k\}. \tag{10.1}$$

Let us consider the formation of an idealized prototype of a planned movement—its deep structure. We will not consider influence of external conditions and corrections necessary during movement execution.

Postulate:

Learned dependencies of U_k on an abstract time x (Bernstein's engrams) are stored in the memory. Any "everyday" movement or a movement consisting of familiar components connected in time is modeled on the basis of $U_k(x)$ functions.

The formation of a deep structure of a movement starts with the choice of some function corresponding to a motor task function $U_{ki}(x)$. Index i

denotes the number of the given $U_k(x)$. When one describes the beginning of a movement, $i = 1$ and the first function looks like $U_{k1}(x)$. It is necessary to choose a point at which playback of the function begins—x_i and a coefficient of the transformation of the function from the abstract time x to real time t, $a_i = x/t$. Playback of the function $U_{ki}(x)$ will take place until some moment x_{i+1} (t_{i+1}), when it will be substituted by a function $U_{ki+1}(x)$.

Thus, ideal representation of a k-joint movement (its deep structure) can be expressed as:

$$U_k(t) = \sum_{i=1}^{n} U_{ki}^{0}(t) \tag{10.2}$$

where

$$U_{ki}^{0}(t) = U_{ki}(x/a_i) \text{ if } x_i < x < x_{i+1};$$
$$= 0 \text{ if } x < x_i \text{ or } x > x_{i+1}.$$

In certain situations, $U_{ki}(x)$ can be expressed in its single-joint components (see Equation 10.1),

$$U_{ki}(x) = \{V_{1i}(x), V_{2i}(x), \ldots, V_{ki}(x),$$

and for functions describing movement in only one joint (in the first one):

$$V_1(t) = V_{1i}^{0}(t),$$

where $V_{1i}^{0}(t)$ is defined similarly to $U_{ki}^{0}(t)$ in Equation 10.2.

Before discussing a process of transformation of a movement idealization (its deep structure) into real changes in joint torques and angles (surface structure), it is worthwhile to say a few words about the parameter a_i, which is a coefficient for transforming abstract time functions into real time. This parameter can change speed of movement execution, accelerating or decelerating the $U_{ki}(x)$ playback (cf. a similar parameter τ_{mov} introduced in a kinematic model; "Variability in a Kinematic Model" in chapter 6).

Our ability to perform different everyday movements at different speeds is self-evident. There are also experimental data demonstrating that execution of a learned sequence of movements at different velocity is performed on the basis of only one pattern, whose temporal structure does not alter when executed at different speeds (e.g., Terzuolo & Viviani, 1979; Carter & Shapiro, 1984).

However, the parameter a_i can also be controlled by means other than the central structures, as was demonstrated by Severin, Orlovsky, and Shik (1967) in experiments on decerebrate cat locomotion when velocity of treadmill motion (peripheral information) could change velocity of induced cat locomotion. The parameter a_i is probably very general because it can govern seemingly independent movements: For example, it is hard to simultaneously tap different rhythms with a hand and a foot (except for harmonics of the same frequency).

Transformation

For the execution of a certain motor task, a function $U_k(t)$ has been synthesized representing an ideal image of the planned movement without taking into account actual conditions in which the movement is going to be performed. To get a function describing real changes of joint torques and angles, $R_k(t)$ (surface structure of the movement), it is necessary to transform the ideal function taking into account conditions of the movement execution. This transformation is in many respects similar to the one introduced by Chomsky. Let us term it T. T depends on time because conditions of movement execution change continuously during the movement. So, $T, = T(t)$. Thus,

$$R_k(t) = T\{t; U_k(t)\}. \tag{10.3}$$

What is included in T?

- Force field, including forces of gravity, inertia, and perhaps others, as, for example, force of water resistance while executing movements in water.
- Geometry of the body.
- Geometry of the adjacent space, including the coordinate system.

For many movements performed in familiar (everyday) conditions the $T(t)$ functions are already stored in the subject's memory. The retrieval of a convenient transformation function should be performed by taking into account the motor task, information on body position in space, and information on conditions of movement execution. Therefore, $T(t)$ should be synthesized on the basis of sensory information, memory, subject's knowledge, and ongoing motor command (or its efference copy) because realization of the motor command will change conditions of the movement execution. A special role is probably played by vision, which gives us the most complete information on the conditions of movement execution and on their possible changes. A significant role can also be played by afferent input from proprioceptors and from the vestibular system. The role of these systems in the formation of T will be discussed later in this chapter.

However, the presence of sensory information is not necessary for the formation of T and its dependence on time. From memory and our knowledge about the surrounding world, we can in certain conditions synthesize $T(t)$ independently of sensory information, as demonstrated by the following example. Note that this and the following examples do not represent results of controlled experiments but rather my own observations. In most of the cases, I was both the experimenter and the subject.

First Escalator Experiment

A person used to stepping on a moving escalator in everyday life does not experience noticeable difficulties, although this change of external conditions for the execution of the walking program requires a complex set of coordinated muscle reactions. If such a person knows that during his fifth step he is going to enter an escalator, he can perform this task without visible motor disruptions even with his eyes closed. The same experiment performed in the absence of advance information, or in the absence of knowledge of what an escalator is, or without previous experience of stepping on an escalator would lead to significant movement and posture disruption.

This experiment clearly demonstrates the essential role of memory and knowledge in the transformation of a motor program according to different external conditions. In everyday life, the function $T(t)$ is certainly significantly based on sensory information. However, memory and knowledge play a considerable role here as well: Compare stepping on an escalator for an experienced and inexperienced subject.

For some unusual conditions, the $T(t)$ function can be absent in our memory. In these cases, we either begin performing familiar semiautomatic movements slowly, attentively controlling their execution, or we experience considerable disruptions in movement patterns due to usage of an inadequate T-function.

First Water Experiment

If a person without special experience tries to run in the water, he usually falls down due to usage of an inadequate $T(t)$, although running itself is a very familiar semiautomatic pattern. A series of attempts of running in the water leads to the formation of a new T-function, and running, with a considerably changed pattern, can be performed without falling down.

First Badminton Experiment

Another example of T-function synthesis can be seen when learning to play badminton. In this case, the geometry of the limb is altered by the requirement to hold a racket. It is interesting that it takes less time for an adult than for a child to adapt to this geometry change and to be able to hit a shuttlecock with the racket. Perhaps it is due to the adult ability to abstractly imagine the changes in system geometry, that is, to synthesize $T(t)$ with a considerable role played by intellectual abilities.

Note that the formation of a T-function on the basis of certain classes of movements [$U(t)$ patterns] lets one execute other movements that were memorized under different conditions (i.e., apply $T(t)$ to other deep structures).

Second Escalator Experiment

An experienced subject may have obtained the experience of ascending on an escalator through walking and, perhaps, running. As such, this subject's T-function was synthesized only on the basis of these highly automatic movements. However, most people can step on an escalator while jumping on one leg, that is, using a motor pattern that never coincided with stepping on an escalator before.

The last example supports the existence of a separate group of transforming functions T. One might a priori presume that motor patterns $U_k(t)$ are memorized already with conditions in which they are performed. This may not necessarily be the case.

Surface Structure

Taking into account Equation 10.3, one can express commands leading to a real movement as

$$R_k(t) = T\{t; \ U_{ki}^0(t)\}, \tag{10.4}$$

where $U_{ki}^0(t)$ functions are given in Equation 10.2.

Thus, setting an ongoing command includes setting of

- a function $U_{ki}(x)$ that represents an ideal basis for a real command $R_k(t)$;
- a moment of this function playback start, x_i;
- a coefficient of $U_{ki}(x)$ transfer into real time, a_i; and
- a function T transforming the ideal command according to actually existing conditions of movement execution.

Equation 10.4 corresponds to a k-joint movement formation when the memory contains the necessary $U_k(x)$ patterns. If there is no pattern for a movement or if it is necessary to modify some pattern in one of its components, the surface structure $R_k(t)$ can be based on stored-in-memory n-joint patterns ($n < k$), or, in an extreme case, on the basis of single-joint patterns. This means that the transformation of a function $U_k(t)$ is replaced by a parallel transformation of several functions. One can expect that in this case the movement will become less "automatic" or more "consciously controlled."

In general,

$$R_k(t) = T \left\{ t; \begin{array}{c} U_{mi}^{0}(t) \\ U_{pi}^{0}(t) \\ \ldots \\ U_{si}^{0}(t) \end{array} \right\}, \tag{10.5}$$

where $m + p + \ldots + s = k$.

Equation 10.5 reflects control of a k-joint movement on a basis of separate $U(x)$ patterns for different $m,p, \ldots ,s$-joint groups.

The choice of patterns forming the deep structure can be ambiguous. It is performed, as noted previously, on the basis of coordination laws applied to the given motor task. Coordination laws generate a variety of possible different pattern sequences in the deep structure (sets of U_m, U_p, $\ldots$ U_s—see Equation 10.5). The choice of a certain deep structure is performed on the basis of many factors. Thus, one may assume that normal variability of voluntary movements is due to at least two factors: the variability of the deep structure and application of different T-functions dependent on irreproducible external conditions of movement execution (cf. chapter 6).

By repeating a movement, controlled as in Equation 10.5, one can store it in the memory as a new k-joint pattern, and its control will be described by Equation 10.4.

First Piano Experiment

A musician playing a familiar piece "automatically" can probably be described as using control of the type in Equation 10.4. If the musician wants to stress a certain note, he should focus his attention on a particular movement, causing a harder pressing of the key. This can be interpreted as transition to a two-pattern control as in Equation 10.5. If the musician repeats this piece with the stress many times, the stress becomes automatic, which can be described as returning to a control corresponding to Equation 10.4.

If a movement is absolutely new and unusual, and it cannot be controlled on the basis of single-joint patterns, it becomes necessary to learn this movement with continuous correction of control signals making use of sensory feedback information. However, if the accuracy demands of a movement are low as compared with the demands of the movement velocity, the movement can be approximated by inadequate patterns stored in the memory.

Many of the considerations concerning the learning of new movement patterns can be applied to the synthesis of new functions T. When a movement is performed in familiar and predictable external conditions, the $T(t)$ function is extracted from the memory according to the motor task and the information regarding conditions of its execution. $T(t)$ adequately transforms deep structure into surface structure according to Equation 10.4. If there is no adequate function T in the memory, it is approximated on the basis of existing information and after that modified according to feedback information on the results of the movement execution. Our knowledge can also play a considerable role in the T-function synthesis as has been demonstrated by the first escalator experiment and the first badminton experiment.

Second Water Experiment

Imagine a subject walking swiftly on a plain surface in total darkness. If he suddenly walks into a pool, he will get sensory information from his feet and probably from his ears that his walking program should be modified. He may try to minimize distortions of his walking pattern by adjusting the T-function. In this case, he will be able to continue walking without reducing speed dramatically even when the water is over his knees. A subject who tries to walk swiftly in water over his knees without previous adjustment of T-function may fall down or experience considerable distortions of the walking pattern.

So, every movement performed without conscious tracking has an analogue in the memory representing an idealized motor pattern and a function reflecting the existing external conditions. One can disagree at this point by giving an example of a complex coordinated semiautomatic movement performed by a person slipping on ice. Every movement of this class looks different, creating an impression that each movement is accomplished by solving a complex dynamic motor problem, rather than by the creation of a movement on the basis of stored memory patterns. However, someone slipping on ice for the first time in his or her life (e.g., a baby or a person who spent all of his or her life on the equator) does not perform any complex coordinated movements but simply falls down, sometimes performing movements that can even exert adverse effects. It seems that the variety of slipping movements are based on a restricted number of patterns stored in memory, and the individual appearance of these movements is due to the uniqueness of conditions in which they are realized (i.e., to different T-functions dependent, in particular, on body geometry at the moment of slipping).

Role of Different Sources of Information

The relative role of information from different sources in retrieving patterns necessary for the formation of a deep structure and in the choice of a transforming function can be considerably different depending on the type of movement (how familiar it is), the conditions of its execution, and also on the level (conscious or not) at which formation of a motor program takes place.

In the case of the conscious execution of a movement consisting of learned patterns in familiar conditions, visual information and our knowledge about the surrounding world probably play the main role in retrieving patterns necessary for the formation of a deep structure and in the choice of a transforming function. The role of proprioception is presumably small because one can plan a movement before its realization.

Third Escalator Experiment

When an experienced subject sees an escalator, he is ready to modify motor patterns to avoid disruptions of movement and/or posture (i.e., to change the T-function). If the escalator is not working (motionless), the subject experiences dramatic movement perturbations although both visual and proprioceptive information would lead to simple adjustment of the walking motor program, as if he were stepping on regular stairs. The movement disruption starts during the first step or two (my own observations) when the different height of the steps cannot yet start playing a role. This example demonstrates the predominant role of our knowledge in T-function formation in certain situations. Note that a person without any "escalator experience" can ascend a motionless escalator without any problems.

The first escalator experiment demonstrates that T-function adjustment can take place even without any visual information.

Vision plays a leading role in the choice of necessary deep structure patterns and transforming functions, but it always does that in cooperation with knowledge. For example, based on visual information and memory, we make a prognosis of surface stiffness during walking. Divergence of real qualities of the surface from the predicted ones leads to movement disturbances in spite of adequate proprioceptive information: Imagine that during walking you unexpectedly place your foot on a wobbling sandy surface. This suggests that proprioceptive information cannot adequately modify general movement patterns, which could be done with participation of memory if you had information on the sandy surface in advance. Automatic corrections induced by proprioceptive information are usually inadequate to totally compensate for a change in movement execution conditions and can even lead to adverse effects (see "Preprogramming in Motor Control" in chapter 1).

Sensory feedback loops start playing an important role while learning new movement patterns or adjusting to new conditions of movement execution. Besides that, the proprioceptors play an important role in creating an "image of the body," that is, the body geometry that is necessary for synthesis of a transforming function T.

Proprioceptive information also takes part in urgent corrections of movements in the case of unexpected changes in the conditions of movement execution. One can assume that these corrections represent existing memory resident patterns $U_k(x)$, whose playback is initiated by corresponding triggering peripheral information (see chapter 1). This means that the basic scheme of correction realization does not differ from the scheme for a voluntary movement (deep structure—transformation surface structure) but is performed automatically in response to a triggering peripheral signal. In this sense, the corrections are preprogrammed movements.

One example of such types of corrections is the "corrective stumbling reaction" (e.g., Forssberg et al., 1977; Forssberg, 1979a,b), demonstrated in cases of an unexpected obstacle during locomotion. This reaction consists of lifting the hindlimb over the obstacle. Note that the idea that afferent signals take part in providing a signal about peripheral changes to modify the ongoing motor command, rather than to participate in automatic compensation of motor disturbances, was expressed by Bizzi and Polit (1979) and Jung (1979).

Vestibular signals probably provide some background against which small alterations in signal level are not interpreted as signals to change a motor command. However, if these changes exceed a certain threshold, one can observe in an inexperienced subject (not an acrobat) global motor disruptions leading to the necessity to elaborate new T-functions.

The same can be said about proprioceptive information. Its changes, which are inevitable during any movement, are unlikely to significantly influence the choice of the deep structure and transformation functions, because reaction to this information should be immediate. Proprioceptive signals are likely to provide a background necessary for normal synthesis of a deep structure and a transformation function. Phasic changes of this background can play a role only by giving rise to preprogrammed reactions such as the corrective stumbling reaction. Overall changes of this background (or lack of proprioception) can also bring about dramatic movement disorders.

One can say that the choice of a deep structure is connected to a choice of a "range of insensitivity" to peripheral signals when corrections of the motor program do not take place. When peripheral signals exceed certain thresholds, motor corrections take place; that is, time-restricted modifications of the motor program (probably in a limited number of joints—transfer to control pattern described by Equation 10.5), or overall changes of the deep structure can occur.

Note that changes in the influence of peripheral information on central structures depending on motor task (gating of afferent signals) has been demonstrated for both eye (Adey & Noda, 1973) and limb movements (Rushton et al., 1981). In particular, it has been shown that rubrospinal collaterals, which can tentatively be assumed to convey the "efference copy" signals (e.g., Ghez, 1975; Kuypers, 1981), can modulate peripheral inputs to the rostral dorsal accessory olive (rDAO) (Gray & Dostrovsky, 1983). Inhibition of the sensory responsiveness of the rDAO was observed during electrical stimulation of the rubrospinal pathways (Weiss et al., 1990) and during cortical stimulation (Leight et al., 1973). Activity of rDAO has also been shown to be strongly modulated in certain phases of movement (Bauswein et al., 1983; Gellman et al., 1985). Taking into account the hypothesized role of corticospinal and rubrospinal pathways in control of voluntary movements (for review see Kyupers, 1981) and of the cerebellum in sensorimotor processing (Houk & Gibson, 1987), these observations provide a tentative basis for gating of proprioceptive information during voluntary movements.

First Walking Experiment

When a person is descending unknown stairs in darkness his or her steps are unconfident and slow, demonstrating that the motor control system is not well suited for functioning on the basis of only proprioceptive information. After several trials, the person gets used to these particular stairs and is able to run up and down, generating the deep structure and the T-function on the basis of memory. This leads to a considerable improvement in the motor pattern. Proprioceptive signals are not taken into consideration if they do not exceed a certain level: Imagine that one stair is unexpectedly destroyed.

This experiment demonstrates that the role of proprioceptive information can considerably increase during unfamiliar movements or movements performed in unfamiliar conditions. Note that presence of sensory corrections can even be undesirable in certain situations, as demonstrated by the following example.

Second Walking Experiment

If it is necessary to cross an abyss via a narrow plank, people usually prefer to walk quickly (or even to run), planning both the deep structure and the transformation function in advance on the basis of visual information and memory. Proprioceptive corrections can only hinder accomplishing the task. Imagine that during this crossing one unexpectedly steps on a nail in the plank. It is quite clear that it would be desirable not to react to the evoked proprioceptive signals. This can be done by making the range of insensitivity wider prior to the movement.

An essential role in the structure of a motor program is played by the "efference copy" (cf. von Holst, 1954), that is, by signals corresponding to ongoing motor command $U_k(t)$. Since motor commands inevitably lead to a change in muscle state giving rise to a movement and/or to force that changes the conditions of movement execution, the efference copy should influence the transforming function $T(t)$. Besides that, every movement automatically changes the target of a successive one, and this fact should be taken into account in choosing movement patterns and (x_i, a_i) parameters. There remains the problem of the interrelation between efference copies taking part in motor control and in kinesthetic perception (cf. Feldman & Latash, 1982a,b). These two information flows can be identical, related, or totally independent.

Tonic Stretch Reflex—An Example of Transformation

According to definition, $R_k(t)$ describes changes in joint torques and angles. However, this definition does not say anything about the control variables independently supplied by the central control structures. To consider this issue, let us analyze an example of transformation, namely, the transformation carried out by the mechanism of the tonic stretch reflex, which is supposed to form the basis of the equilibrium-point hypothesis (λ-model; see chapters 1 and 3).

Remember that the hypothetical tonic stretch reflex defines the dependence of active force upon muscle length (e.g., Matthews, 1959). Registration of its force-length characteristics in human subjects in a special paradigm (requiring the subjects "not to change voluntarily" the motor commands to their muscles) has demonstrated that these characteristics for a given muscle do not intersect. They differ only in the value of the threshold (i.e., muscle length at which the motor unit recruitment takes place). Stimulation of different descending systems in animal experiments has also demonstrated practically parallel transfer of these characteristics (Feldman & Orlovsky, 1972).

The lack of intersections of the force-length curves let Feldman (1966a,b, 1974, 1979, 1986) introduce a monoparametric description of the central regulation of muscle state with changes of λ as the centrally supplied variable. Actual force and length of the muscle are defined by interaction of the centrally defined characteristic and external load giving rise to an equilibrium state of the system (equilibrium point).

Joint position is usually controlled by at least a pair of antagonist muscles. In this case, one may say that the central command regulating joint position and torque can be described by a pair of variables (λ_f and λ_e) representing the reflex thresholds for flexors and extensors, respectively. In a more common case, any motor command can be described by changes in time of a set of λ-variables $\{\lambda_a, \lambda_b, \ldots, \lambda_z\}$, that is, represent a trajectory in a "λ space" (Feldman et al., 1990b).

In other words, the mechanism of the tonic stretch reflex interacting with load (which depends on effector geometry, force fields, and adjacent space geometry) leads, given the values of λ-parameters, to certain values of joint torques and angles. An idealized command (virtual trajectory), expressed via the λ changes in time, is transformed by the reflex mechanism into actual joint angles and torques depending on existing conditions of movement execution.

The hypothetical tonic stretch reflex represents the last step of the transformation that is probably used during control of practically all voluntary movements. The universal nature of this transformation lets one express the surface structure of a movement with changes in λ:

$$R_k(t) = T_\lambda \{R'(t)\},$$

where $R'(t)$ is a command at the previous step expressed with time functions $\lambda(t)$, which can be associated with the notion of virtual trajectory for single-joint movements (cf. Chapters 2 and 5), and T_λ is a transformation corresponding to the λ-mechanism. Note that the effects of the T_λ transformation can be quite substantial and far from trivial. As suggested in chapter 2, smooth single-joint movements can result from a nonmonotonic N-shaped virtual trajectory. Actual peripheral patterns of muscle activation are also dependent on the peripheral conditions of movement execution (taken into account by T_λ) as, for example, during isotonic and isometric contractions (see chapter 5).

Concluding Comments

Let me repeat the main points of the described general scheme for generating motor commands. They look more like axioms:

• All familiar movements or movements consisting of familiar components (the components can represent familiar single-joint movements) are based on memorized patterns of changes in an abstract time of a regulated parameter. Movement plan (or its deep structure) is formed by choosing necessary patterns in a certain sequence and transferring them into real time.

• Functions used for transformation of the deep structure into real motor commands (into surface structure) taking into account real conditions of movement execution (force fields, effector, and adjacent space geometry) are also memorized. These functions are time dependent because conditions of movement execution change in time.

• Choice of necessary pattern sequence, transferring it into real time, and choice of a transforming function are usually based on motor task, visual information, and subject's memory and knowledge.

• Choice of a motor program defines a range of insensitivity to peripheral sensory signals where they do not influence execution of the movement. Sensory corrections start playing an important role during urgent reactions to unexpected changes in conditions of movement execution (getting beyond the range of insensitivity). Tonic function of vestibular and proprioceptive signals is mainly in delivering information on body geometry and its position in space.

• Sensory corrections play an important role in learning new movements or movements in unfamiliar conditions.

• Reflex connections of peripheral receptors can subserve automatic realization of certain transformations as in an example of the hypothetical tonic stretch reflex (λ-mechanism). This mechanism can be considered a step in transformation of the deep structure of a motor command.

Essential features of the described scheme include (a) lack of solution of complex dynamic problem and (b) lack of continuous corrections of ongoing motor program—of the deep structure; changes of the surface structure certainly take place because the transforming function depends on changing external conditions of movement execution.

This scheme can be applied at different levels of motor hierarchy. The given description has dealt primarily with consciously controlled voluntary movements. Depending on the control level, the relative importance of different sources of information on different stages of surface structure command formation can change (cf. the example of the λ transformation).

Getting back to language-movement relations, one can rephrase the statement of Gelfand et al. (1971) presented at the beginning of this chapter:

One may say that words of the language of movements represent commands to particular muscles and that synergies unite these words into sentences. An important feature of coordinated movements is grammaticality of "synergy phrases."

In the evolution process, movements appeared much earlier than speech. However, the border between these two phenomena is not well defined. Movements play a crucial role in information transmission between animals. Any usage of language by humans is also associated with organized coordinated movements of vocalization organs (during speech) or of hand (during writing). This means that all the vocabulary of a given person, all sequences of words, all peculiarities of grammar can be expressed by sequences of motor commands leading to writing or speaking the phrases. Phrase construction can thus be associated with construction of a synergy, say a synergy necessary for writing the phrase.

These considerations suggest that, theoretically speaking, language could be based on movements. Therefore, a hypothesis that the basic laws of these phenomena can in essence be common looks not improbable. One can tentatively hypothesize that the basic logics of central structures controlling coordinated movements coincides with the logics of similar language structures. Perhaps the second ones evolved in the evolution process on the basis of the first ones (cf. MacNeilage et al., 1984; Ojemann, 1984).

Chapter 11

What to Do Next?

I would like now to speculate on what I consider the most promising future directions of motor control studies. My opinions are certainly conditioned by my experience and beliefs, in particular by a strong attachment to the λ-model. These thoughts were provoked by several nights of discussion with Gregor Schoner, to whom I feel deeply indebted.

I believe that motor control studies should be at least two dimensional. First is a "horizontal" direction that refers to complexity of the moving object. The complexity increases from single-joint to multijoint to multilimb movements (Figure 11.1). The "vertical" direction refers to steps or levels of processing the motor task and the generation of peripheral

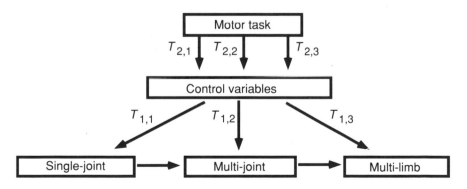

Figure 11.1 Two dimensional motor control studies. The horizontal direction corresponds to the increasing complexity of a moving object, while the vertical direction refers to different steps in the processing of motor tasks leading to the generation of peripheral movement patterns. $T_{i,j}$ represents a transformation of a control pattern for step i and object j.

movement patterns that are likely to be object-independent. Figure 11.1 suggests three levels (which I believe represent the minimal number), namely, motor task, control variables, and peripheral patterns. To me, the processes of transformation ($T_{i,j}$ in Figure 11.1, where i refers to the level of control and j to the object) of a control pattern at one level into a pattern on a lower level are the most exciting objects of study: How does a motor task generate control-variable patterns? How does a control-variable pattern generate peripheral movement patterns? These laws of transformation (or laws of coordination; see chapter 10) can be studied separately for single-joint, multijoint, and multilimb movements. Both questions require at least hypothetical control variables at the intermediate level; presently such variables are available only for single-joint movements (λ-model).

Most studies have addressed the relations between motor task and peripheral patterns of movements, jumping over the intermediate level of control variables. Theories and models have been created to try to express the hypothetical control patterns via indices of performance of the peripheral apparatus (kinematics, dynamics, and electromyograms). However, according to the scheme in Figure 11.1, control variables should be searched for at a different level. How can one study such a system with an explicitly controlled input (motor task), explicitly observable output (motor performance), and an assumed concealed layer? Recent experience suggests two fruitful approaches.

The dynamic pattern generation approach has been gaining support of late. This approach, although based on observations of the relations between motor task and peripheral patterns, need not be restricted to these relations. In particular, one may assume that patterns of control variables are generated based on motor task using the mechanism of dynamic pattern generation and that peripheral patterns are generated based on control variable patterns using the same mechanism. To be productive within the suggested scheme, the dynamic pattern generation approach needs to specify at least hypothetical control variables at the intermediate level. We have recently illustrated the dynamic properties of the $T_{1,1}$ transformation (Figure 11.1) by observing typical phase jumps between the virtual joint trajectories (control patterns) and actual joint trajectories during oscillatory single-joint movements (Latash, 1992).

The second approach involves changes in external conditions for the moving effectors. It has been used to reconstruct joint compliant characteristics in both the absence and the presence of voluntary changes in the motor command (see chapter 2). This group of experiments may be termed "iso-command" because they require noninteraction by the subject, that is, preserving a constant motor command or a time pattern of motor command when external conditions change (e.g., when perturbations occur). Until now, this approach has been used exclusively in single-joint movement studies. It has allowed reconstruction of control variable patterns during a variety of single-joint movements, although certainly under a number of assumptions and simplifications. The horizontal direction in Figure 11.1 suggests applying the same method to analyze multijoint, and eventually multilimb, movements. With a bit of luck and intuition, this may lead to a good guess concerning the control variables (analogous to λ for single-joint movements), creation of a good model for control of multijoint movements, and eventually to reconstruction of control variable patterns.

The other direction (vertical in Figure 11.1) suggests studying the relations between motor task and control variable patterns in "iso-task" experiments with perturbations. Small changes in the force field may lead to minor, quantitative changes in the control patterns that preserve the basic features of the patterns, such as when moving along a straight path at slightly different speeds. After a certain threshold, however, a further change in the force field may require a qualitative change in the control patterns corresponding to a global change in the system's dynamics that is likely to be reflected in an abrupt change in an appropriate order parameter. Iso-task experiments will, I hope, help elucidate the relations between the motor task level and the control-variable level.

Neither suggested direction explicitly suggests which experiments to perform, nor does either promise success. But when we deal with such a complex system as motor control, direct induction and deduction are likely to bring limited success. Intuition has played a crucial role in all motor control studies that have led to a qualitative jump in our understanding of the system (e.g., those by N. A. Bernstein). Let us hope that it will not fail us when the time comes.

Epilogue

Writing a book is like telling a story. This applies to scientific books as well. I'm not sure if I have failed or succeeded in telling my version of the story of the control of human voluntary movements. However, I would like another chance, so I'll substitute a boring and repetitive summary of the last eleven chapters with a short story written to capture the essence of this book.

The Story of the Dead Rat

When Huckleberry Finn found a dead rat, he was both pleasantly surprised and puzzled. The dead rat certainly promised a lot of entertainment. He knew from his buddy Tom Sawyer that it could be tied to a rope and swung through the air. Two months ago, Huckleberry himself played with Tom's rat at the cost of a cigarette puff per minute. That certainly was fun but not too much. Somehow, Huckleberry felt that the rat had more to offer.

"Well," said Huckleberry, thinking out loud, "the rope either pulls the rat with the force I apply or hangs slack without doing anything. So, the rat is either as far from my hand as the rope allows or it is out of control. This does not leave much room for exercising one's imagination. . . .

"What will happen if I take a metal rod and tie the rat to one of its ends? Then, I can at least both push and pull. But this is too simple and will soon be boring, because the rat will exactly follow the movements of my hand. And also if the rat hits something, I am going to hurt my hand and smash the precious animal. This is too crude.

"Well, what is in between?"

At that moment, Tom Sawyer was walking home from school.

"Hi, Huckleberry!" said Tom. "What's cooking? Oh, this is a marvel! Such a pretty one!" And Tom closely examined the unfortunate animal. "What are you going to do with it? Tie it to a rope?"

"I don't know yet. Just thinking about something more exciting."

"I have an idea," said Tom. "They've been talking today at school about how you can touch your nose without poking yourself in the eye. A guy from Russia, I forgot his name, has suggested that your muscles act like springs. He thinks this is more fun than just telling them to push or pull with the force you like. He also thinks this is how you can still touch your nose even if I kick your elbow. So, let's try the rat on a spring."

"Sounds good to me, but where shall we get a spring?"

"No problem. I have a long thick elastic band. It should do."

The boys tied the dead rat to an end of the elastic band.

"Well, what's next?" asked Huckleberry with the rat hanging just above the ground.

"Say you want the rat to swing quickly from here to the tree and stop there. What will you do?"

"Why, move my hand quickly from here to there."

"And smash the poor rodent against the tree. Don't forget that the elastic band stretches."

"So what?"

And Huckleberry did what he had promised. The rat hit the tree and started to look even more unfortunate.

"Well," admitted Huckleberry, "I need to quickly move there and then quickly back, so that the rat will not touch the bark."

He tried that, and the rat swung forward, then backward, and stopped a couple of feet from the tree.

"You see," said Tom Sawyer, "the rat is swinging there and back because the piece of rubber is elastic, but it will always eventually stop where your hand is. That's how the equilibrium control works."

"Well then, I guess I will move my hand quickly to the tree, and when the rat is halfway there, I will move my hand quickly back to brake it, and then, a little bit more smoothly, to the tree again."

Huckleberry tried this strategy and after a couple of attempts, he was able to move the rat quickly to the target and smoothly stop it there.

"Great!" said Tom. "This is, I believe, what they called today the N-shaped virtual trajectory."

"Virtual trajectory? What is so virtual about it?"

"This is simple. Say there is a guy behind that fence who cannot see you or your hand but just sees the rat and the elastic band. He sees the rat moving quickly to the tree and stopping there. He knows that there should be some Mr. Finn handling the other end of the elastic band. Can he guess what your hand is doing?"

"Never! Unless he is as smart as you are, Tom."

Tom looked pleased.

"And can he guess what you wanted to do?"

"This is simple. If the rat swings to the tree and stops there, this is probably what I wanted it to do."

"Exactly! That guy behind the fence looks at the rat and thinks: There is someone who wants to move the rat quickly to the tree. First, he should decide where and how fast he wants it to go. He plans the future rat movement in his mind. For that he must have had a lot of practice and know how the rat behaves, what the wind is, and all that stuff. And, probably, he will not want to calculate what tension will be in the elastic band during the movement and all that . . . A long time ago, there was a guy who looked through the fence and thought just like that. Another Russian. So, when you have decided where and how the rat should go, you need to tell your hand how to move, and this is not that simple, as you have just seen. The movements of your hand can be quite different from what you want from the dead rat but they are what cause the future rat movement. These hand movements are the virtual trajectory of the rat. Got it?"

"Yes, sir! But can that guy guess what my virtual trajectory is just by looking at the rat?"

"Probably this is not enough. But he can study the virtual trajectories step by step. First, he needs to know something about the elastic band you have in your hand, how stiff it is, how long it is, and all that. For example, he can start throwing stones at the rat while it hangs. If he hits the rat, it will move sideways and then return to the same point. Then, if you know some physics, which I don't, you can guess something about the elastic band."

"I got it! And then he will throw the stones while the rat is moving . . ."

"Not that simple. If a stone hits the rat while it is moving, you are likely to make some jerky hand movement, aren't you? What will you do if I hit the rat while you are moving it to the tree?"

"Hit you over the head! This is my rat, Tom! I did not spoil your fun when you were painting this very fence, did I?"

"Okay, cool down, I am not going to hit it. But this is exactly the point. If something nasty happens, you will probably forget about moving the hand smoothly to where it should move but rather pull the cord from me and hit me over the head. These are called preprogrammed reactions. Also, if the stone hits the rat too hard, the elastic band can start playing nasty jokes, jumping and prancing. That guy behind the fence would never guess what you actually wanted to do with your hand."

"Tom, I have an idea. You need to do something so that the rat will move differently, but you need to do it in a kind of sneaky way so that the elastic band does not prance too much and I do not start those preprogrammed hits over the head. Okay? Say that guy has a kind of wind-blower, so that he can make wind. Then, he can quietly puff at the rat, it will move differently, and he will guess what my hand was doing. How does that sound?"

"Sounds fine to me. This is really clever. Let's talk it over with our teacher tomorrow."

Huckleberry Finn gleamed with pleasure and gently caressed the dead rat.

"Okay," continued Tom, "let us think now what will happen if I grab the rat and you do not notice and still move the hand as if the rat were free to move."

"Well, the elastic band will pull differently because the rat will be at the wrong place all the time. This means that I can move my hand the same way but the tension on the elastic band may be quite different if you grab the rat or you don't."

"So the teacher was right when he told us that we could not control our muscle forces or their electrical signals because they both depend on what happens around them."

"But what is so special about this particular piece of elastic? Can we use something different? Say, a long metal spring?"

"Sure we can. But if the spring is too long and jumpy, the rat will jump all the time even if you do not want it to move because of wind and other things. And if we take a rod which is too short and rigid, the rat will move in a kind of stiff way, it will be hard to get around things like trees and poles, and you will hurt your hand if the rat hits something. In both cases, it will look more like a disease. So the rod should be elastic but not too much, long but also not too much, just like the one we have."

"Looks like we are lucky, Tom. Hey, and what if we take your rat as well, and tie both rats to a short metal rod. Then let us tie the bands to each rat and I . . ."

"Oh no, Huckleberry. You are now talking about multijoint movements. The teacher said today that no one knows how we control them. Even those Russians. It is interesting, however, that no matter how many elastic pieces you use, the equilibrium control will work anyway. I mean, if you move your hand somewhere, and you know that the rats are free to move, they will all come to the same new places, no matter how you move your hand to that new position and how many stones are thrown at the rats while they are moving. But let us, for the time being, try to get the maximum fun from one rat and see. Maybe tomorrow you will invent some kind of air puffs for a couple of rats as well."

"I sure will, Tom. See you tomorrow, and don't forget to bring your rat."

The two boys left in different directions. Huck looked around hoping to find another dead rat or, even better, a cat. Tom tried to invent an explanation for Aunt Polly why he was late, which looked like a much more complicated problem than all the rat tricks taken together.

References

Abbs, JH, Hartman, DE, & Vishwanat, B (1987). Orofacial motor control impairment in Parkinson's disease. *Neurology* 37: 394-398.

Abdusamatov, RM, & Feldman, AG (1986). Description of the electromyograms with the aid of a mathematical model for single joint movements. *Biophysics* 31: 549-552.

Abdusamatov, RM, Adamovich, SV, & Feldman, AG (1987). A model for one-joint motor control in man. In GN Gantchev, B Dimitrov, & P Gatev (Eds.) *Motor control* (pp. 183-187). New York: Plenum Press.

Abend, W, Bizzi, E, & Morasso, P (1982). Human arm trajectory formation. *Brain* 105: 331-348.

Abraham, LD, & Loeb, GE (1986). The distal hindlimb musculature of the cat. Patterns of normal use. *Exp Brain Res* 58: 580-593.

Abraham, RH, & Shaw, CD (1982). *Dynamics—The geometry of behavior.* Santa Cruz: Aerial Press.

Accornero, N, Berardelli, A, Argenta, M, & Manfredi, M (1984). Two joint ballistic arm movements. *Neurosci Lett* 46: 91-95.

Accornero, N, Berardelli, A, Argenta, M, & Manfredi, M (1985). Two-joint fast arm movements in normal subjects and in patients with Parkinson's disease. In PJ Delwaide & A Agnoli (Eds.) *Clinical neurophysiology in Parkinsonism* (pp. 83-89). Amsterdam: Elsevier.

Adamovitch, SV, Burlachkova, NI, & Feldman, AG (1984). Wave nature of the central process of formation of the trajectories of change in the joint angle in man. *Biophysics* 29: 130-134.

Adamovitch, SV, & Feldman, AG (1984). Model of the central regulation of the parameters of motor trajectories. *Biophysics* 29: 338-342.

Adamovitch, SV, & Feldman, AG (1989). The prerequisites for one-joint motor control theories. *Behav Brain Sci* 12: 210-211.

Adams, JA (1971). A closed-loop theory of motor learning. *J Mot Behav* 3: 111-150.

Adey, WR, & Noda, H (1973). Influence of eye movements on geniculostriate excitability in the cat. *J Physiol* 235: 805-821.

Agarwal, GC, & Gottlieb, GL (1977). Compliance of the human ankle joint. *J Biomech Eng* 99: 166-170.

Agarwal, GC, & Gottlieb, GL (1980). Effect of vibration on the ankle stretch reflex in man. *Electroencephalog Clin Neurophysiol* 49: 81-92.

Agarwal, GC, & Gottlieb, GL (1982). Mathematical modeling and simulation of the postural control loop: Part I. *CRC Crit Rev Biomed Eng* 8: 93-134.

Agarwal, GC, & Gottlieb, GL (1986). Complexity in control of movements. *Behav Brain Sci* 9: 599-600.

Akazawa, K, Aldridge, JW, Steeves, JD, & Stein, RB (1982). Modulation of stretch reflexes during locomotion in the mesencephalic cat. *J Physiol* 329: 553-567.

Akazawa, K, Milner, TE, & Stein, RB (1983). Modulation of reflex EMG and stiffness in response to stretch of human finger muscle. *J Neurophysiol* 49: 16-27.

Aleshinsky, SY (1986). An energy "sources" and "fraction" approach to the mechanical energy expenditure problem.—1. Basic concepts, descriptions of the model, analysis of a one-link system movement. *J Biomechanics* 19: 287-293.

Alexander, GE (1987). Selective neuronal discharge in monkey putamen reflects intended direction of planned limb movements. *Exp Brain Res* 67: 623-634.

Alexander, GE, & Crutcher, MD (1990a). Preparation for movement: Neural representations of intended direction in three motor areas of the monkey. *J Neurophysiol* 64: 133-150.

Alexander, GE, & Crutcher, MD (1990b). Neural representations of the target (goal) of visually guided arm movements in three motor areas of the monkey. *J Neurophysiol* 64: 164-178.

Alexander, GE, & DeLong, MR (1985). Microstimulation of the primate neostriatum. II. Somatotopic organization of microexcitable zones and their relation to neuronal response properties. *J Neurophysiol* 53: 1401-1416.

Allan, G (1979). The perception of time. *Percept Psychophys* 26: 340-354.

Allum, JHJ (1975). Response to load disturbances in human shoulder muscles: The hypothesis that one component is a pulse test information signal. *Exp Brain Res* 22: 307-326.

Allum, JHJ (1983). Organization of stabilizing reflex responses in tibialis anterior muscles following ankle flexion perturbations of standing man. *Brain Res* 264: 297-301.

Allum, JHJ, Honneger, F, & Pfaltz, CR (1989). The role of stretch and vestibulospinal reflexes in the generation of human equilibrating reactions. *Prog Brain Res* 80: 399-409.

Allum, JHJ, & Keshner, EA (1986). Vestibular and proprioceptive control of sway stabilization. In W Bles & T Brandt (Eds.) *Disorders of posture and gait* (pp. 19-40). Amsterdam: Elsevier.

Almeida, GL, Corcos, DM, & Latash, ML (1991). Effects of practice of fast voluntary movements in Down syndrome individuals. *Abstr Soc Neurosci* 17: 1027.

Alstermark, B, Gorska, T, Lundberg, A, & Petersson, L-G (1990). Integration in descending motor pathways controlling the forelimb in the cat. 16. Visually guided switching of target-reaching. *Exp Brain Res* 80: 1-11.

Ammon, K, & Gandevia, SC (1990). Transcranial magnetic stimulation can influence the selection of motor programmes. *J Neurol Neurosurg Psychiat* 53: 705-707.

An, CH, Atkeson, CG, & Hollerbach, JM (1988). *Model-based control of a robot manipulator.* Cambridge: MIT Press.

Anden, N-E, Jukes, MGM, Lundberg, A, & Vyklicky, L (1966). The effect of DOPA on the spinal cord. I. Influence on transmission from primary afferents. *Acta Physiol Scand* 67: 373-386.

Andersson, O, Forssberg, H, Grillner, S, & Lindquist, M (1978). Phasic gain control of the transmission in cutaneous reflex pathways to motoneurones during "fictive" locomotion. *Brain Res* 149: 503-507.

Andrews, CJ, Burke, D, & Lance, JW (1972). The response to muscle stretch and shortening in Parkinsonian rigidity. *Brain* 95: 795-812.

Angel, RW (1983). Muscular contractions elicited by passive shortening. *Adv Neurol* 39: 555-563.

Angel, RW, Alston, W, & Higgins, JR (1970). Control of movement in Parkinson's disease. *Brain* 93: 1-14.

Anson, JG (1989). Down syndrome: Neuromotor programming and fractionated reaction time. In ML Latash (Ed.) *Motor control in Down syndrome* (pp. 6-13). Chicago: Rush Medical Center.

Anson, JG, & Davis, SA (1988). Neuromotor programming and Down syndrome. *Int J Neurosci* 40: 82.

Arbib, MA (1980). Interacting schemas for motor control. In GE Stelmach & J Requin (Eds.) *Tutorials in motor behavior* (pp. 71-81). Amsterdam: North-Holland.

Armstrong, DM, Campbell, NC, Edgley, SA, Schild, RF, & Trott, JR (1982). Investigations of the olivocerebellar and spinoolivary pathways. In SL Palay & V Chan-Palay (Eds.) *The cerebellum — New vistas* (pp. 195-232). Berlin: Springer-Verlag.

Arnt, K, & Scheel-Kruger, G (1980). Intracranial GABA antagonists produce dopamine-independent biting in rats. *Eur J Pharmacol* 62: 51-61.

Asanuma, H (1973). Cerebral cortical control of movements. *Physiologist* 16: 143-166.

Asanuma, H, Babb, RS, Mori, A, & Waters, RS (1981). Input-output relationship in cat's motor cortex after pyramidal section. *J Neurophysiol* 46: 694-703.

Asatryan, DG, & Feldman, AG (1965). Functional tuning of the nervous system with control of movements or maintenance of a steady posture. I. Mechanographic analysis of the work of the limb on execution of a postural task. *Biophysics* 10: 925-935.

Ashby, P, Andrews, C, Knowles, L, & Lance, JW (1972). Pyramidal and extrapyramidal control of tonic mechanisms in the cat. *Brain* 95: 21-30.

Ashby, P, & McCrea, DA (1987). Neurophysiology of spinal spasticity. In RA Davidoff (Ed.) *Handbook of the spinal cord* (pp. 119-143). New York: Dekker.

Ashby, P, & Verrier, M (1976). Neurophysiological changes in hemiplegia, possible explanation for initial disparity between muscle tone and tendon reflexes. *Neurology* 26: 1145-1151.

Athans, M, & Falb, PL (1966). *Optimal control: An introduction to the theory and its applications.* New York: McGraw-Hill.

Atkeson, CG (1989). Learning arm kinematics and dynamics. *Ann Rev Neurosci* 12: 157-183.

Atkeson, CG, & Hollerbach, JM (1985). Kinematic features of unrestrained vertical arm movements. *J Neurosci* 5: 2318-2320.

Baillieul, J, Hollerbach, JM, & Brockett, R (1984). Programming and control of kinematically redundant manipulators. *Proc 23rd Conf on Decision and Control*, pp. 768-774, Las Vegas.

Baldissera, F, Cavallari, P, & Civaschi, P (1982). Preferential coupling between voluntary movements of ipsilateral limbs. *Neurosci Lett* 34: 95-100.

Baldissera, F, Hultborn, H, & Illert, M (1981). Integration in spinal neuronal systems. In VB Brooks (Ed.) *Handbook of physiology II. Motor control* (pp. 509-595). Bethesda: Amer Physiol Soc.

Bankhead, I, & Mackay, DN (1982). Fine motor performance in subjects of subnormal, normal, and superior intelligence. I. Reaction time and task complexity. *J Ment Defic Res* 26: 73-89.

Bassler, U (1976). Reversal of a reflex to a single motoneuron in the stick insect Carausius morosus. *Biol Cybern* 24: 47-49.

Bauswein, E, Kolb, FP, Leimbeck, B, & Rubia, FJ (1983). Simple and complex spike activity of cerebellar Purkinje cells during active and passive movements in the awake monkey. *J Physiol* 339: 379-394.

Bawa, P, & McKenzie, DC (1981). Contribution of joint and cutaneous afferents to longer-latency reflexes in man. *Brain Res* 211: 185-189.

Baxendale, RH, & Ferrell, WR (1981). The effect of knee joint afferent discharge on transmission in flexor reflex pathways in decerebrate cat. *J Physiol* 315: 231-242.

Bazalgette, D, Zattara, M, Bathien, N, Bouisset, S, & Rondot, P (1986). Postural adjustments associated with rapid voluntary arm movements in patients with Parkinson's disease. *Adv Neurol* 45: 371-374.

Becker, W, & Jurgens, R (1979). An analysis of the saccadic system by means of double step stimuli. *Vision Res* 19: 967-983.

Beggs, WDA, & Howart, CI (1970). Movement control in a repetitive motor task. *Nature* 225: 752-753.

Belen'kii, VY, Gurfinkel, VS, & Pal'tsev, YI (1967). Elements of control of voluntary movements. *Biofizika* 10: 135-141.

Bellman, KL, & Goldberg, LJ (1984). Common origin of linguistic and movement abilities. *Amer J Physiol* 15: R915-R921.

Benecke, R, Meinck, HM, & Conrad, B (1985). Rapid goal-directed elbow flexion movements: Limitations of speed control system due to neural constraints. *Exp Brain Res* 59: 470-477.

Benecke, R, Rothwell, JC, Dick, JPR, Day, BL, & Marsden, CD (1986). Performance of simultaneous movements in patients with Parkinson's disease. *Brain* 109: 739-757.

Benecke, R, Rothwell, JC, Dick, JPR, Day, BL, & Marsden, CD (1987). Disturbance of sequential movements in patients with Parkinson's disease. *Brain* 110: 361-379.

Bennett, DJ, Xu, Y, Hollerbach, JM, & Hunter, IW (1989). Identifying the mechanical impedance of the elbow joint during posture and movement. *Abstr Soc Neurosci* 15: 396.

Berardelli, A, Dick, JPR, Rothwell, JC, Day, BL, & Marsden, CD (1986). Scaling of the size of the first agonist EMG burst during rapid wrist movements in patients with Parkinson's disease. *J Neurol Neurosurg Psychiat* 49: 1273-1279.

Berardelli, A, & Hallett, M (1984). Shortening reaction of human tibialis anterior. *Neurology* 34: 242-246.

Berardelli, A, Rothwell, JC, Day, BL, & Marsden, CD (1984). Movements not involved in posture are abnormal in Parkinson's disease. *Neurosci Lett* 47: 47-50.

Berardelli, A, Sabra, AF, Hallett, M, Berenberg, W, & Simon, SR (1983). Stretch reflexes of triceps surae in patients with upper motor neuron syndromes. *J Neurol Neurosurg Psychiat* 46: 54-60.

Berger, W, Horstmann, D, & Dietz, V (1984). Tension development and muscle activation in the leg during gait in spastic hemiparesis: Independence of muscle hypertonia and exaggerated stretch reflexes. *J Neurol Neurosurg Psychiat* 47: 1029-1033.

Bergmans, J, & Grillner, S (1968). Changes in dynamic sensitivity of primary endings of muscle spindle afferents induced by DOPA. *Acta Physiol Scand* 74: 618-639.

Berkinblit, MB, Zharkova, IS, Feldman, AG, & Fukson, OI (1984). Biomechanical singularities of the wiping reflex cycle. *Biofizika* 29: 483-488.

Berkinblit, MB, & Feldman, AG (1988). Some problems of motor control. *J Mot Behav* 20: 369-373.

Berkinblit, MB, Feldman, AG, & Fukson, OI (1986a). Adaptability of innate motor patterns and motor control mechanisms. *Behav Brain Sci* 9: 585-638.

Berkinblit, MB, Gelfand, IM, & Feldman, AG (1986b). A model for the control of multijoint movements. *Biofizika* 31: 128-138.

Berkinblit, MB, Gelfand, IM, & Feldman, AG (1986c). A model for the aiming phase of the wiping reflex. In S Grillner, PSG Stein, D Stuart, H Forssberg, & RM Herman (Eds.) *Neurobiology of vertebrate locomotion.* Wenner-Gren International Symposium Series, 45: 217-227.

Berkson, G (1960). An analysis of reaction time in normal and mentally deficient young men. I, II, III. *J Ment Defic Res* 4: 51-77.

Bernstein, NA (1926). *General biomechanics.* Moscow: Medgiz (in Russian).

Bernstein, NA (1935). The problem of interrelation between coordination and localization. *Arch Biol Sci* 38: 1-35 (in Russian).

Bernstein, NA (1947). *On the construction of movements.* Moscow: Medgiz (in Russian).

Bernstein, NA (1967). *The co-ordination and regulation of movements.* Pergamon Press, Oxford.

Bernstein, NA (1991). *On dexterity and its development.* Moscow: Physical Culture and Sport Press (in Russian).

Berthoz, A, Lacour, M, Soechting, JF, & Vidal, PP (1979). The role of vision in the control of posture during linear motion. In R Granit & O Pompeiano (Eds.) *Reflex control of posture and movement* (pp. 197-209). Amsterdam, New York, Oxford: Elsevier.

Beuter, A, Milton, JG, Labrie, C, Glass, L, & Gauthier, S (1990). Delayed visual feedback and movement control in Parkinson's disease. *Exp Neurol* 110: 228-235.

Bigland, B, & Lippold, O (1954). The relation between force, velocity and integrated electrical activity in human muscles. *J Physiol* 123: 214-224.

Bingham, GP, Schmidt, RC, Turvey, MT, & Rosenblum, LD (1991). Task dynamics and response dynamics in the assembly of a coordinated rhythmic movement. *J Exp Psychol: Hum Percept Perform* 17: 359-381.

Bizzi, E (1980). Central and peripheral mechanisms in motor control. In GE Stelmach & J Requin (Eds.) *Tutorials in motor behavior* (pp. 131-143). Amsterdam: North-Holland.

Bizzi, E, Accornero, N, Chapple, W, & Hogan, N (1982). Arm trajectory formation in monkeys. *Exp Brain Res* 46: 139-143.

Bizzi, E, Accornero, N, Chapple, W, & Hogan, N (1984). Posture control and trajectory formation during arm movements. *J Neurosci* 4: 2738-2744.

Bizzi, E, Dev, P, Morasso, P, & Polit, A (1978a). Effect of load disturbances during centrally initiated movements. *J Neurophysiol* 41:542-556.

Bizzi, E, Dev, P, Morasso, P, & Polit, A (1978b). Role of neck proprioceptors during visually triggered head movements. *Prog Clin Neurophysiol* 4: 141-152.

Bizzi, E, Mussa-Ivaldi, FA, & Giszter, S (1991). Computations underlying the execution of movement: A biological perspective. *Science* 253: 287-291.

Bizzi, E, & Polit, A (1979). Characteristics of the motor programs underlying visually evoked movements. In RE Talbott & DR Humphrey (Eds.) *Posture and movement* (pp. 169-176). New York: Raven Press.

Bizzi, E, Polit, A, & Morasso, P (1976). Mechanisms underlying achievement of final head position. *J Neurophysiol* 39: 435-444.

Bloxham, CA, Mindel, TA, & Frith, CD (1984). Initiation and execution of predictable and unpredictable movements in Parkinson's disease. *Brain* 107: 371-384.

Bock, O (1990). Load compensation in human goal-directed arm movements. *Behav Brain Res* 41: 167-177.

Bock, O, & Eckmiller, R (1986). Goal-directed arm movements in absence of visual guidance: Evidence for amplitude rather than position control. *Exp Brain Res* 62: 451-458.

Bonnet, M (1983). Anticipatory changes of long-latency stretch responses during preparation for directional hand movements. *Brain Research,* 280: 51-62.

Bonnet, M, Requin, J, & Stelmach, GE (1991). Changes in electromyographic responses to muscle stretch, related to the programming of movement parameters. *Electroencephalog Clin Neurophysiol* 81: 135-151.

Bouisset, S, & Lestienne, F (1974). The organization of simple voluntary movement as analyzed from its kinematic properties. *Brain Res* 71: 451-457.

Bouisset, S, & Zattara, M (1981). A sequence of postural movements precedes voluntary movement. *Neurosci Lett* 22: 263-270.

Bouisset, S, & Zattara, M (1983). Anticipatory postural movements related to a voluntary movement. In *Physiologie spatiale* (pp. 137-141). Toulouse: Cepadues Editions.

Bouisset, S, & Zattara, M (1987). Biomechanical study of the programming of anticipatory postural adjustments associated with voluntary movement. *J Biomech* 20: 735-742.

Bouisset, S, & Zattara, M (1990). Segmental movement as a perturbation to balance? Facts and concepts. In JM Winters & SL-Y Woo (Eds.) *Multiple muscle systems. Biomechanics and movement organization* (pp. 498-506). New York: Springer-Verlag.

Bowery, NG, Hill, DR, Hudson, AL, Doble, AL, Middlemiss, A, Shaw, J, & Turnbull, M (1980). Baclofen decreases neurotransmitter release in the mammalian CNS by an action at a novel GABA receptor. *Nature* 283: 92-94.

Bowery, NG, Price, GW, Hudson, AL, Hill, DR, Wilkin, GP, & Turnbull, MJ (1984). GABA receptor multiplicity. *Neuropharmacol* 23: 219-231.

Brodie, EE, & Ross, HE (1984). Sensorimotor mechanisms in weight discrimination. *Percept Psychophys* 36: 477-481.

Bronstein, AM, Hood, JD, Gresty, MA, & Panagi, C (1990). Visual control of balance in cerebellar and parkinsonian syndromes. *Brain* 113: 767-779.

Brookhart, JM (1979). Convergence of an understanding of motor control. In RE Talbott & DR Humphrey (Eds.) *Posture and movement* (pp. 295-303). New York: Raven Press.

Brooks, VB (1979). Motor programs revisited. In RE Talbott & DR Humphrey (Eds.) *Posture and movement* (pp. 13-49). New York: Raven Press.

Brooks, VB (1981). Task-related cell assemblies. In O Pompeiano & CA Marsan (Eds.) *Brain mechanisms and perceptual awareness* (pp. 295-309). New York: Raven Press.

Brooks, VB (1982). Optimal path generation for cooperating or redundant manipulators. Proc 2 Intern Computer Engineering Conf, San Diego, pp. 119-122.

Brown, JE, & Frank, FS (1987). Influence of event anticipation of postural actions accompanying voluntary movement. *Exp Brain Res* 67: 645-650.

Brown, MC, Engberg, I, & Matthews, PBC (1967). The relative sensitivity to vibration of muscle receptors of the cat. *J Physiol* 192: 773-800.

Brown, RG, & Marsden, CD (1991). Dual task performance and processing resources in normal subjects and patients with Parkinson's disease. *Brain* 114: 215-231.

Brown, SH, & Cooke, JD (1981). Responses to force perturbations preceding voluntary human arm movements. *Brain Res* 220: 350-355.

Brown, SH, & Cooke, JD (1984). Initial agonist burst duration depends on movement amplitude. *Exp Brain Res* 55: 523-527.

Brown, SH, & Cooke, JD (1990). Movement-related phasic muscle activation. I. Relations with temporal profile of movement. *J Neurophysiol* 63: 455-464.

Brown, TIH, Rack, PMH, & Ross, HF (1982). A range of different stretch reflex responses in the human thumb. *J Physiol* 332: 101-112.

Bruwer, M, & Cruse, H (1990). A network model for the control of the movement of a redundant manipulator. *Biol Cybern* 62: 549-555.

Buchanan, TS, Almdale, DPJ, Lewis, JL, & Rymer, WZ (1986). Characteristics of synergetic relations during isometric contractions of human elbow muscles. *J Neurophysiol* 56: 1225-1241.

Buchanan, TS, Rovai, GP, & Rymer, WZ (1989). Strategies for muscle activation during isometric torque generation at the human elbow. *J Neurophysiol* 39: 925-935.

Buchtal, F (1942). The mechanical properties of the single striated muscle fibre at rest and during contraction and their structural interpretation. *Dan Biol Medd Kbh* 17: 1-140.

Buchwald, NA, Hull, CD, Levine, MS, & Villablanca, JR (1975). The basal ganglia and the regulation of response and cognitive sets. In MAB Brazier (Ed.) *Growth and development of the brain: Nutritional, genetic, and environmental factors* (pp. 171-189). New York: Raven Press.

Bullock, D (1989). Saturation is not an evolutionary stable strategy. *Behav Brain Sci* 12: 212-214.

Bullock, D, & Grossberg, S (1988). Neural dynamics of planned arm movements: Emergent invariants and speed-accuracy properties during trajectory formation. *Psychol Rev* 95: 49-90.

Burbaud, P, Garanx, F, Gross, CH, & Bioulac, B (1988). Postural adjustments in the monkey: Effects of velocity on EMG sequence. *Neurosci Lett* 84: 51-56.

Burke, D (1988). Spasticity as an adaptation to pyramidal tract injury. In SG Waxman (Ed.) *Functional recovery in neurological disease* (pp. 401-423). New York: Raven Press.

Burke, D, Andrews, CJ, & Ashby, P (1971). Autogenic effects of static muscle stretch in spastic man. *Arch Neurol* 25: 367-372.

Burke, D, Gandevia, SC, & McKeon, B (1983). The afferent volleys responsible for spinal proprioceptive reflexes in man. *J Physiol* 339: 535-552.

Burke, D, Gandevia, SC, & McKeon, B (1984). Monosynaptic and oligosynaptic contributions to human ankle jerk and H-reflex. *J Neurophysiol* 52: 435-448.

Cannon, SC, & Zahalak, GI (1982). The mechanical behavior of active human skeletal muscle in small oscillations. *J Biomech* 15: 111-121.

Capaday, C, Forget, R, Fraser, R, & Lamarre, Y (1991). Evidence for a contribution of the motor cortex to the long-latency stretch reflex of the human thumb. *J Physiol* 440: 243-255.

Carlton, LG (1981). Processing visual feedback information for movement control. *J Exp Psychol: Hum Percept Perform* 7: 1019-1030.

Carlton, LG, Robertson, RN, Carlton, LG, & Newell, KM (1985). Response timing variability: Coherence of kinematic and EMG patterns. *J Mot Behav* 17: 301-321.

Carter, MC, & Shapiro, DC (1984). Control of sequential movements: Evidence for generalized motor programs. *J Neurophysiol* 52: 787-796.

Carter, RR, Crago, PE, & Keith, MW (1990). Stiffness regulation by reflex action in the normal human hand. *J Neurophysiol* 64: 105-118.

Chan, CWY, & Kearney, RE (1982). Is the functional stretch reflex servo controlled or preprogrammed? *Electroencephalog Clin Neurophysiol* 53: 310-324.

Chan, CWY, Melvill Jones, G, Kearney, RE, & Watt, DGD (1979a). The late electromyographic response to limb displacement in man. I. Evidence for supraspinal contribution. *Electroencephalog Clin Neurophysiol* 46: 173-181.

Chan, CWY, Melvill Jones, G, & Catchlove, RFH (1979b). The late electromyographic response to limb displacement in man. II. Sensory origin. *Electroencephalog Clin Neurophysiol* 46: 182-188.

Chapanis, A, Garner, WR, & Morgan, CT (1949). *Applied experimental psychology: Human factors in engineering design*. New York: Wiley.

Chapman, CE, Ruegg, DG, & Wiesendanger, M (1983). Effects of dorsal cord stimulation on stretch reflexes. *Brain Res* 258: 211-215.

Chapman, CE, & Wiesendanger, M (1982). Recovery of function following unilateral lesions of the bulbar pyramid in the monkey. *Electroencephalog Clin Neurophysiol* 53: 374-387.

Cheney, PD, & Fetz, EE (1984). Corticomotoneuronal cells contribute to long-latency stretch reflexes in the rhesus monkey. *J Physiol* 349: 249-272.

Cheron, G, & Godaux, E (1986). Self-terminated fast movement of the forearm in man: Amplitude dependence of the triple burst pattern. *J Biophys Biomec* 10: 109-117.

Chomsky, N (1971). *Selected readings.* JPB Allen & P van Buren (Eds.). London: Oxford University Press.

Chomsky, N (1975). *Reflections on language.* New York: Pantheon Books.

Christakos, CN, Wolf, H, & Meyer-Lohmann, J (1983). The "M2" electromyographic response to random perturbations of arm movements is missing in long-trained monkeys. *Neurosci Lett* 41: 295-300.

Cody, FWJ, MacDermott, N, Matthews, PBC, & Richardson, HC (1986). Observations on the genesis of the stretch reflex in Parkinson's disease. *Brain* 109: 229-249.

Cole, KJ, Abbs, JH, & Turner, GS (1988). Deficits in the production of grip force in Down Syndrome. *Devel Med Child Neurol* 30: 752-758.

Colebatch, JG, Gandevia, SC, McCloskey, DI, & Potter, EK (1979). Subject instruction and long-latency reflex responses to muscle stretch. *J Physiol* 292: 527-534.

Connor, NP, Abbs, JH, Cole, KJ, & Gracco, VL (1989). Parkinsonian deficits in serial multiarticular movements for speech. *Brain* 112: 997-1009.

Conrad, B (1978). The motor cortex as a primary device for fast adjustment of programmed motor patterns to afferent signals. *Prog Clin Neurophysiol* 4: 123-140.

Cooke, JD (1980). The organization of simple skilled movements. In GE Stelmach & J Requin (Eds.) *Tutorials in motor behavior* (pp. 199-211). Amsterdam: North-Holland.

Cooke, JD, & Brown, SH (1990). Movement-related phasic muscle activation. II. Generation and functional role of the triphasic pattern. *J Neurophysiol* 63: 465-472.

Cooke, JD, Brown, S, Forget, R, & Lamarre, Y (1985). Initial agonist burst changes with movement amplitude in a deafferented patient. *Exp Brain Res* 60: 184-187.

Corcos, DM, Gottlieb, GL, & Agarwal, GC (1988). Accuracy constraints upon rapid elbow movements. *J Mot Behav* 20: 255-272.

Corcos, DM, Gottlieb, GL, & Agarwal, GC (1989). Organizing principles for single joint movements. II. A speed-sensitive strategy. *J Neurophysiol* 62: 358-368.

Corcos, DM, Gottlieb, GL, Agarwal, GC, & Flaherty, BP (1990a). Organizing principles for single joint movements. IV. Implications for isometric contractions. *J Neurophysiol* 64: 1033-1042.

Corcos, DM, Gottlieb, GL, Jaric, S, Cromwell, RL, & Agarwal, GC (1990b). Organizing principles underlying motor skill acquisition. In JM Winters & SL-Y Woo (Eds.) *Multiple muscle systems. Biomechanics and movement organization* (pp. 251-267). New York: Springer-Verlag.

Corcos, DM, Gottlieb, GL, Penn, RD, Myklebust, B, & Agarwal, GC (1986). Movement deficits caused by hyperexcitable stretch reflexes in spastic humans. *Brain* 109: 1043-1058.

Cordo, PJ (1990). Kinesthetic control of a multijoint movement sequence. *J Neurophysiol* 63: 161-172.

Cordo, PJ, Horak, FB, & Moore, SP (1989). On to real-life movements. *Behav Brain Sci* 12: 214-215.

Cordo, PJ, & Nashner, LM (1982). Properties of postural adjustments associated with rapid arm movements. *J Neurophysiol* 47: 287-302

Cordo, PJ, & Rymer, WZ (1982). Contribution of motor-unit recruitment and rate modulation to compensation for muscle yielding. *J Neurophysiol* 47: 797-809.

Cowie, V (1970). *A study in the early development of mongols*. London: Pergamon.

Crago, PE, Houk, JC, & Hasan, Z (1976). Regulatory actions of human stretch reflex. *J Neurophysiol* 39: 925-935.

Craik, RL, Cozzens, BA, & Freedman, W (1982). The role of sensory conflict on stair descent performance in humans. *Exp Brain Res* 45: 399-409.

Crane, HD, & Ostrem, JS (1983). Automatic signature verification using a three-axis force-sensitive pen. *IEEE Trans Syst Man Cybern* 13: 329-337.

Crenna, P, Frigo, C, Giovanni, P, & Piccolo, I (1990). The initiation of gait in Parkinson's disease. In A Berardelli, R Benecke, M Manfredi, & CD Marsden (Eds.) *Motor disturbances II* (pp. 161-173). London: Academic Press.

Crenna, P, Frigo, C, Massion, J, & Pedotti, A (1987). Forward and backward axial synergies in man. *Exp Brain Res* 65: 538-548.

Crossman, ERFW, & Goodeve, PJ (1983). Feedback control of hand-movement and Fitts' law. *Quart J Exp Psychol* 35A: 251-278.

Cruse, H, & Bruwer, M (1987). The human arm as a redundant manipulator: The control of path and joint angles. *Biol Cybern* 57: 137-144.

Cruse, H, Wischmeyer, E, Bruwer, M, Brockfield, P, & Dress, A (1990). On the cost functions for the control of the human arm movement. *Biol Cybern* 62: 519-528.

Crutcher, MD, & Alexander, GE (1990). Movement-related neuronal activity selectively coding either direction or muscle pattern in three motor areas of the monkey. *J Neurophysiol* 64: 151-163.

Crutcher, MD, & DeLong, MR (1984). Single cell studies of the primate putamen. II Relation to direction of movement and pattern of muscular activity. *Exp Brain Res* 53: 244-258.

Currie, SN, & Stein, PSG (1990). Cutaneous stimulation evokes long-lasting excitation of spinal interneurons in the turtle. *J Neurophysiol* 64: 1134-1148.

Curtis, DR, & Malik, R (1985). The differential effects of baclofen on segmental and descending excitation of spinal interneurones in the cat. *Exp Brain Res* 58: 333-337.

Darling, WG, Cole, KJ, & Abbs, JH (1988). Kinematic variability of grasp movements as a function of practice and movement speed. *Exp Brain Res* 73: 225-235.

Darling, WG, & Cooke, JD (1987). A linked muscular activation model for movement generation and control. *J Mot Behav* 19: 333-354.

Darton, K, Lippold, OCJ, Shahani, M, & Shahani, U (1985). Long-latency spinal reflexes in humans. *J Neurophysiol* 53: 1604-1618.

Davidoff, RA (1978). Pharmacology of spasticity. *Neurology (Minneap)* 28 (Suppl): 46-51.

Davidoff, RA (1985). Antispasticity drugs: Mechanisms of action. *Ann Neurol* 17: 107-116.

Davies, J (1981). Selective depression of synaptic excitation in cat spinal neurones by baclofen: An iontophoretic study. *Br J Pharmacol* 72: 373-384.

Davis, WR, & Kelso, JAS (1982). Analysis of "invariant characteristics" in the motor control of Down's syndrome and normal subjects. *J Mot Behav* 14: 194-212.

Davis, WE, & Sinning, WE (1987). Muscle stiffness in Down Syndrome and other mentally handicapped subjects: A research note. *J Mot Behav* 19: 130-144.

Day, BL, & Marsden, CD (1981). Movement of a human thumb is not achieved by resetting its "spring" constants. *J Physiol* 317: 60P-61P.

Day, BL, & Marsden, CD (1982). Accurate repositioning of the human thumb against unpredictable dynamic loads is dependent upon peripheral feedback. *J Physiol* 327: 393-407.

Day, BL, Dick, JPR, & Marsden, CD (1984a). Patients with Parkinson's disease can employ a predictive motor strategy. *J Neurol Neurosurg Psychiat* 47: 1299-1306.

Day, BL, Marsden, CD, Obeso, JA, & Rothwell, JC (1984b). Reciprocal inhibition between the muscles of the human forearm. *J Physiol* 349: 519-534.

Day, BL, Riescher, H, Struppler, A, Rothwell, JC, & Marsden, CD (1991). Changes in the response to magnetic and electrical stimulation of the motor cortex following muscle stretch in man. *J Physiol* 433: 41-57.

De Gail, P, Lance, JW, & Neilson, PD (1966). Differential effects on tonic and phasic reflex mechanisms produced by vibration of muscles in man. *J Neurol Neurosurg Psychiat* 29: 1-11.

deGuzman, GC, & Kelso, JAS (1991). Multifrequency behavioral patterns and the phase attractive circle map. *Biol Cybern* 64: 485-495.

DeLong, MR, Crutcher, MD, & Georgopoulos, AP (1987). Primate globus pallidus and subthalamus nucleus: Functional organization. *J Neurophysiol* 53: 530-543.

Delwaide, PJ (1973). Human monosynaptic reflexes and presynaptic inhibition. An interpretation of spastic hyperreflexia. In JE Desmedt (Ed.) *New developments in electromyography and clinical neurophysiology, v 3* (pp. 508-522). Basel: Karger.

Dengler, R, Konstanzer, A, Gillespie, J, Argenta, M, Wolf, W, & Struppler, A (1990). Behavior of motor units in parkinsonism. *Adv Neurol* 53: 167-173.

Denier van der Gon, JJ, & Turing, J (1965). The guiding of human writing movements. *Kybernetik* 2: 145-148.

Denier van der Gon, JJ, & Wadman, WJ (1977). Control of fast ballistic human arm movements. *J Physiol* 271: 28-29P.

Denny-Brown, D (1929). On the nature of postural reflexes. *Proc Roy Soc B* 104: 252-301.

Destcherevsky, VI (1977). *Mathematical models of muscle contraction.* Moscow: Nauka.

Desmedt, JE, & Godaux, E (1978). Ballistic skilled movements: Load compensation and patterning of the motor commands. In JE Desmedt (Ed.) *Cerebral motor control in man: Long-loop mechanisms. Prog Clin Neurophysiol* 4: 21-55. Basel: Karger.

Desmedt, JE, & Godaux, E (1981). Spinal motoneuron recruitment in man: Rank deordering with direction but not with speed of voluntary movements. *Science* 214: 933-936.

Diener, HC, Horak, FB, & Nashner, LM (1988). Influence of stimulus parameters on human postural responses. *J Neurophysiol* 59: 1888-1905.

Dietz, V, & Berger, W (1983). Normal and impaired regulation of muscle stiffness in gait: A new hypothesis about muscle hypertonia. *Exp Neurol* 79: 680-687.

Dietz, V, Horstmann, GA, Trippel, M, & Gollhofer, A (1989). Human postural reflexes and gravity—An under water simulation. *Neurosci Lett* 106: 350-355.

Dietz, V, Noth, J, & Schmidtbleicher, D (1981a). Interaction between pre-activity and stretch reflex in human triceps brachii during landing from forward falls. *J Physiol* 311: 113-125.

Dietz, V, Quintern, J, & Berger, W (1981b). Electrophysiological studies of gait in spasticity and rigidity. Evidence that altered mechanical properties of muscle contribute to hypertonia. *Brain* 104: 431-449.

Dietz, V, Quintern, J, & Berger, W (1984). Corrective reactions to stumbling in man: Functional significance of spinal and transcortical reflexes. *Neurosci Lett* 44: 131-135.

Di Fabio, RP, Badke, MB, McEvoy, A, & Breunig, A (1990). Influence of local sensory afference in the calibration of human balance responses. *Exp Brain Res* 80: 591-599.

Dimitrijevic, MR (1988). Residual motor function in spinal cord injury. *Adv Neurol* 47: 139-155.

Dinnerstein, AJ, Frigyesi, T, & Lowenthal, M (1962). Delayed feedback as a possible mechanism in parkinsonism. *Percept Mot Skills* 15: 667-680.

Dormont, JF, Farin, D, Schmied, A, & Amalric, M (1989). Cat red nucleus activity preceding movement depends on initiation conditions. *Exp Brain Res* 77: 271-282.

Dralle, D, Muller, H, & Zierski, J (1988). A short historical review of spasticity and its therapy. In H Muller, J Zierski, & RD Penn (Eds.) *Local-spinal therapy of spasticity* (pp. 3-16). Berlin: Springer-Verlag.

Dufresne, JR, Gurfinkel, VS, Soechting, JF, & Terzuolo, CA (1978). Response to transient disturbances during intentional forearm flexion in man. *Brain Res* 150: 103-115.

Duysens, J, & Loeb, GE (1980). Modulation of ipsi- and contralateral reflex responses in unrestrained walking cats. *J Neurophysiol* 44: 1024-1037.

Duysens, J, Loeb, GE, & Weston, BJ (1980). Crossed flexor reflex responses and their reversal in freely walking cats. *Brain Res* 197: 538-542.

Duysens, J, & Pearson, KG (1976). The role of cutaneous afferents from the distal hindlimb in the regulation of the stepcycle in thalamic cats. *Exp Brain Res* 24: 245-255.

Duysens, J, Trippel, M, Horstmann, GA, & Dietz, V (1990). Gating and reversal of reflexes in ankle muscles during human walking. *Exp Brain Res* 82: 351-358.

Eccles, JC (1964). Presynaptic inhibition in the spinal cord. In JC Eccles & JP Schade (Eds.) *Physiology of spinal neurons* (pp. 65-89). Amsterdam: Elsevier.

Eccles, JC (1969). *The inhibitory pathways of the central nervous system.* Springfield: Thomas.

Eccles, JC (1981). Physiology of motor control in man. *Appl Neurophysiol* 44: 5-15.

Edelman, S, & Flash, T (1987). A model of handwriting. *Biol Cybern* 57: 25-36.

Eldred, E, Granit, R, & Merton, PA (1953). Supraspinal control of the muscle spindles and its significance. *J Physiol* 122: 498-523.

Engelhorn, R (1988). EMG and motor performance changes with practice of a forearm movement by children. *Percep Motor Skills* 67: 523-529.

Enoka, RM (1983). Muscular control of a learned movement: The speed control system hypothesis. *Exp Brain Res* 51: 135-145.

Enoka, RM (1988). *Neuromechanical basis of kinesiology.* Champaign: Human Kinetics.

Ertekin, C, & Akcali, D (1978). Effect of continuous vibration on nociceptive flexor reflexes. *J Neurol Neurosurg Psychiat* 41: 532-537.

Evarts, EV (1973). Motor cortex reflexes associated with learned movement. *Science* 179: 501-503.

Evarts, EV (1974). Precentral and postcentral cortical activity in association with visually triggered movement. *J Neurophysiol* 37: 373-381.

Evarts, EV, & Granit, R (1976). Relations of reflexes and intended movements. *Progr Brain Res* 44: 1-14.

Evarts, EV, & Tanji, J (1974). Gating of motor cortex reflexes by prior instruction. *Brain Res* 71: 479-494.

Evarts, EV, Teravainen, H, Beuchert, D, & Calne, DB (1979). Pathophysiology of motor performance in Parkinson's disease. In K Fuxe, D Calne (Eds.) *Ergot derivatives and motor function* (pp. 45-59). London: Pergamon Press.

Evarts, EV, Teravainen, H, & Calne, DB (1981). Reaction time in Parkinson's disease. *Brain* 104: 167-186.

Fahn, S (1990). Akinesia. In A Berardelli, R Benecke, M Manfredi, & CD Marsden (Eds.) *Motor disturbances II* (pp. 142-150). London: Academic Press.

Feldman, AG (1966a). Functional tuning of the nervous system with control of movement or maintenance of a steady posture. II. Controllable parameters of the muscle. *Biophysics* 11: 565-578.

Feldman, AG (1966b). Functional tuning of the nervous system with control of movement or maintenance of a steady posture. III. Mechanographic analysis of execution by man of the simplest motor task. *Biophysics* 11: 667-675.

Feldman, AG (1974). Control of the length of a muscle. *Biophysics* 19: 776-771.

Feldman, AG (1979). Central and reflex mechanisms of motor control. Moscow: Nauka (in Russian).

Feldman, AG (1980a). Superposition of motor programs. I. Rhythmic forearm movements in man. *Neuroscience* 5: 81-90.

Feldman, AG (1980b). Superposition of motor programs. II. Rapid flexion of forearm in man. *Neuroscience* 5: 91-95.

Feldman, AG (1981). The composition of central programs subserving horizontal eye movements in man. *Biol Cybern* 42: 107-116.

Feldman, AG (1986). Once more on the equilibrium-point hypothesis (λ model) for motor control. *J Mot Behav* 18: 17-54.

Feldman, AG, Adamovitch, SV, Ostry, DJ, & Flanagan, JR (1990a). The origin of electromyograms—Explanations based on the equilibrium point hypothesis. In JM Winters & SL-Y Woo (Eds.) *Multiple muscle systems. Biomechanics and movement organization* (pp. 195-213). New York: Springer-Verlag.

Feldman, AG, Flanagan, JR, & Ostry, DJ (1990b). Equilibrium vector spaces for the control of multi-muscle systems. *Abstr Soc Neurosci* 16: 1088.

Feldman, AG, & Latash, ML (1982a). Afferent and efferent components of joint position sense: Interpretation of kinaesthetic illusions. *Biol Cybern* 42: 205-214.

Feldman, AG, & Latash, ML (1982b). Interaction of afferent and efferent signals underlying joint position sense: Empirical and theoretical approaches. *J Mot Behav* 14: 174-193.

Feldman, AG, & Latash, ML (1982c). Inversions of vibration-induced senso-motor events caused by supraspinal influences in man. *Neurosci Lett* 31: 147-151.

Feldman, AG, & Orlovsky, GN (1972). The influence of different descending systems on the tonic stretch reflex in the cat. *Exp Neurol* 37: 481-494.

Feldman, AG, & Orlovsky, GN (1975). Activity of interneurons mediating reciprocal Ia inhibition during locomotion in cats. *Brain Res* 84: 181-194.

Ferrell, WR (1965). Remote manipulation with transmission delay. *IEEE Trans Hum Fact Electr* 6: 24-32.

Fetters, L, & Todd, J (1987). Quantitative assessment of infant reaching movements. *J Mot Behav* 19: 147-166.

Fitts, PM (1954). The information capacity of the human motor system in controlling the amplitude of movements. *J Exp Psychol* 47: 381-391.

Fitts, PM, & Peterson, JR (1964). Information capacity of discrete motor responses. *J Exp Psychol* 67: 103-112.

Flament, D, Hore, J, & Vilis, T (1984). Braking of fast and accurate elbow flexions in the monkey. *J Physiol* 349: 195-203.

Flanagan, JR, & Ostry, DJ (1990). Trajectories of human multi-joint arm movements: Evidence of joint level planning. In V Hayward (Ed.) *Experimental robotics*. New York: Springer-Verlag.

Flanders, M, & Soechting, JF (1990). Arm muscle activation for static forces in three-dimensional space. *J Neurophysiol* 64: 1818-1837.

Flash, T (1987). The control of hand equilibrium trajectories in multi-joint arm movements. *Biol Cybern* 57: 257-274.

Flash, T (1990). The organization of human arm trajectory control. In JM Winters & SL-Y Woo (Eds.) *Multiple muscle systems. Biomechanics and movement organization* (pp. 282-301). New York: Springer-Verlag.

Flash, T, & Hogan, N (1985). The coordination of arm movements: An experimentally confirmed mathematical model. *J Neurosci* 5: 1688-1703.

Flash, T, & Mussa-Ivaldi, FA (1984). Inferring movement and muscle synergies from multi-joint arm posture. *Neuroscience* 10: 635.

Flash, T, & Mussa-Ivaldi, FA (1990). Human arm stiffness characteristics during the maintenance of posture. *Exp Brain Res* 82: 315-326.

Flowers, KA (1975). Ballistic and corrective movements on an aiming task. *Neurol (Minneap)* 25: 413-421.

Flowers, KA (1976). Visual "closed-loop" and "open-loop" characteristics of voluntary movement in patients with parkinsonism and intention tremor. *Brain* 99: 269-310.

Flowers, KA (1978a). Some frequency response characteristics of parkinsonism on pursuit tracking. *Brain* 101: 19-34.

Flowers, KA (1978b). Lack of prediction in the motor behavior of parkinsonism. *Brain* 101: 35-52.

Forget, R, & Lamarre, Y (1987). Rapid elbow flexion in the absence of proprioceptive and cutaneous feedback. *Hum Neurobiol* 6: 27-37.

Forssberg, H (1979a). Stumbling corrective reaction: A phase dependent compensatory reaction during locomotion. *J Neurophysiol* 42: 936-953.

Forssberg, H (1979b). On integrative motor functions in the cat's spinal cord. *Acta Physiol Scand* 107 (Suppl. 474), 56 pp.

Forssberg, H, Grillner, S, & Rossignol, S (1975). Phase dependent reflex reversal during walking in chronic spinal cat. *Brain Res* 85: 103-107.

Forssberg, H, Grillner, S, & Rossignol, S (1977). Phasic gain control of reflexes from the dorsum of the paw during spinal locomotion. *Brain Res* 132: 121-139.

Forssberg, H, Grillner, S, Rossignol, S, & Wallen, P (1976). Phasic control of reflexes during locomotion in vertebrates. In R Herman (Ed.) *Neural control of locomotion* (pp. 647-674). New York, London: Plenum Press.

Fournier, E, Katz, R, & Pierrot-Deseilligny, E (1983). Descending control of reflex pathways in the production of voluntary isolated movements in man. *Brain Res* 288: 375-377.

Freund, HJ, & Budingen, HJ (1978). The relationship between speed and amplitude of the fastest voluntary contractions of human arm muscles. *Exp Brain Res* 55: 167-171.

Frith, U, & Frith, CD (1974). Specific motor disabilities in Down's syndrome. *J Child Psychol Psychiat* 15: 293-301.

Fromm, C (1983). Changes of steady-state activity in motor cortex consistent with the length-tension relation of muscle. *Pflug Arch* 398: 318-323.

Fukson, OI, Berkinblit, MB, & Feldman, AG (1980). The spinal frog takes into account the scheme of its body during the wiping reflex. *Science* 209: 1261-1263.

Gan, K, & Hoffman, ER (1988). Geometrical conditions for ballistic and visually controlled movements with one mechanical degree of freedom. *Ergonomics* 31. 829-839.

Garfinkel, A (1983). A mathematics for physiology. *Amer J Physiol* 14: R455-R466.

Garnett, R, & Stephens, JA (1981). Changes in the recruitment thresholds of motor units produced by cutaneous stimulation in man. *J Physiol* 311: 463-473.

Gelfand, IM, Gurfinkel, VS, Tsetlin, ML, & Shik, ML (1971a). Some problems in the analysis of movements. In IM Gelfand, VS Gurfinkel, SV Fomin, & ML Tsetlin (Eds.) *Models of the structural-functional organization of certain biological systems*. Cambridge: MIT Press.

Gelfand, IM, Gurfinkel, VS, Fomin, SV, & Tsetlin, ML (Eds.) (1971b). *Models of the structural-functional organization of certain biological systems*. Cambridge: MIT Press.

Gellman, R, Gibson, AR, & Houk, JC (1985). Inferior olivary neurons in the awake cat: Detection of contact and passive body displacement. *J Neurophysiol* 54: 40-60.

Georgopoulos, AP (1986). On reaching. *Ann Rev Neurosci* 9: 147-170.

Georgopoulos, AP, Kalaska, JF, & Caminiti, R (1985). Relations between two-dimensional arm movements and single-cell discharge in motor cortex and area 5: Movement direction versus movement end point. *Exp Brain Res* Suppl 10: 175-183.

Georgopoulos, AP, Kalaska, JF, Caminiti, R, & Massey, JT (1982). On the relations between the direction of two-dimensional arm movements and cell discharge in primate motor cortex. *J Neurosci* 2: 1527-1537.

Georgopoulos, AP, Kalaska, JF, Caminiti, R, Massey, JT (1983). Interruption of motor cortical discharge subserving aimed arm movements. *Exp Brain Res* 49: 327-340.

Georgopoulos, AP, Kalaska, JF, & Massey, JT (1981). Spatial trajectories and reaction times of aimed movements: Effects of practice, uncertainty, and change in target location. *J Neurophysiol* 46: 725-743.

Georgopoulos, AP, Lurito, JT, Petrides, M, Schwartz, AB, & Massey, JT (1989). Mental rotation of the neuronal population vector. *Science* 243: 234-236.

Georgopoulos, AP, Schwartz, AB, & Kettner, RE (1986). Neural population coding of movement direction. *Science* 233: 1416-1419.

Getty, DJ (1975). Discrimination of short temporal intervals: A comparison of two models. *Percept Psychophys* 18: 1-8.

Ghez, C (1975). Input-output relations of the red nucleus in the cat. *Brain Res* 98: 93-108.

Ghez, C, & Gordon, J (1987). Trajectory control in targeted force impulses. I. Role of opposing muscles. *Exp Brain Res* 67: 225-240.

Ghez, C, & Martin, JH (1982). The control of rapid limb movement in the cat. III. Agonist-antagonist coupling. *Exp Brain Res* 45: 115-125.

Ghez, C, & Shinoda, Y (1978). Spinal mechanisms of the functional stretch reflex. *Exp Brain Res* 32: 55-68.

Ghez, C, & Vicario, D (1978). The control of rapid limb movement in the cat. I. Response latency. *Exp Brain Res* 33: 173-189.

Gibson, AR, Houk, JC, Kohlerman, NJ (1985). Relation between red nucleus discharge and movement parameters in trained macaque monkeys. *J Physiol* 358: 551-570.

Gibson, JJ (1966). *The senses considered as perceptual systems.* Boston: Houghton Mifflin.

Gielen, CCAM, & Denier van der Gon, JJ (1989). If a particular strategy is used, what aspects of the movement are controlled? *Behav Brain Sci* 12: 218.

Gielen, CCAM, & Houk, JC (1984). Nonlinear viscosity of human wrist. *J Neurophysiol* 52: 553-569.

Gielen, CCAM, van der Heuvel, PJM, & Denier van der Gon, JJ (1984). Modification of muscle activation patterns during fast goal-directed movements. *J Mot Behav* 16: 2-19.

Gielen, S, van Ingen Schenau, GJ, Tax, T, & Theeuwen, M (1990). The activation of mono- and bi-articular muscles in multi-joint movements. In JM Winters & SL-Y Woo (Eds.) *Multiple muscle systems. Biomechanics and movement organization* (pp. 302-311). New York: Springer-Verlag.

Gillies, JD, Lance, JW, Neilson, PD, & Tassinari, CA (1969). Presynaptic inhibition of the monosynaptic reflex by vibration. *J Physiol* 205: 329-339.

Gilman, S, Bloedel, JR, & Lechtenberg, R (Eds.) (1981). *Disorders of the cerebellum.* Philadelphia. F.A. Davis.

Giszter, SF, McIntyre, J, & Bizzi, E (1989). Kinematic strategies and sensorimotor transformations in the wiping movements of frogs. *J Neurophysiol* 62: 750 767.

Giszter, SF, Mussa-Ivaldi, FA, & Bizzi, E (1990). The organization of limb motor space in the spinal cord. *Abstr Soc Neurosci* 16: 117.

Glencross, DJ (1973). Temporal organization in a repetitive speed skill. *Ergonomics* 16: 765-776.

Gordon, J, & Ghez, C (1987a). Trajectory control in targeted force impulses. II. Pulse height control. *Exp Brain Res* 67: 241-252.

Gordon, J, & Ghez, C (1987b). Trajectory control in targeted force impulses. III. Compensatory adjustments for initial errors. *Exp Brain Res* 67: 253-269.

Gottlieb, GL, Agarwal, GC (1980a). Response to sudden torque about ankle in man. II. Postmyotatic reactions. *J Neurophysiol* 43: 86-101.

Gottlieb, GL, & Agarwal, GC (1980b). Response to sudden torque about ankle in man. III. Suppression of stretch-evoked responses during phasic contraction. *J Neurophysiol* 44: 233-246.

Gottlieb, GL, & Agarwal, GC (1986). The invariant characteristic isn't. *Behav Brain Sci* 9: 608-609.

Gottlieb, GL, & Agarwal, GC (1988). Compliance of single joints: Elastic and plastic characteristics. *J Neurophysiol* 59: 937-951.

Gottlieb, GL, Agarwal, C, & Tanavde, AS (1986). A method of estimating human joint compliance which is insensitive to reflex, triggered or voluntary reactions. *Abstr Soc Neurosci* 12: 468.

Gottlieb, GL, Corcos, DM, & Agarwal, GC (1989a). Strategies for the control of voluntary movements with one mechanical degree of freedom. *Behav Brain Sci* 12: 189-250.

Gottlieb, GL, Corcos, DM, & Agarwal, GC (1989b). Organizing principles for single joint movements. I: A speed-insensitive strategy. *J Neurophysiol* 62: 342-357.

Gottlieb, GL, Corcos, DM, Agarwal, GC, & Latash, ML (1990a). Organizing principles for single joint movements. III: Speed-insensitive strategy as a default. *J Neurophysiol* 63: 625-636.

Gottlieb, GL, Corcos, DM, Agarwal, GC, & Latash, ML (1990b). Principles underlying single-joint movement strategies. In JM Winters & SL-Y

Woo (Eds.) *Multiple muscle systems. Biomechanics and movement organization* (pp. 236-250). New York: Springer-Verlag.

Gottlieb, GL, Corcos, DM, Jaric, S, & Agarwal, GC (1988). Practice improves even the simplest movements. *Exp Brain Res* 73: 435-440.

Gottlieb, GL, Myklebust, BM, Penn, RD, & Agarwal, GC (1982). Reciprocal excitation of muscle antagonists by the primary afferent pathway. *Exp Brain Res* 46: 454-456.

Graham Brown, T (1911). Studies in the physiology of the nervous system. VIII. Neural balance and reflex reversal with a note on progression in the decerebrate guinea-pig. *Quart J Exp Physiol* 4: 273-288 (cited after Magnus, 1924).

Granit, R (1955). *Receptors and sensory perception*. New Haven: Yale University Press.

Granit, R (1970). *The basis of motor control*. London: Academic Press.

Gray, BG, & Dostrovsky, JO (1983). Modulation of the sensory responses of cat trigeminal and cuneate neurons by electrical stimulation of the red nucleus. *Abstr Soc Neurosci* 9: 247.

Graybiel, AM, Hirsch, EC, & Agid, Y (1990). The nigrostriatal system in Parkinson's disease. *Adv Neurol* 53: 17-29.

Greene, PH (1971). Introduction. In IM Gelfand, VS Gurfinkel, SV Fomin, & ML Tsetlin (Eds.) *Models of the structural-functional organization of certain biological systems*. Cambridge: MIT Press.

Greene, PH (1972). Problems of organization of motor systems. *Prog Theor Biol* 2: 303-338.

Greene, PH (1982). Why is it easy to control your arms? *J Mot Behav* 4: 260-286.

Gregoric, M, Dimitrijevic, MD, Trontelj, JV, & Sherwood, A (1979). Reflex response of paraspinal muscles to tapping. *Acta Neurol Scand* 60: (Suppl. 73), 169.

Griffiths, RI (1991). Shortening of muscle fibres during stretch of the active cat medial gastrocnemius muscle: The role of tendon compliance. *J Physiol* 436: 219-236.

Grillner, S (1975). Locomotion in vertebrates: Central mechanisms and reflex interaction. *Physiol Rev* 55: 247-304.

Grillner, S (1979). Interaction between central and peripheral mechanisms in the control of locomotion. In R Granit & O Pompeiano (Eds.) *Reflex control of posture and movement* (pp. 227-235). Amsterdam: Elsevier.

Gurfinkel, VS, & Latash, ML (1978). Motor reversals in calf muscles. *Phyziologiya Cheloveka (Human Physiol)* 4: 30-35.

Gurfinkel, VS, & Latash, ML (1979). Segmental postural mechanisms and reversal of muscle reflexes. *Agressologie* 20B: 145-146.

Gurfinkel, VS, & Latash, ML (1980). The reversals of muscle reflexes. In A Ovsyannikov & A Talyshev (Eds.) *Physiological basis of motor control* (pp. 19-29). Moscow: Nauka.

Gurfinkel, VS, Lipshits, MI, & Lestienne, FG (1988). Anticipatory neck muscle activity associated with rapid arm movements. *Neurosci Lett* 94: 104-108.

Gurfinkel, VS, Lipshits, MI, & Popov, KE (1974). Is the stretch reflex the main mechanism in the system of regulation of the vertical posture in man? *Biophysics* 19: 744-748.

Gurfinkel, VS, & Osovets, SM (1972). Dynamics of equilibrium of the vertical posture in man. *Biophysics* 17: 496-506.

Gutman, SR, & Gottlieb, GL (1989). A solution for the problem of multi-joint redundancy by minimizing joint angle increments. *Abstr Soc Neurosci* 15: 606.

Gutman, SR, & Gottlieb, GL (1990). Non-linear "inner" time in reaching movement trajectory formation. *Abstr First World Congr Biomech* I: 190, San Diego, CA.

Gutman, SR, Gottlieb, GL, & Corcos, DM (1992). Exponential model of a reaching movement trajectory with nonlinear time. *Comments Theor Biol* 2: 357-383.

Haar ter Romeny, BM, Denier van der Gon, JJ, & Gielen, CCAM (1982). Changes in recruitment order of motor units in the human biceps muscle. *Exp Neurol* 78: 360-368.

Hagbarth, K-E, & Eklund, G (1968). The effect of muscle vibration in spasticity, rigidity, and cerebellar disorders. *J Neurol Neurosurg Psychiat* 31: 207-213.

Haken, H (1983). *Synergetics. An introduction.* Berlin: Springer-Verlag.

Haken, H, Kelso, JAS, & Bunz, H (1985). A theoretical model of phase transitions in human hand movements. *Biol Cybern* 51: 347-356.

Halbertsma, JM, Miller, S, & van der Meche, FGA (1976). Basic programs for the phasing of flexion and extension movements of the limbs during locomotion. In R Herman (Ed.) *Neural control of locomotion* (pp. 489-517). New York, London: Plenum Press.

Hallett, M (1989). Experiment and reality. *Behav Brain Sci* 12: 219.

Hallett, M, & Khoshbin, S (1980). A physiological mechanism of bradykinesia. *Brain* 103: 301-314.

Hallett, M, & Marsden, CD (1979). Ballistic flexion movements of the human thumb. *J Physiol* 294: 33-50.

Hallett, M, Shahani, B, & Young, R (1975). EMG analysis of stereotyped voluntary movements in man. *J Neurol Neurosurg Psychiatr* 38: 1154-1162.

Halsband, U, Homberg, V, & Lange, HJ (1990). Slowing of different types of voluntary movement in extrapyramidal disease: Fitts' law and idiographic writing. In A Berardelli, R Benecke, M Manfredi, & CD Marsden (Eds.) *Motor disturbances II* (pp. 182-190). London: Academic Press.

Hammond, PH (1954). Involuntary activity in biceps following the sudden application of velocity to the abducted forearm. *J Physiol* 127: 23P-25P.

Hancock, PA, & Newell, KM (1985). The movement speed-accuracy relationship in space-time. In H Heuer, U Kleinbeck, & KH Schmidt (Eds.) *Motor behavior: Programming, control, and acquisition* (pp. 153-188). Berlin: Springer-Verlag.

Hannaford, B, & Winters, JM (1990). Actuator properties and movement control: Biological and technological models. In JM Winters & SL-Y Woo (Eds.) *Multiple muscle systems. Biomechanics and movement organization* (pp. 101-119). New York: Springer-Verlag.

Harrington, DL, & Haaland, KY (1991). Sequencing in Parkinson's disease. Abnormalities in programming and controlling movement. *Brain* 114: 99-115.

Hasan, Z (1986a). Optimized movement trajectories and joint stiffness in unperturbed, inertially loaded movements. *Biol Cybern* 53: 373-382.

Hasan, Z (1986b). Do subprograms for movement always seek equilibrium? *Behav Brain Sci* 9: 609-610.

Hasan, Z, & Enoka, RM (1985). Isometric torque-angle relationship and movement-related activity of human elbow flexors: Implications for the equilibrium-point hypothesis. *Exp Brain Res* 59: 441-450.

Hasan, Z, & Karst, GM (1989). Task variables and the saturation of the excitation pulse. *Behav Brain Sci* 12: 219-220.

Hattab, JR (1980). Review of European clinical trials with baclofen. In RG Feldman, RR Young, & WP Koella (Eds.) *Spasticity: Disordered motor control* (pp. 71-85). Chicago, London: Year Book Medical.

Hatze, H, & Buys, JD (1977). Energy-optimal controls in the mammalian neuromuscular system. *Biol Cybern* 27: 9-20.

Hayashi, A, Kagamihara, Y, Nakajima, Y, Narabayashi, H, Okuma, Y, & Tanaka, R (1988). Disorder in reciprocal innervation upon initiation of voluntary movement in patients with Parkinson's disease. *Exp Brain Res* 70: 437-440.

Hayes, KC (1982). Biomechanics of postural control. *Exer Sport Sci Rev* 10: 363-391.

Heilman, KM, Bowers, D, Watson, RT, & Greer, M (1976). Reaction time in Parkinson's disease. *Arch Neurol* 33: 139-140.

Hemami, H, & Stokes, BT (1983). Four neural circuit models and their role in the organization of voluntary movement. *Biol Cybern* 49: 69-77.

Henderson, SE (1985). Motor skill development. In D Lane & B Stratford (Eds.) *Current approaches to Down syndrome* (pp. 187-218). London: Holt, Rhinehart & Winston.

Henderson, SE, Morris, J, & Frith, V (1981). The motor deficit in Down's syndrome children: A problem of timing? *J Child Psychol Psychiat* 22: 233-245.

Hendrie, A, & Lee, RG (1978). Selective effects of vibration on human spinal and long-loop reflexes. *Brain Res* 157: 369-375.

Henneman, E, Somjen, G, & Carpenter, DO (1965). Excitability and inhibitibility of motoneurones of different sizes. *J Neurophysiol* 28: 599-620.

Henry, JL (1980). Pharmacologic studies on baclofen in the spinal cord of the cat. In RG Feldman, RR Young, & WP Koella (Eds.) *Spasticity: Disordered motor control* (pp. 437-452). Chicago and London: Year Book Medical.

Herman, R (1970). The myotatic reflex: Clinico-physiological aspects of spasticity and contracture. *Brain* 93: 273-312.

Herman, R, Freedman, W, & Meeks, SM (1973). Physiological aspects of hemiplegic and paraplegic spasticity. In JE Desmedt (Ed.) *New developments in EMG and clinical neurophysiology* 3: 463-468.

Heuer, H (1989). Movement strategies as points on equal-outcome curves. *Behav Brain Sci* 12: 220-221.

Hill, AV (1938). The heat of shortening and the dynamic constants of muscle. *Proc Roy Soc London Ser B* 126: 136-195.

Hill, AV (1953). The mechanics of active muscle. *Proc Roy Soc London Ser B* 141: 104-117.

Hoffer, JA, & Andreassen, S (1981). Regulation of soleus muscle stiffness in premammillary cats: Intrinsic and reflex components. *J Neurophysiol* 45: 267-285.

Hoffmann, P (1922). *Untersuchungen uber die eigenreflexe (sehnenreflexe) menschlicher muskeln*. Berlin: Springer.

Hoffmann, DS, & Strick, PL (1986). Steptracking movements of the wrist in humans: Kinematic analysis. *J Neurosci* 6: 3309-3318.

Hogan, N (1984). An organizational principle for a class of voluntary movements. *J Neurosci* 4: 2745-2754.

Hogan, N (1985). The mechanics of multi-joint posture and movement control. *Biol Cybern* 52: 315-331.

Hogan, N (1990). Mechanical impedance of single- and multi-articular systems. In JM Winters & SL-Y Woo (Eds.) *Multiple muscle systems. Biomechanics and movement organization* (pp. 149-164). New York: Springer-Verlag.

Hollerbach, JM (1981). An oscillation theory of handwriting. *Biol Cybern* 39: 139-156.

Hollerbach, JM (1982). Computers, brain, and the control of movements. *Trends in Neuroscience* 5: 189-192.

Hollerbach, JM, & Atkeson, CG (1987). Deducing planning variables from experimental arm trajectories: Pitfalls and possibilities. *Biol Cybern* 56: 279-292.

Hollerbach, JM, & Flash, T (1982). Dynamic interaction between limb segments during planar arm movements. *Biol Cybern* 44: 67-77.

Hollerbach, JM, & Suh, KC (1985). Redundancy resolution of manipulators through torque optimization. *Proc IEEE Intern Conf on Robotics and Automation*, St. Louis.

Holst, E von (1954). Relation between the central nervous system and the peripheral organs. *Brit J Anim Behav* 2: 89-94.

Horak, FB, & Nashner, LM (1986). Central program of postural movements: Adaptation to altered support-surface configurations. *J Neurophysiol* 55: 1369-1381.

Horak, FB, Nashner, LM, & Diener, HC (1990). Postural strategies associated with somatosensory and vestibular loss. *Exp Brain Res* 82: 167-177.

Hore, J, McCloskey, DI, & Taylor, JL (1990). Task-dependent changes in gain of the reflex response to imperceptible perturbations of joint position in man. *J Physiol* 429: 309-321.

Horne, DJ de L (1983). Sensorimotor control in Parkinsonians. *J Neurol Neurosurg Psychiat* 36: 742-746.

Horstink, MWIM, Berger, HJC, van Spaendonck, KPM, van der Bercken, JHL, & Cools, AR (1990). Bimanual simultaneous motor performance and impaired ability to shift attention in Parkinson's disease. *J Neurol Neurosurg Psychiat* 53: 685-690.

Horstmann, GA, & Dietz, V (1990). A basic posture control mechanism: The stabilization of the centre of gravity. *Electroencephalog Clin Neurophysiol* 76: 165-176.

Houk, JC (1976). An assessment of stretch reflex function. *Prog Brain Res* 44: 303-314.

Houk, JC (1979). Regulation of stiffness by skeletomotor reflexes. *Ann Rev Physiol* 41: 99-114.

Houk, JC (1989). Cooperative control of limb movements by the motor cortex, brainstem, and cerebellum. In RMJ Cotterill (Ed.) *Models of brain function* (pp. 309-325). Cambridge: University Press.

Houk, JC, Crago, PE, & Rymer, WZ (1981). Function of the spindle dynamic response in stiffness regulation—A predictive mechanism provided by non-linear feedback. In A Taylor & A Prochazka (Eds.) *Muscle receptors and movement*. New York: McMillan.

Houk, JC, & Gibson, AR (1987). Sensorimotor processing through the cerebellum. In JS King (Ed.) *New concepts in cerebellar neurobiology* (pp. 387-416). New York: Liss.

Houk, JC, & Rymer, WZ (1981). Neural control of muscle length and tension. In VB Brooks (Ed.) *The nervous system. Handbook of physiology, vol. II* (pp. 257-324). Bethesda: American Physiological Society.

Houk, JC, Singer, JJ, & Goldman, MR (1970). An evaluation of length and force feedback to soleus muscles of decerebrate cats. *J Neurophysiol* 33: 784-811.

Hoyle, G (1983). *Muscles and their motor control*. New York: Wiley-Interscience.

Hufschmidt, A, & Mauritz, K-H (1985). Chronic transformation of muscle in spasticity: A peripheral contribution to increased tone. *J Neurol Neurosurg Psychiat* 48: 676-685.

Hughlings Jackson, J (1889). On the comparative study of diseases of the nervous system. *Br Med J* 2: 355-362.

Hugon, M, Massion, J, & Wiesendanger, M (1982). Anticipatory postural changes induced by active unloading and comparison with passive unloading in man. *Pflug Arch* 393: 292-296.

Hultborn, H (1972). Convergence of interneurons in the reciprocal Ia inhibitory pathway to motoneurones. *Acta Physiol Scand* (Suppl 375): 1-42.

Hultborn, H, Lindstrom, S, & Wigstrom, H (1979). On the function of recurrent inhibition in the spinal cord. *Exp Brain Res* 37: 399-403.

Hultborn, H, & Malmsten, J (1977). Late adjustments in segmental reflex transmission following experimental spinal cord lesions. *Intern Symp on Acute Manag and Rehab after Spinal Cord Inj*, Warsaw, May 15-24, 1977, pp. 149-179.

Hultborn, H, & Pierrot-Deseilligny, E (1979). Changes in recurrent inhibition during voluntary soleus contractions in man studied by an H-reflex technique. *J Physiol* 297: 229-251.

Humphrey, DR (1982). Separate cell systems in the motor cortex of the monkey for the control of joint movement and of joint stiffness. In PA Buser, VA Cobb, & M Okuma (Eds.) *Kyoto Symp, Electroencephalog Clin Neurophysiol* (Suppl 36), pp. 393-408. Amsterdam: Elsevier.

Hunt, CC (1952). The effect of stretch receptors from muscle on the discharge of motoneurones. *J Physiol* 117: 359-379.

Hunter, IW, & Kearney, RE (1982). Dynamics of human ankle stiffness: Variation with mean ankle torque. *J Biomech* 15: 747-752.

Hunter, JP, Ashby, P, & Lang, AE (1988). Afferents contributing to the exaggerated long latency reflex response to electrical stimulation in Parkinson's disease. *J Neurol Neurosurg Psychiat* 51: 1405-1410.

Iles, JF (1977). Responses in human pretibial muscle to sudden stretch and to nerve stimulation. *Exp Brain Res* 30: 451-470.

Inzelberg, R, Flash, T, & Korczyn, AD (1990). Kinematic properties of upper-limb trajectories in Parkinson's disease and idiopathic torsion dystonia. *Adv Neurol* 53: 183-189.

Jacobson, SC, Iverson, EK, Knutyti, DF, Johnson, RT, & Biggers, KB (1986). Design of the Utah-MIT dextrous hand. *Proc IEEE Robot Autom*, pp. 1520-1532.

Jankowska, E (1979). New observations on neuronal organization of reflexes from tendon organ afferents and their relation to reflexes evoked from muscle spindle afferents. In R Granit & O Pompeiano (Eds.) *Reflex control of posture and movement* (pp. 29-36). Amsterdam: Elsevier.

Jankowska, E, Lundberg, A, & Stuart, D (1983). Propriospinal control of interneurons in spinal reflex pathways from tendon organs in the cat. *Brain Res* 261: 317-320.

Jankowska, E, & McCrea, DA (1983). Shared reflex pathways from Ib tendon organ afferents and Ia muscle spindle afferents in the cat. *J Physiol* 338: 99-111.

Jankowska, E, McCrea, D, & Mackel, R (1981a). Pattern of "non-reciprocal" inhibition of motoneurones by impulses in group Ia muscle spindle afferents in the cat. *J Physiol* 316: 393-409.

Jankowska, E, McCrea, D, & Mackel, R (1981b). Oligosynaptic excitation of motoneurones by impulses in group Ia muscle spindle afferents in the cat. *J Physiol* 316: 411-425.

Jankowska, E, & Roberts, WJ (1972). Synaptic action of single interneurons mediating reciprocal Ia inhibition of motoneurons. *J Physiol* 222: 623-642.

Jeannerod, M (1988). *The neural and behavioral organization of goal-directed movements*. Oxford: Clarendon Press.

Jellinger, K (1986). Overview of morphological changes in Parkinson's disease. *Adv Neurol* 45: 1-19.

Johnson, MTV, Kipnis, AN, Lee, MC, Loewenson, RB, & Ebner, TJ (1991). Modulation of the stretch reflex during volitional sinusoidal tracking in Parkinson's disease. *Brain* 114: 443-460.

Johnston, GAR, Hailstone, MH, & Freeman, CG (1980). Baclofen: Stereoselective inhibition of excitant amino acid release. *J Pharm Pharmacol* 32: 230.

Jongen, HAH, Denier van der Gon, JJ, Gielen, CCAM (1989). Activation of human arm muscles during flexion/extension and supination/pronation tasks: A theory of muscle coordination. *Biol Cybern* 61: 1-9.

Jordan, H (1905). Untersuchungen zur physiologie des nervensystems bei pulmonalen tonus und erregbarkeit. *Pflugers Archiv d ges Physiologie* 110: 533 (cited after Magnus, 1924).

Josin, G (1988). Neural-space generalization of a topological transformation. *Biol Cybern* 59: 283-290.

Joyce, GC, Rack, PMH, & Ross, HF (1974). The forces generated in the human elbow joint in response to imposed sinusoidal movements of the forearm. *J Physiol* 240: 351-374.

Joyce, GC, Rack, PMH, & Westbury, DR (1969). The mechanical properties of cat soleus muscle during controlled lengthening and shortening movements. *J Physiol* 204: 461-474.

Jung, R (1979). Two functions of reflexes in human movement: Interaction of preprograms and gain of force. In R Granit & O Pompeiano (Eds.) *Reflex control of posture and movement* (pp. 237-241). Amsterdam, New York, Oxford: Elsevier.

Kalaska, JF, Caminiti, R, & Georgopoulos, AP (1983). Cortical mechanisms related to the direction of two-dimensional arm movements: Relations in parietal area 5 and comparison with motor cortex. *Exp Brain Res* 51: 247-260.

Kaminski, T, & Gentile, AM (1986). Joint control strategies and hand trajectories in multijoint pointing movements. *J Mot Behav* 18: 261-278.

Karst, GM, & Hasan, Z (1987). Antagonist muscle activity during human forearm movements under varying kinematic and loading conditions. *Exp Brain Res* 67: 391-401.

Karst, GM, & Hasan, Z (1989). Timing and magnitude of EMG activity at the shoulder and elbow for initiation of planar arm movements. *Soc Neurosci Abstr* 15: 48.

Karst, GM, & Hasan, Z (1990). Direction-dependent strategy for control of multi-joint arm movements. In JM Winters & SL-Y Woo (Eds.) *Multiple muscle systems. Biomechanics and movement organization* (pp. 268-281). New York: Springer-Verlag.

Kathib, O (1983). Dynamic control of manipulators in operational space. *6 IFTMM Congress on Theory of Machines and Mechanisms*. New Delhi.

Katz, B (1950). Depolarization of sensory terminals and the initiation of impulses in the muscle spindle. *J Physiol* 111: 261-282.

Katz, R, & Pierrot-Desseilligny, E (1984). Facilitation of soleus-coupled Renshaw cells during voluntary contraction of pretibial flexor muscle in man. *J Physiol* 335: 587-603.

Kawato, M, Maeda, Y, Uno, Y, & Suzuki, R (1990). Trajectory formation of arm movement by cascade neural network model based on minimum torque-change criterion. *Biol Cybern* 62: 275-288.

Kay, BA, Kelso, JAS, Saltzman, EL, & Schoner, G (1987). Space-time behavior of single and bimanual rhythmical movements: Data and limit cycle model. *J Exp Psychol: Hum Percept Perform* 13: 178-192.

Kay, BA, Saltzman, EL, & Kelso, JAS (1991). Steady-state and perturbed rhythmical movements: A dynamical analysis. *J Exp Psychol: Hum Percept Perform* 17: 183-197.

Keele, SW (1968). Movement control in skilled motor performances. *Psychol Bull* 70: 387-403.

Keele, SW (1986). Motor control. In JK Boff, L Kaufman, & JP Thomas (Eds.) *Handbook of human perception and performance, vol. II* 30: 1-60. New York: Wiley & Sons.

Keele, SW, & Ells, JG (1972). Memory characteristics of kinesthetic information. *J Mot Behav* 4: 127-134.

Keele, SW, Ivry, RI, & Pokorny, RA (1987). Force control and its relation to timing. *J Mot Behav* 19: 96-114.

Keele, SW, & Posner, MI (1968). Processing of visual feedback in rapid movements. *J Exp Psychol* 77: 155-158.

Kelso, JAS (1977). Motor control mechanisms underlying human movement reproduction. *J Exp Psychol: Hum Percep Perform* 3: 529-543.

Kelso, JAS (1981). On the oscillatory basis of movement. *Bull Psychonom Soc* 18: 63.

Kelso, JAS (1984). Phase transitions and critical behavior in human bimanual coordination. *Amer J Physiol* 246: R1000-R1004.

Kelso, JAS, Buchanan, JJ, & Wallace, SA (1991). Order parameters for the neural organization of single, multijoint limb movement patterns. *Exp Brain Res* 85: 432-444.

Kelso, JAS, & Holt, KG (1980). Exploring a vibratory system analysis of human movement production. *J Neurophysiol* 43: 1183-1196.

Kelso, JAS, Holt, KG, Kugler, PN, & Turvey, MT (1980). On the concept of coordinative structures as dissipative structures. II. Empirical lines of convergence. In GE Stelmach & J Requin (Eds.) *Tutorials in motor behavior* (pp. 49-70). Amsterdam: North-Holland.

Kelso, JAS, Holt, KG, Rubin, P, & Kugler, PN (1981). Patterns of human interlimb coordination emerge from the properties of non-linear, limit cycle oscillatory processes: Theory and data. *J Mot Behav* 13: 226-261.

Kelso, JAS, Putnam, CA, & Goodman, D (1983). On the spacetime structure of human interlimb coordination. *Quart J Exp Psychol* 35: 347-375.

Kelso, JAS, & Schoner, G (1988). Self-organization of coordinative movement patterns. *Hum Mov Sci* 7: 27-46.

Kelso, JAS, Southard, DL, & Goodman, D (1979a). On the coordination of two-handed movements. *J Exp Psychol: Hum Percept Perform* 5: 529-543.

Kelso, JAS, Southard, DL, & Goodman, D (1979b). On the nature of human interlimb coordination. *Science* 203: 1029-1031.

Kelso, JAS, & Tuller, B (1984). Converging evidence in support of common dynamical principles for speech and movement coordination. *Amer J Physiol* 15: R928-R935.

Kennedy, PR, Gibson, AR, Houk, JC (1986). Functional and anatomic differentiation between parvocellular and magnocellular regions of red nucleus in the monkey. *Brain Res* 364: 124-136.

Kerr, R (1978). Diving adaptation, and Fitts' law. *J Mot Behav* 10: 255-260.

Kerr, R, & Blais, C (1985). Motor skill acquisition by individuals with Down syndrome. *Amer J Ment Defic* 90: 313-318.

Kerr, R, & Blais, C (1987). Down syndrome and extended practice of a complex motor task. *Amer J Ment Defic* 91: 591-597.

Keshner, EA, & Allum, JHJ (1990). Muscle activation patterns coordinating postural stability from head to foot. In JM Winters & SL-Y Woo (Eds.) *Multiple muscle systems. Biomechanics and movement organization* (pp. 481-497). New York: Springer-Verlag.

Keshner, EA, Woollacott, MH, & Debu, B (1988). Neck and trunk muscle responses during postural perturbations in humans. *Exp Brain Res* 71: 455-466.

Klapp, ST (1979). Doing two things at once: The role of temporal compatibility. *Memory Cognition* 7: 375-381.

Klapp, ST (1981). Temporal compatibility in dual motor tasks II: Simultaneous articulation and hand movements. *Memory Cognition* 9: 398-401.

Kniffki, K-D, Schomburg, ED, Steffens, H (1981). Convergence in segmental reflex pathways from fine muscle afferents and cutaneous or group II muscle afferents to α-motoneurons. *Brain Res* 218: 342-346.

Knight, AA, & Dagnall, PR (1967). Precision in movements. *Ergonomics* 10: 321-330.

Knutsson, E (1983). Analysis of gait and isokinetic movements for evaluation of antispastic drugs or physical therapies. In JE Desmedt (Ed.)

Motor control mechanisms in health and disease (pp. 1013-1034). New York: Raven Press.

Kohonen, T (1982). Self-organized formation of topologically correct feature maps. *Biol Cybern* 43: 59-69.

Kornhuber, HH (1971). Motor functions of cerebellum and basal ganglia: The cerebellocortical saccadic (ballistic) clock, the cerebellonuclear hold regulator, and the basal ganglia ramp (voluntary speed smooth movement) generator. *Kybernetik* 8: 157-162.

Kornhuber, HH (1974). The vestibular system and general motor system. In HH Kornhuber (Ed.) *Handbook of sensory physiology, vol. 6* (pp. 581-620). Berlin: Springer.

Kudina, LP (1980). Reflex effects of muscle afferents on antagonists studied on single firing motor units in man. *Electroencephalog Clin Neurophysiol* 50: 214-221.

Kugler, PN, Kelso, JAS, & Turvey, MT (1980). On the concept of coordinative structures as dissipative structures. I. Theoretical lines of convergence. In GE Stelmach & J Requin (Eds.) *Tutorials in motor behavior* (pp. 3-45). Amsterdam: North-Holland.

Kugler, PN, & Turvey, MT (1987). *Information, natural law, and the self-assembly of rhythmic movement.* Hillsdale, NJ: Erlbaum.

Kugler, PN, Turvey, MT, Schmidt, RC, & Rosenblum, LD (1990). Investigating a nonconservative invariant of motion in coordinated rhythmic movements. *Ecol Psychol* 2: 151-189.

Kuperstein, M (1987). Adaptive visual-motor coordination in multijoint robots using parallel architecture. *IEEE Conf Robot Automat* (pp. 1595-1602). Raleigh, NC.

Kuypers, HGJM (1981). Anatomy of the descending pathways. In *Handbook of physiology. The nervous system. Motor control, vol. II* (pp. 597-666). Bethesda: Am Physiol Soc.

Kvalseth, TO (1980). An alternative to Fitts' law. *Bull Psychonom Soc* 16: 371-373.

Lacquaniti, F (1989). Central representations of human limb movement as revealed by studies of drawing and handwriting. *Trends Neurosci* 12: 287-291.

Lacquaniti, F, Borghese, NA, & Carrozzo, M (1991). Transient reversal of the stretch reflex in human arm muscles. *J Neurophysiol* 66: 939-949.

Lacquaniti, F, Ferrigno, G, Pedotii, A, Soechting, JF, & Terzuolo, C (1987). Changes in spatial scale in drawing and handwriting: Kinematic contributions by proximal and distal joints. *J Neurosci* 7: 819-828.

Lacquaniti, F, Licata, F, & Soechting, JF (1982a). The mechanical behavior of the human forearm in response to transient perturbations. *Biol Cybern* 44: 67-77.

Lacquaniti, F, Soechting, JF, & Terzuolo, CA (1982b). Some factors pertinent to the organization and control of arm movements. *Brain Res* 252: 394-397.

Lacquaniti, F, Soechting, JF, & Terzuolo, CA (1986). Path constraints on point to point arm movements in three-dimensional space. *Neurosci* 17: 313-324.

Lance, JW (1980a). Symposium synapsis. In RG Feldman, RR Young, & WP Koella (Eds.) *Spasticity: Disordered motor control* (pp. 485-494). Chicago, London: Year Book Med.

Lance, JW (1980b). Pathophysiology of spasticity and clinical experience with baclofen. In RG Feldman, RR Young & WP Koella (Eds.) *Spasticity: Disordered motor control* (pp. 185-204). Chicago, London: Year Book Med Publ.

Landau, WM (1974). Spasticity: The fable of a neurological demon and the emperor's new therapy. *Arch Neurol* 31: 217-219.

Langolf, GD, Chaffin, DB, Foulke, JA (1976). An investigation of Fitts' law using a wide range of movement amplitudes. *J Mot Behav* 8: 113-128.

Latash, ML (1986). Coordination, grammar, and spasticity. *Behav Brain Sci* 9: 612.

Latash, ML (1989a). Direct pattern-imposing control or dynamic regulation? *Behav Brain Sci* 12: 226-227.

Latash, ML (1989b). Implications of the equilibrium-point hypothesis for the variability of aimed hand movements. In CJ Worringham (Ed.) *Spatial, temporal and electromyographical variability in human motor control* (pp. 16-17). Ann Arbor: University of Michigan.

Latash, ML (1989c). *Dynamic regulation of single-joint voluntary movements.* Doctoral dissertation, Rush University.

Latash, ML (1992). Virtual trajectories, joint stiffness, and changes in natural frequency during single-joint oscillatory movements. *Neurosci* 49: 209-220.

Latash, ML, Almeida, GL, & Corcos, DM (1992). Pre-programmed reactions in individuals with Down syndrome: The effects of instruction and predictability of the perturbation. *Arch Phys Med Rehab* (in press).

Latash, ML, & Corcos, DM (1991). Kinematic and electromyographic characteristics of single-joint movements in Down syndrome. *Amer J Ment Retard* 96: 189-201.

Latash, ML, Corcos, DM, & Gottlieb, GL (1989a). Kinematic and electromyographic characteristics of single-joint elbow movements in Down syndrome subjects. In ML Latash (Ed.) *Motor control in Down syndrome* (pp. 22-29). Chicago: Rush-Presbyterian St. Luke's Medical Center.

Latash, ML, & Gottlieb, GL (1990a). Compliant characteristics of single joints: Preservation of equifinality with phasic reactions. *Biol Cybern* 62: 331-336.

Latash, ML, & Gottlieb, GL (1990b). Virtual trajectories of single-joint movements show two basic strategies. *Abstr Soc Neurosci* 20: 152.

Latash, ML, & Gottlieb, GL (1990c). Equilibrium-point hypothesis and variability of the amplitude, speed, and time of single-joint movements. *Biofizika* 35: 870-874.

Latash, ML, & Gottlieb, GL (1991a). An equilibrium-point model of dynamic regulation for fast single-joint movements: I. Emergence of strategy-dependent EMG patterns. *J Mot Behav* 23: 163-177.

Latash, ML, & Gottlieb, GL (1991b). An equilibrium-point model of dynamic regulation for fast single-joint movements. II. Similarity of isometric and isotonic programs. *J Mot Behav* 23: 179-191.

Latash, ML, & Gottlieb, GL (1991c). Reconstruction of joint compliant characteristics during fast and slow movements. *Neurosci* 43: 697-712.

Latash, ML, & Gottlieb, GL (1992). Virtual trajectories of single-joint movements performed under two basic strategies. *Neurosci* 47: 357-365.

Latash, ML, & Gurfinkel, VS (1976). Tonic vibration reflex and position of the body. *Physiologiya Cheloveka (Hum Physiol)* 2: 593-598.

Latash, ML, & Gutman, SR (1992). Variability of fast single-joint movements and the equilibrium-point hypothesis. In K Newell & DM Corcos (Eds.) *Variability and motor control*. Champaign, IL: Human Kinetics (in press).

Latash, ML, Gutman, SR, & Gottlieb, GL (1991). Relativistic effects in single-joint voluntary movements. *Biological Cybernetics* 65: 401-406.

Latash, ML, Penn, RD, Corcos, DM, & Gottlieb, GL (1989b). Short-term effects of intrathecal baclofen in spasticity. *Exp Neurol* 103: 165-172.

Latash, ML, Penn, RD, Corcos, DM, & Gottlieb, GL (1990). Effects of intrathecal baclofen on voluntary motor control in spastic paresis. *J Neurosurg* 72: 388-392.

Lee, RG, & Hendrie, A (1977). Selective modification of human spinal and long-loop reflexes by vibration. *Electroencephalog Clin Neurophysiol* 43: 606-610.

Lee, RG, Lucier, GE, Mustard, BE, & White, DG (1986). Modification of motor output to compensate for unanticipated load conditions during rapid voluntary movements. *Canad J Neurol Sci* 13: 97-102.

Lee, RG, Murphy, JT, & Tatton, WG (1983). Long-latency myotatic reflexes in man: Mechanisms, functional significance, and changes in patients with Parkinson's disease. *Adv Neurol* 39: 489-508.

Lee, RG, & Tatton, WG (1975). Motor responses to sudden limb displacements in primates with specific CNS lesions and in human patients with motor system disorders *Canad J Neurol Sci* 2: 285-293.

Lee, RG, & Tatton, WG (1978). Long loop reflexes in man: Clinical application. *Prog Clin Neurophysiol* 4: 320-333.

Lee, RG, & Tatton, WG (1982). Long latency reflexes to imposed displacements of the human wrist: Dependence on duration of movement. *Exp Brain Res* 45: 207-216.

Lee, WA (1980). Anticipatory control of posture and task muscles during rapid arm flexion. *J Mot Behav* 12: 185-196.

Lee, WA, Boughton, A, & Rymer, WZ (1987). Absence of stretch reflex gain enhancement in voluntarily activated spastic muscle. *Exp Neurol* 98: 317-335.

Leicht, R, Rowe, MJ, & Schmidt, RF (1973). Cortical and peripheral modification of cerebellar climbing fiber activity arising from cutaneous mechanoreceptors. *J Physiol* 228: 619-635.

Lennard, RP (1985). Afferent perturbations during "monopodal" swimming movements in the turtle: Phase-dependent cutaneous modulation and proprioceptive resetting of the locomotor rhythm. *J Neurosci* 5: 1434-1445.

Lennerstrand, G (1968). Position and velocity sensitivity of muscle spindles in the cat. I. Primary and secondary endings deprived of fusimotor activation. *Acta Physiol Scand* 73: 281-299.

Lestienne, F (1979). Effects of inertial loads and velocity on the braking process of voluntary limb movements. *Exp Brain Res* 35: 407-418.

Liddell, EGT, & Sherrington, CS (1924). Reflexes in response to stretch (myotatic reflexes). *Proc Roy Soc London Ser B* 96: 212-242.

Lippold, OCJ (1971). Physiological tremor. *Sci Amer* 224: 65-73.

Lisberger, SG, Fuchs, AF, King, WM, & Evinger, LC (1975). Effect of mean reaction time on saccadic responses to two-step stimuli with horizontal and vertical components. *Vision Res* 15: 1021-1025.

Lisin, VV, Frankstein, SI, & Rechtman, MB (1973). The influence of locomotion on flexor reflex of the hind limb in cat and man. *Exp Neurol* 38: 180-183.

Litvintsev, AI (1972). Vertical posture control mechanisms in man. *Automat Remote Contr* 33: 590-600.

Lloyd, DPC (1943). Conduction and synaptic transmission of reflex response to stretch in spinal cat. *J Neurophysiol* 6: 317-326.

Lloyd, DPC (1946). Facilitation and inhibition of spinal motoneurons. *J Neurophysiol* 9: 421-438.

Loeb, GE (1989). Strategies for the control of studies of voluntary movements with one mechanical degree of freedom. *Behav Brain Sci* 12: 227.

Loo, CKC, & McCloskey, DI (1985). Effects of prior instruction and anaesthesia on long-latency responses to stretch in the long flexor of the human thumb. *J Physiol* 365: 285-296.

Lou, J-S, & Bloedel, JR (1986). The response of simultaneously recorded Purkinje cells in perturbations of the step cycle in the walking ferret: A study using a new analytical method—The real time postsynaptic response (RTPR). *Brain Res* 365: 340-344.

Lundberg, A (1966). Integration in reflex pathway. In R Granit (Ed.) *Muscular afferents and motor control. Nobel symposium, I,* (pp. 275-305). Stockholm: Almqvist & Wiksell.

Lundberg, A (1975). Control of spinal mechanisms from the brain. In DB Tower (Ed.) *The nervous system, vol. 2.* New York: Raven Press.

Lundberg, A (1979). Multisensory control of spinal reflex pathways. In R Granit & O Pompeiano (Eds.) *Reflex control of posture and movement* (pp. 11-128) Amsterdam: Elsevier.

Ma, S-P, & Zahalak, GI (1985). The mechanical response of the active human triceps brachii muscle to very rapid stretch and shortening. *J Biomech* 18: 585-598.

MacDonell, RAL, Talalla, A, Swash, M, & Grundy, D (1989). Intrathecal baclofen and the H-reflex. *J Neurol Neurosurg Psychiat* 52: 1110-1112.

MacKay, WA, Crammond, DJ, Kwan, HC, & Murphy, JT (1986). Measurements of human forearm viscoelasticity. *J Biomech* 19: 231-238.

MacKenzie, CL, Marteniuk, RG, Dugas, C, Liske, D, & Bickmeier, B (1987). Three-dimensional movement trajectories in Fitts' task: Implications for control. *Quart J Exp Psychol* 39A: 629-647.

MacNeilage, PF, Studdert-Kennedy, MG, & Lindblom, B (1984). Functional precursors to language and its lateralization. *Amer J Physiol* 15: R912-R914.

Macpherson, JM, Rushmer, DS, & Dunbar, DC (1986). Postural responses in the cat to unexpected rotations of the supporting surface: Evidence for a centrally generated synergic organization. *Exp Brain Res* 62: 152-160.

McClelland, JL, & Rumelhart, DE (1988). *Explorations in parallel distributed processing*. Cambridge: MIT Press.

McCloskey, DI (1978). Kinaesthetic sensibility. *Physiol Rev* 58: 763-820.

McGovern, DE (1974). *Factors affecting control allocation for augmented remote manipulation*. Doctoral dissertation, Stanford University.

McGrain, P (1980). Trends in selected kinematic and myoelectric variables associated with learning a novel motor task. *Res Quart Exer Sport* 51: 509-520.

McKinley, PA, & Smith, JL (1983). Visual and vestibular contributions to prelanding EMG during jump-downs in cats. *Exp Brain Res* 52: 439-448.

McKinley, PA, Smith, JL, & Gregor, RJ (1983). Responses of elbow extensors to landing forces during jump downs in cats. *Exp Brain Res* 49: 218-228.

McMahon, TA (1984). *Muscles, reflexes, and locomotion*. Princeton: Princeton University Press.

Magnus, R (1909a). Zur regelung der bewegungen durch das zentralnervensystem. I. Mitteilung. *Pflugers Archiv d ges Physiologie* 130: 219-252 (cited after Magnus, 1924).

Magnus, R (1909b). Zur regelung der bewegungen durch des zentralnervensystem. II. Mitteilung. *Pflugers Archiv d ges Physiologie* 130: 253 (cited after Magnus, 1924).

Magnus, R (1910). Zur regelung der bewegungen durch des zentralnervensystem. III. Mitteilung. *Pflugers Archiv d ges Physiologie* 134: 545-583 (cited after Magnus, 1924).

Magnus, R (1924). *Korperstellung*. Berlin: Springer.

Mailis, A, & Ashby, P (1990). Alterations in group Ia projections to motoneurons following spinal lesions in humans. *J Neurophysiol* 64: 637-647.

Marsden, CD (1982). The mysterious motor function of the basal ganglia. The Robert Wartenberg lecture. *Neurology* 32: 514-539.

Marsden, CD (1984). Which motor disorder in Parkinson's disease indicates the true motor function of the basal ganglia? In D Evered & M O'Connor (Eds.) *Functions of the basal ganglia. Ciba foundation symposium N 107* (pp. 225-237). London: Pitman.

Marsden, CD, Merton, RA, & Morton, HB (1972). Servo action in human voluntary movement. *Nature* 238: 140-143.

Marsden, CD, Merton, RA, & Morton, HB (1975). Behaviour of short and long latency components of the stretch reflex in the human muscle. *J Physiol* 246: 43P-44P.

Marsden, CD, Merton, RA, & Morton, HB (1976). Stretch reflex and servo action in a variety of human muscles. *J Physiol* 259: 531-560.

Marsden, CD, Merton, RA, & Morton, HB (1977). The sensory mechanism of servo action in human muscles. *J Physiol* 265: 521-535.

Marsden, CD, Merton, RA, Morton, HB, Adam, JER, & Hallett, M (1978). Automatic and voluntary responses to muscle stretch in man. *Prog Clin Neurophysiol* 4: 167-177.

Marsden, CD, Merton, RA, Morton, HB, Rothwell, JC, & Traub, MM (1981). Reliability and efficacy of the long-latency stretch reflex in the human thumb. *J Physiol* 316: 47-60.

Marsden, CD, Obeso, JA, & Rothwell, JC (1983). The function of the antagonist muscle during fast limb movements in man. *J Physiol* 335: 1-13.

Marsden, CD, Rothwell, JC, & Traub, M (1979). Long latency stretch reflex of the human thumb can be reversed if the task is changed. *J Physiol* 293: 41P-42P.

Marteniuk, RG (1973). Retention characteristics of motor short-term memory cues. *J Mot Behav* 5: 249-259.

Marteniuk, RG, Shields, KW, & Campbell, S (1972). Amplitude, position, timing, and velocity as cues in reproduction of movement. *Percept Motor Skills* 35: 51-58.

Marteniuk, RG, MacKenzie, CL, & Baba, DM (1984). Bimanual movement control: Information processing and interaction effects. *Quart J Exp Psychol* 36A: 335-365.

Martin, DP, Baillieul, J, & Hollerbach, JM (1989). Resolution of kinematic redundancy using optimization techniques. *IEEE Trans Robot Automat* 5: 529-533.

Martin, JP (1967). *The basal ganglia and posture*. London: Pitman.

Matthews, BHC (1933). Nerve endings in mammalian muscle. *J Physiol* 78: 1-33.

Matthews, PBC (1959). The dependence of tension upon extension in the stretch reflex of the soleus of the decerebrate cat. *J Physiol* 47: 521-546.

Matthews, PBC (1972). *Mammalian muscle receptors and their central actions*. Baltimore: Williams & Wilkins.

Matthews, PBC (1981). Muscle spindles: Their messages and their fusimotor supply. In *Handbook of physiology, sect. I: The nervous system* (vol. 2, part 1), American Physiological Society.

Matthews, PBC, Cody, FWJ, Richardson, HC, & MacDermott, N (1990). Observations on the reflex effects seen in Parkinson's disease on terminating a period of tendon vibration. *J Neurol Neurosurg Psychiat* 53: 215-219.

Megaw, ED (1974). Possible modification to a rapid on-going programmed manual response. *Brain Res* 71: 425-441.

Meinck, H, Benecke, R, Meyer, W, Hohne, J, & Conrad, B (1984). Human ballistic finger flexion: Uncoupling of the three-burst pattern. *Exp Brain Res* 55: 127-133.

Melvill Jones, G, & Watt, DGD (1971a). Observation on the control of stepping and hopping movements in man. *J Physiol* 219: 709-727.

Melvill Jones, G, & Watt, DGD (1971b). Muscular control of landing from unexpected falls in man. *J Physiol* 219: 729-737.

Merritt, JL (1981). Management of spasticity in spinal cord injury. *Mayo Clin Proc* 56: 614-622.

Merton, PA (1953). Speculations on the servo-control of movements. In JL Malcolm, JAB Gray, & GEW Wolstenholm (Eds.) *The spinal cord* (pp. 183-198). Boston: Little, Brown.

Merton, PA, & Morton, HB (1980). Stimulation of the cerebral cortex in the intact human subject. *Nature* 285: 227.

Meulen, JHP van der, Gooskens, RHJM, Denier van der Gon, JJ, Gielen, CCAM, & Wilhelm, K (1990). Mechanisms underlying accuracy in fast goal-directed arm movements in man. *J Mot Behav* 22: 67-84.

Meyer, DE, Abrams, RA, Kornblum, S, Wright, CE, & Smith, JE (1988). Optimality in human motor performance. Ideal control of rapid aimed movements. *Psychol Rev* 95: 340-370.

Meyer, DE, Smith, JE, Kornblum, S, Abrams, RA, & Wright, CE (1990). Speed-accuracy tradeoffs in aimed movements: Toward a theory of rapid voluntary action. In M Jeannerod (Ed.) *Attention and performance XIII* (pp. 173-225).

Meyer, DE, Smith, JEK, & Wright, CE (1982). Models for the speed and accuracy of aimed movements. *Psychol Rev* 89: 449-482.

Miller, AD, & Brooks, VB (1981). Late muscular responses to arm perturbations persist during supraspinal dysfunctions in monkeys. *Exp Brain Res* 41: 146-158.

Miller, S, Ruit, JB, & van der Meche, FGA (1977). Reversal of sign of long spinal reflexes dependent on the phase of the step cycle in the high decerebrate cat. *Brain Res* 128: 447-459.

Miller, S, & van der Meche, FGA (1976). Coordinated stepping of all four limbs in the high spinal cat. *Brain Res* 109: 395-398.

Milner, TE (1986). Controlling velocity in rapid movements. *J Mot Behav* 18: 147-161.

Milner, TE, & Ijaz, MM (1990). The effect of accuracy constraints on three-dimensional movement kinematics. *Neurosci* 35: 365-374.

Mitchell, SJ, Richardson, RT, Baker, FH, & DeLong, MR (1987). The primate globus pallidus: Neuronal activity related to direction of movement. *Exp Brain Res* 68: 491-505.

Mizrahi, EM, & Angel, RW (1979). Impairment of voluntary movement by spasticity. *Ann Neurol* 5: 594-595.

Morasso, P (1981). Spatial control of arm movements. *Exp Brain Res* 42: 223-227.

Morasso, P (1983a). Coordination aspects of arm trajectory formation. *Hum Mov Sci* 2: 197-210.

Morasso, P (1983b). Three-dimensional arm trajectories. *Biol Cybern* 48: 187-194.

Morasso, P, & Mussa-Ivaldi, FA (1982). Trajectory formation and handwriting: A computational model. *Biol Cybern* 45: 131-142.

Morris, AF, Vaughan, SE, & Vaccaro, P (1982). Measurements of neuromuscular tone and strength in Down's syndrome children. *J Ment Defic Res* 26: 41-46.

Mortimer, JA, & Webster, DD (1979). Evidence for a quantitative association between EMG stretch responses and Parkinsonian rigidity. *Brain Res* 162: 169-173.

Mortin, LI, Keifer, J, & Stein, PSG (1985). Three forms of scratch reflex in the spinal turtle: Movement analysis. *J Neurophysiol* 53: 1501-1516.

Mpitsos, GJ (1990). Chaos in brain function and the problem of nonstationarity: A commentary. In E Basar (Ed.) *Chaos in brain function* (pp. 162-176) Springer-Verlag: Berlin e.a.

Muir, RB, & Lemon, RN (1983). Corticospinal neurons with a special role in precision grip. *Brain Res* 261: 312-316.

Muller, F, & Abbs, JH (1990). Precision grip in parkinsonian patients. *Adv Neurol* 53: 191-195.

Muller, H, Zierski, J, Dralle, D, Borner, U, & Hoffmann, O (1987). The effect of intrathecal baclofen on electrical muscle activity in spasticity. *J Neurol* 234: 348-352.

Muller, H, Zierski, J, Dralle, D, Hoffmann, O, & Michaelis, G (1988a). Intrathecal baclofen in spasticity. In H Muller, J Zierski, & RD Penn (Eds.) *Local-spinal therapy of spasticity* (pp. 155-214). Berlin: Springer-Verlag.

Muller, H, Zierski, J, & Penn, RD (Eds.) (1988b). *Local-spinal therapy of spasticity*. Berlin: Springer-Verlag.

Mussa-Ivaldi, FA (1988). Do neurons in the motor cortex encode movement direction? An alternative hypothesis. *Neurosci Lett* 91: 106-111.

Mussa-Ivaldi, FA, Hogan, N, & Bizzi, E (1985). Neural, mechanical, and geometric factors subserving arm posture in humans. *J Neurosci* 5: 2732-2743.

Mussa-Ivaldi, FA, Morasso, P, & Zaccaria, R (1989). Kinematic networks. A distributed model for representing and regularizing motor redundancy. *Biol Cybern* 60: 1-16.

Mustard, BE, & Lee, RG (1987). Relationship between EMG patterns and kinematic properties for flexion movements at the human wrist. *Exp Brain Res* 66: 247-256.

Myklebust, BM, Gottlieb, GL, Agarwal, GC, & Penn, RD (1982). Reciprocal excitation of antagonist muscles as a differentiating feature in spasticity. *Ann Neurol* 12: 367-374.

Nagasaki, H (1989). Asymmetric velocity and acceleration profiles of human arm movements. *Exp Brain Res* 74: 319-326.

Nashner, LM (1976). Adapting reflexes controlling human posture. *Exp Brain Res* 26: 59-72

Nashner, LM (1979). Organization and programming of motor activity during posture control. In R Granit & O Pompeiano (Eds.) *Reflex control of posture and movement* (pp. 177-184). Amsterdam: Elsevier.

Nashner, LM (1980). Balance adjustments of humans perturbed while walking. *J Neurophysiol* 44: 650-664.

Nashner, LM, & Cordo, PJ (1981). Relation of automatic postural responses and reaction-time voluntary movements of human leg muscles. *Exp Brain Res* 43: 395-405.

Nashner, LM, & Grimm, RJ (1978). Analysis of multiloop dyscontrols in standing cerebellar patients. In JE Desmedt (Ed.) *Prog Clin Neurophysiol vol. 4* (pp. 300-319). Basel: Karger.

Nashner, LM, & McCollum, G (1985). The organization of human postural movements: A formal basis and experimental synthesis. *Brain Behav* 8: 135-172.

Nashner, LM, & Woollacott, M (1979). The organization of rapid postural adjustments of standing humans: An experimental-conceptual model. In RE Talbott & DR Humphrey (Eds.) *Posture and movement* (pp. 243-257). New York: Raven Press.

Nashner, LM, Woollacott, M, & Tuma, G (1979). Organization of rapid responses to postural and locomotor-like perturbations of standing man. *Exp Brain Res* 36: 463-479.

Nashner, LM, Shupert, CL, Horak, FB, & Black, FO (1989). Organization of posture controls: An analysis of sensory and mechanical constraints. *Prog Brain Res* 80: 411-418.

Nativ, J, & Abbs, JH (1989). Goal-directed arm movements in Down syndrome. In ML Latash (Ed.) *Motor control in Down syndrome* (pp. 36-42). Chicago: Rush Medical Center.

Neilson, PD, & McCaughey, J (1981). Effect of contraction level and magnitude of stretch on tonic stretch reflex transmission characteristics. *J Neurol Neurosurg Psychiat* 44: 1007-1012.

Nelson, W (1983). Physical principles for economies of skilled movements. *Biol Cybern* 46: 135-147.

Newell, KM, & Carlton, LG (1985). On the relationship between peak force and peak force variability in isometric tasks. *J Mot Behav* 17: 230-241.

Newell, KM, & Carlton, LG (1988). Force variability in isometric responses. *J Exp Psychol: Hum Percept Perform* 14: 24-36.

Newell, KM, Carlton, LG, & Carlton, MJ (1982). The relationship of impulse to timing error. *J Mot Behav* 12: 47-56.

Newell, KM, Carlton, LG, & Hancock, PA (1984). Kinetic analysis of response variability. *Psychol Bull* 96: 133-151.

Newell, KM, van Emmerik, REA, & McDonald, PV (1989). On simple movements and complex theories (and vice versa). *Behav Brain Sci* 12: 229-230.

Newsom Davis, J, & Sears, TA (1970). The proprioceptive reflex control of the intercostal muscles during their voluntary activation. *J Physiol* 209: 711-738.

Nichols, TR (1974). *Soleus muscle stiffness and its reflex control*. Ph.D. thesis, Harvard University.

Nichols, TR (1989). The organization of heterogenic reflexes among muscles crossing the ankle joint in the decerebrate cat. *J Physiol* 410: 463-477.

Nichols, TR, & Houk, JC (1976). Improvement in linearity and regulation of stiffness that results from actions of stretch reflex. *J Neurophysiol* 39: 119-142.

Nichols, TR, & Steeves, JD (1986). Resetting of resultant stiffness in ankle flexor and extensor muscles in the decerebrate cat. *Exp Brain Res* 62: 401-410.

Noguchi, T, Homma, S, & Nakajima, Y (1979). Measurement of excitatory postsynaptic potentials in the stretch reflex of normal subjects and spastic patients. *J Neurol Neurosurg Psychiat* 42: 1100-1105.

Oguztoreli, MN, & Stein, RB (1983). Optimal control of antagonistic muscles. *Biol Cybern* 48: 91-99.

Oguztoreli, MN, & Stein, RB (1990). Optimal task performance of antagonistic muscles. *Biol Cybern* 64: 87-94.

Ohye, C, Tsukahara, N, & Narabayashi, H (1965). Rigidity and disturbance of reciprocal innervation. *Confin Neurol* 26: 24-40 (cited after Hayashi et al., 1988).

Ojakangas, CL, & Ebner, TJ (1991). Scaling of the metrics of visually-guided arm movements during motor learning in primates. *Exp Brain Res* 85: 314-323.

Ojemann, GA (1984). Common cortical and thalamic mechanisms for language and motor function. *Amer J Physiol* 15: R901-R903.

Ongeboer de Visser, BW, Bour, LJ, Koelman, JHTM, & Speelman, JD (1989). Cumulative vibratory indices and the H/M ratio of the soleus H-reflex: A quantitative study in control and spastic patients. *Electroencephalog Clin Neurophysiol* 73: 162-166.

Osovets, SM, Ginsburg, DA, Gurfinkel, VS, Zenkov, LP, Latash, LP, Malkin, VB, Mel'nichuk, PV, & Pasternak, EB (1983). Electrical activity of the brain: Mechanisms and interpretation. *Adv Phys Sci (USSR)* 141: 103-150.

Ostry, DJ, Cooke, JD, & Munhall, KG (1987). Velocity curves of human arm and speech movements. *Exp Brain Res* 68: 37-46.

Ostry, DJ, Feldman, AG, & Flanagan, JR (1991). Kinematics and control of frog hindlimb movements. *J Neurophysiol* 65: 547-562.

Paillard, J (1955). *Reflexes et regulations d'origine proprioceptive chez l'homme. Etude neurophysiologique et psychophysiologique.* Paris: Lib Arnette.

Paintal, AS, & Walsh, EG (1981). Inhibition of tonic stretch reflex by J receptor activity. *J Physiol* 316: 22P-23P.

Pal'tsev, YI, & El'ner, AM (1967). Preparatory and compensatory period during voluntary movement in patients with involvement of the brain of different localization. *Biofizika* 12: 142-147.

Parker, AW, & Bronks, RB (1980). Gait of children with Down syndrome. *Arch Phys Med Rehab* 61: 345-351.

Partridge, LD (1965). Modifications of neural output signals by muscles: A frequency response study. *J Appl Physiol* 20: 150-156.

Partridge, LD, & Benton, LA (1981). Muscle, the motor. In VB Brooks (Ed.) *The nervous system. Handbook of physiology, vol. II, part I* (pp. 43-106). Bethesda: American Physiological Society.

Partridge, LD, & Glaser, GH (1960). Adaptation in regulation of movement and posture. A study of stretch responses in spastic animals. *J Neurophysiol* 23: 257-268.

Pedersen, E (1974). Clinical assessment of pharmacological therapy of spasticity. *Arch Phys Med Rehab* 55: 344-355.

Pedotti, A, & Crenna, P (1990). Individual strategies of muscle recruitment in complex natural movements. In JM Winters & SL-Y Woo (Eds.) *Multiple muscle systems. Biomechanics and movement organization* (pp. 542-548). New York: Springer-Verlag.

Penn, RD (1988). Chronic intrathecal baclofen for severe rigidity and spasms. In H Muller, J Zierski, & RD Penn (Eds.) *Local-spinal therapy of spasticity* (pp. 151-153). Berlin: Springer-Verlag.

Penn, RD, & Kroin, JS (1985). Continuous intrathecal baclofen for severe spasticity. *Lancet* 2: 125-127.

Penn, RD, & Kroin, JS (1987). Long-term intrathecal baclofen infusion for treatment of spasticity. *J Neurosurg* 66: 181-185.

Penn, RD, Savoy, SM, Corcos, D, Latash, M, Gottlieb, G, Parke, B, & Kroin, J (1989). Intrathecal baclofen for severe spinal spasticity. *New Engl J Med* 320: 1517-1521.

Peters, M (1977). Simultaneous performance of two motor activities: The factor of timing. *Neuropsychol* 15: 461-464.

Pew, RW (1966). Acquisition of hierarchical control over the temporal organization of a skill. *J Exp Psychol* 71: 764-771.

Phillips, CG (1969). Motor apparatus of the baboon's hand. *Proc Roy Soc London Ser B* 173: 141-174.

Pierau, FK, & Zimmermann, P (1973). Action of a GABA-derivative on post-synaptic potentials and membrane properties of cats' spinal motoneurones. *Brain Res* 54: 376-380.

Pierrot-Deseilligny, E (1983). Pathophysiology of spasticity. *Triangle* 22: 165-174.

Plamondon, R (1989). Handwriting control: A functional model. In RMJ Cotterill (Ed.) *Models of brain function* (pp. 563-574). Cambridge: Cambridge University Press.

Plamondon, R, & Lorette, G (1989). Automatic signature verification and writer identification: The state of the art. *Pattern Recogn* 22: 107-131.

Plamondon, R, Stelmach, GE, & Teasdale, N (1990). Motor program coding representation from a handwriting generator model: The production of line responses. *Biol Cybern* 63: 443-451.

Polit, A, & Bizzi, E (1978). Processes controlling arm movements in monkey. *Science* 201: 1235-1237.

Polit, A, & Bizzi, E (1979). Characteristics of motor programs underlying arm movement in monkey. *J Neurophysiol* 42: 183-194.

Pompeiano, O (1960). Alpha types of "release" studied in tension-extension diagrams from cat's forelimb triceps muscle. *Arch Ital Biol* 98: 92-117.

Poppele, RE, & Terzuolo, CA (1968). Myotatic reflex: Its input-output relations. *Science* 159: 743-745.

Pozzo, T, Berthoz, A, Lefort, L (1990). Head stabilization during various locomotor tasks in humans. I. Normal subjects. *Exp Brain Res* 82: 97-106.

Prablanc, C, Pellison, D, & Goodale, MA (1986). Visual control of reaching movements without vision of the limb. I. Role of retinal feedback of target position in guiding the hand. *Exp Brain Res* 62: 293-302.

Price, GW, Wilkin, GP, Turnbull, MJ, & Bowery, NG (1984). Are baclofen-sensitive $GABA_B$ receptors present on primary afferent terminals of the spinal cord? *Nature* 307: 71-73.

Pullman, SL, Watts, RL, Juncos, JL, & Sanes, JN (1990). Movement amplitude choice reaction time performance in Parkinson's disease may be independent of dopaminergic status. *J Neurol Neurosurg Psychiat* 53: 279-283.

Rack, PMH, Ross, HF, & Brown, TIH (1978). Reflex responses during sinusoidal movement of human limbs. In JE Desmedt (Ed.) *Cerebral motor control in man: Long loop mechanisms* (pp. 216-218) Basel: Karger.

Rack, PMH, Ross, HF, & Thilmann, AF (1984). The ankle stretch reflexes in normal and spastic subjects: The response to sinusoidal movement. *Brain* 107: 637-654.

Rack, PMH, & Westbury, DR (1969). The effects of length and stimulus rate on tension in the isometric cat soleus muscle. *J Physiol* 204: 443-460.

Rademaker, GGJ (1947). On the lengthening and shortening reactions and their occurrence in man. *Brain* 70: 109-126.

Raibert, MH (1977). *Motor control and learning by the state space model.* Doctoral dissertation, MIT.

Raibert, MH (1978). A model for sensorimotor control and learning. *Biol Cybern* 29: 29-36.

Raibert, MH (1986). *Legged robots that balance.* Cambridge: MIT Press.

Ralston, HJ, Inman, VT, Strait, LA, & Shaffrath, MD (1947). Mechanics of human isolated voluntary muscle. *Am J Physiol* 151: 612-620.

Ramos, CF, & Stark, LW (1990a). Postural maintenance during movement: Simulations of a two joint model. *Biol Cybern* 63: 363-375.

Ramos, CF, & Stark, LW (1990b). Postural maintenance during fast forward bending: A model simulation experiment determines the "reduced trajectory." *Exp Brain Res* 82: 651-657.

Rarick, GL, Dobbins, DA, & Broadhead, GG (1976). *The motor domain and its correlates in education handicapped children.* Englewood Cliffs, NJ: Prentice Hall.

Renshaw, B (1941). Influence of the discharge of motoneurons upon excitation of neighboring motoneurons. *J Neurophysiol* 4: 167-183.

Riccio, GE, & Stoffregen, TA (1988). Affordances as constraints on the control of stance. *Hum Mov Sci* 7: 265-300.

Ritchie, JM, & Wilkie, DR (1958). The dynamics of muscular contraction. *J Physiol* 143: 104-124.

Ritter, HJ, Martinez, TM, & Schulten, KJ (1989). Topology-conserving maps for learning visuo-motor coordination. *Neural Net* 2: 159-168.

Robertson, C, & Flowers, KA (1990). Motor set in Parkinson's disease. *J Neurol Neurosurg Psychiat* 53: 583-592.

Robertson, GA, & Stein, PSG (1988). Synaptic control of hindlimb motoneurons during three forms of the fictive scratch reflex in the turtle. *J Physiol* 404: 101-128.

Rogers, MW, & Pai, Y-C (1990). Dynamic transitions in stance support accompanying leg flexion movements in man. *Exp Brain Res* 81: 398-402.

Rondot, P, & Bathien, N (1986). Movement disorders in patients with coexistent neuroleptic-induced tremor and tardive dyskinesia: EMG and pharmacological study. *Adv Neurol* 45: 361-366.

Rosenthal, JC, McKean, TA, Roberts, WJ, & Terzuolo, CA (1970). Frequency analysis of stretch reflex and its main subsystems in triceps surae muscles of the cat. *J Neurophysiol* 33: 713-749.

Rossignol, S, & Gauthier, L (1980). An analysis of mechanisms controlling the reversal of crossed spinal reflexes. *Brain Res* 182: 31-45.

Rothwell, JC, Day, BL, Berardelli, A, & Marsden, CD (1986). Habituation and conditioning of the human long latency stretch reflex. *Exp Brain Res* 63: 197-204.

Rothwell, JC, Obeso, JA, Traub, MM, & Marsden, CD (1983). The behavior of the long-latency stretch reflex in patients with Parkinson's disease. *J Neurol Neurosurg Psychiat* 46: 35-44.

Rothwell, JC, Traub, MM, Day, BL, Obeso, JA, Thomas, PK, & Marsden, CD (1982a). Manual motor performance in a deafferented man. *Brain* 105: 515-542.

Rothwell, JC, Traub, MM, & Marsden, D (1982b). Automatic and "voluntary" responses compensating for disturbances of human thumb movements. *Brain Res* 248: 33-41.

Rushmer, DS, Roberts, WJ, & Augter, GK (1976). Climbing fiber responses of cerebellar Purkinje cells to passive movement of the cat forepaw. *Brain Res* 106: 1-20.

Rushton, DN, Rothwell, JC, & Craggs, MD (1981). Gating of somatosensory evoked potentials during different kinds of movement in man. *Brain* 104: 465-491.

Sahrmann, SA, & Norton, BJ (1977). The relationship of voluntary movement to spasticity in the upper motor neuron syndrome. *Ann Neurol* 2: 460-465.

Saltzman, EL (1979). Levels of sensorimotor representation. *J Math Psychol* 20: 91-163.

Saltzman, EL, & Kelso, JAS (1985). Synergies: Stabilities, instabilities, and modes. *Behav Brain Sci* 8: 161-163.

Saltzman, EL, & Kelso, JAS (1987). Skilled actions: A task-dynamic approach. *Psychol Rev* 94: 84-106.

Sanes, JN (1985). Information processing deficits in Parkinson's disease during movement. *Neuropsychol* 23: 381-392.

Sanes, JN (1986). Kinematics and end-point control of arm movements are modified by unexpected changes in viscous loading. *J Neurosci* 6: 3120-3127.

Sanes, JN, & Jennings, VA (1984). Centrally programmed patterns of muscle activity in voluntary motor behavior of humans. *Exp Brain Res* 54: 23-32.

Sanes, JN, Mauritz, KH, Dalakas, MC, & Evarts, EV (1985). Motor control in humans with large-fiber sensory neuropathy. *Hum Neurobiol* 4: 101-114.

Sanes, JN, Mauritz, K, Evarts, EV, Dalakas, MC, & Chu, A (1984). Motor deficits in patients with large-fiber sensory neuropathy. *Proc Natl Acad Sci* 81: 979-982.

Schieber, MH, & Thach, Jr, WT (1985). Trained slow tracking. I. Muscular production of wrist movements. *J Neurophysiol* 54: 1213-1227.

Schieppati, M, & Nardone, A (1991). Free and supported stance in Parkinson's disease. The effect of posture and "postural set" on leg muscle responses to perturbation, and its relation to the severity of the disease. *Brain* 114: 1227-1244.

Schmidt, RA (1975). A schema theory of discrete motor skill learning. *Psychol Rev* 82: 225-260.

Schmidt, RA (1980a). On the theoretical status of time in motor-program representations. In GE Stelmach & J Requin (Eds.) *Tutorials in motor behavior* (pp. 145-165). Amsterdam: North-Holland.

Schmidt, RA (1980b). Past and future issues in motor programming. *Res Quart Exer Sport* 51: 122-140.

Schmidt, RA (1988). *Motor control and learning. A behavioral emphasis.* Champaign: Human Kinetics.

Schmidt, RA, & McGown, C (1980). Terminal accuracy of unexpected loaded rapid movements: Evidence for a mass-spring mechanism in programming. *J Mot Behav* 12: 149-161.

Schmidt, RA, & Sherwood, DE (1982) An inverted-U relation between spatial error and force requirements in rapid limb movements: Further evidence for the impulse-variability model. *J Exp Psychol: Hum Percept Perform* 8: 158-170.

Schmidt, RA, Sherwood, DE, & Walter, CB (1988). Rapid movements with reversals in direction. I. The control of movement time. *Exp Brain Res* 69: 344-354.

Schmidt, RA, Zelaznik, HN, & Frank, JS (1978). Sources of inaccuracy in rapid movement. In GE Stelmach (Ed.) *Information processing in motor control and learning*. New York: Academic.

Schmidt, RA, Zelaznik, H, Hawkins, B, Franks, JS, & Quinn, JT (1979). Motor output variability: A theory for the accuracy of rapid motor acts. *Psychol Rev* 86: 415-451.

Schmidt, RC, Beek, PJ, Treffner, PJ, & Turvey, MT (1991). Dynamic substructure of coordinated rhythmic movements. *J Exp Psychol: Hum Percept Perform* 17: 635-651.

Schmidt, RC, Carello, C, & Turvey, MT (1990). Phase transitions and critical fluctuations in the visual coordination of rhythmic movements between people. *J Exp Psychol: Hum Percept Perform* 16: 227-247.

Schmidtbleicher, D, Gollhofer, A, & Frick, U (1988). Effects of a stretch-shortening typed training on the performance capability and innervation characteristics of leg extensor muscles. In G Groot, AP Hollander, PA Huijing, & VI Schenau (Eds.) *Biomechanics XI-A*. Amsterdam: Free University Press.

Scholz, JP, & Kelso, JAS (1990). Intentional switching between patterns of bimanual coordination depends on the intrinsic dynamics of the patterns. *J Mot Behav* 22: 98-124.

Schomaker, LRB, & Plamondon, R (1990). The relation between pen force and pen-point kinematics in handwriting. *Biol Cybern* 63: 277-289.

Schoner, G (1990). A dynamic theory of coordination of discrete movement. *Biol Cybern* 63: 257-270.

Schoner, G, & Kelso, JAS (1988). Dynamic pattern generation in behavioral and neural systems. *Science* 239: 1513-1520.

Schotland, JL, Lee, WA, & Rymer, WZ (1989). Wiping reflex and flexion withdrawal reflexes display different EMG patterns prior to movement onset in the spinalized frog. *Exp Brain Res* 78: 649-653.

Schwab, RS, Chafetz, ME, & Walker, S (1954). Control of two simultaneous voluntary motor acts in normals and in Parkinsonians. *Arch Neurol Psychiat* 72: 591-598.

Schwartz, M, Schulte-Tamburen, A, & Noth, J (1990). Static fusimotor activity in Parkinsonian rigidity: Reconsideration of an old concept. In A Berardelli, R Benecke, M Manfredi, & CD Marsden (Eds.) *Motor disturbances II* (pp. 191-201). London: Academic Press.

Sechenov, IM (1863/1952). *Reflexes of the brain.* Moscow: Medgiz (in Russian).

Seiaky, R, Lacquaniti, F, Terzuolo, C, & Soechting, JF (1987). A note on the kinematics of drawing movements in children. *J Mot Behav* 19: 518-525.

Seif-Naraghi, AH, & Winters, JM (1990). Optimized strategies for scaling goal-directed dynamic limb movements. In JM Winters, & SL-Y Woo (Eds.) *Multiple muscle systems. Biomechanics and movement organization* (pp. 312-334). New York: Springer-Verlag.

Severin, FV, Orlovsky, GN, & Shik, ML (1967). Work of the muscles and single motoneurones during controlled locomotion. *Biofizika* 12: 762-772.

Shannon, CE (1948). A mathematical theory of computation. *Bell Syst Tech J* 27: 379-423.

Shapiro, DC, & Walter, CB (1986). An examination of rapid positioning movements with spatiotemporal constraints. *J Mot Behav* 18: 373-395.

Sheridan, MR, & Flowers, KA (1990). Movement variability and bradykinesia in Parkinson's disease. *Brain* 113: 1149-1161.

Sheridan, MR, Flowers, KA, & Hurrell, J (1987). Programming and execution of movement in Parkinson's disease. *Brain* 110: 1247-1271.

Sherrington, CS (1906). *The integrative action of the nervous system.* London.

Sherrington, CS (1908). Observation on the scratch reflex in the spinal dog. *J Physiol* 34: 1.

Sherrington, CS (1910). Flexion reflex of the limb, crossed extension reflex, and reflex stepping and standing. *J Physiol* 40: 28-121.

Sherwood, DE (1986). Impulse characteristics in rapid movement. Implications for impulse-variability models. *J Mot Behav* 18: 188-214.

Sherwood, DE, & Schmidt, RA (1980). The relationship between force and force variability in minimal and near-maximal static and dynamic contractions. *J Mot Behav* 12: 75-89.

Sherwood, DE, Schmidt, RA, & Walter, CB (1988). Rapid movements with reversals in direction. II. Control of movement amplitude and inertial load. *Exp Brain Res* 69: 355-367.

Shik, ML, & Orlovsky, GN (1976). Neurophysiology of locomotor automatism. *Physiol Rev* 56: 465-501.

Shumway-Cook, A, & Woollacott, MH (1985). Dynamics of postural control in the child with Down syndrome. *Phys Ther* 65: 1315-1322.

Shumway-Cook, A, & Horak, FB (1989). Vestibular rehabilitation: An exercise approach to managing symptoms of vestibular dysfunction. *Seminars in Hearing* 10: 196-208.

Simmons, RW, & Richardson, C (1984). Maintenance of equilibrium point control during an unexpectedly loaded rapid limb movement. *Brain Res* 302: 239-244.

Simonetta, M, Clanet, M, & Rascol, O (1991). Bereitschaftpotential in a simple movement or in a motor sequence starting with the same simple movement. *Electroencephalog Clin Neurophysiol* 81: 129-134.

Sinkjaer, T, Toft, E, Andreassen, S, & Hornemann, BC (1988). Muscle stiffness in human ankle dorsiflexors: Intrinsic and reflex components. *J Neurophysiol* 60: 1110-1121.

Sinkjaer, T, Wu, CH, Barto, A, & Houk, JC (1990). Cerebellar control of endpoint position—A simulation model. *Proc Intern Joint Conf on Neural Networks*. San Diego, vol. II: 705-710.

Smeets, JBJ, Erkelens, CJ, & Denier van der Gon, JJ (1990). Adjustments of fast goal-directed movements in response to an unexpected inertial load. *Exp Brain Res* 81: 303-312.

Smith, JC, Betts, B, Edgerton, VR, & Zernicke, RF (1980). Rapid ankle extension during paw shakes: Selective recruitment of fast ankle extensors. *J Neurophysiol* 43: 612-620.

Soechting, JF (1984). Effect of target size on spatial and temporal characteristics of a pointing movement in man. *Exp Brain Res* 54: 121-132.

Soechting, JF, & Lacquaniti, F (1981). Invariant characteristics of a pointing movement in man. *J Neurosci* 1: 710-720.

Soechting, JF, & Lacquaniti, F (1983). Modification of a trajectory of a pointing movement in response to a change in target location. *J Neurophysiol* 49: 548-564.

Soechting, JF, Lacquaniti, F, & Terzuolo, CA (1986). Coordination of arm movements in three-dimensional space. Sensorimotor mapping during drawing movement. *Neurosci* 17: 295-311.

Soechting, JF, & Terzuolo, CA (1986). An algorithm for the generation of curvilinear wrist motion in an arbitrary plane in three-dimensional space. *Neurosci* 19: 1395-1405.

Soechting, JF, & Terzuolo, CA (1987a). Organization of arm movements. Motion is segmented. *Neurosci* 23: 39-51.

Soechting, JF, & Terzuolo, CA (1987b). Organization of arm movements in three-dimensional space. Wrist motion is piecewise planar. *Neurosci* 23: 53-61.

Sonderen, JF van, Denier van der Gon, JJ, & Gielen, CCAM (1988). Conditions determining early modification of motor programmes in response to changes in target location. *Exp Brain Res* 71: 320-328.

Sonderen, JF van, Gielen, CCAM, & Denier van der Gon, JJ (1989). Motor programmes for goal-directed movements are continuously adjusted according to changes in target location. *Exp Brain Res* 78: 139-146.

Sonderen, JF van, & Denier van der Gon, JJ (1990). A simulation study of a programme generator for centrally programmed fast two-joint arm movements: Responses to single- and double-step target displacements. *Biol Cybern* 63: 35-44.

Stein, JF (1978). Long loop motor control in monkeys. *Prog Clin Neurophysiol* 4: 107-122.

Stein, PSG (1983). The vertebrate scratch reflex. *Symp Soc Exp Biol* 37: 393-403.

Stein, PSG (1984). Central pattern generators in the spinal cord. In RA Davidoff (Ed.) *Handbook of the spinal cord, vol. 2-3: Anatomy and physiology* (pp. 647-672). New York, Basel: Marcel Dekker.

Stein, RB (1974). Peripheral control of movement. *Physiol Rev* 54: 215-243.

Stein, RB (1982). What muscle variable(s) does the nervous system control in limb movements? *Behav Brain Sci* 5: 535-577.

Stein, RB, Cody, FWJ, & Capaday, C (1988). The trajectory of human wrist movements. *J Neurophysiol* 59: 1814-1830.

Stein, RB, & Oguztoreli, MN (1976). Does the velocity sensitivity of muscle spindles stabilize the stretch reflex? *Biol Cybern* 22: 147-157.

Stelmach, GE, & Worringham, CJ (1988a). The control of bimanual aiming movements in Parkinson's disease. *J Neurol Neurosurg Psychiat* 51: 223-231.

Stelmach, GE, & Worringham, CJ (1988b). The preparation and production of isometric force in Parkinson's disease. *Neuropsychol* 26: 93-103.

Stelmach, GE, Worringham, CJ, & Strand, EA (1986). Movement preparation in Parkinson's disease: The use of advance information. *Brain* 109: 1179-1194.

Strick, PL (1978). Cerebellar involvement in "volitional" muscle responses to load changes. *Prog Clin Neurophysiol* 4: 85-93.

Swinnen, SP, Walter, CB, & Shapiro, DC (1988). The coordination of limb movements with different kinematic patterns. II. *Brain Cognition* 8: 326-347.

Swinnen, SP, Young, DE, Walter, CB, & Serrien, DJ (1991). Control of asymmetrical bimanual movements. *Exp Brain Res* 85: 163-173.

Szentagothai, J, & Arbib, MA (1974). *Conceptual models of neural organization.* Boston: Yvonne M. Homsy NRP Writer-Editor.

Talbott, RE (1979). Ferrier, the synergy concept and the study of posture and movement. In RE Talbott & DR Humphrey (Eds.) *Posture and movement* (pp. 1-12). New York: Raven Press.

Tanji, J, & Kurata, K (1982). Comparison of movement-related activity in two cortical motor areas of primates. *J Neurophysiol* 48: 633-653.

Tatton, WG, Bawa, P, Bruce, IC, & Lee, RG (1978). Long loop reflexes in monkeys: An interpretive base for human reflexes. *Prog Clin Neurophysiol* 4: 229-245.

Tatton, WG, & Lee, RG (1975). Evidence for abnormal long-loop reflexes in rigid Parkinsonian patients. *Brain Res* 100: 671-676.

Tax, AAM, Denier van der Gon, JJ, & Erkelens, CJ (1990). Differences in coordination of elbow flexor muscles in force tasks and in movement tasks. *Exp Brain Res* 81: 567-572.

Tax, AAM, Denier van der Gon, JJ, Gielen, CCAM, & Tempel, CMM van den (1989). Differences in the activation of m.biceps brachii in the control of slow isotonic movements and isometric contractions. *Exp Brain Res* 76: 55-63.

Teasdale, N, Phillips, J, & Stelmach, GE (1990). Temporal movement control in patient with Parkinson's disease. *J Neurol Neurosurg Psychiat* 53: 862-868.

Teasdale, N, & Schmidt, RA (1991). Deceleration requirements and the control of pointing movements. *J Mot Behav* 23: 131-138.

Terzuolo, CA, & Viviani, P (1979). The central representation of learned motor patterns. In RE Talbott & DR Humphrey (Eds.) *Posture and movement* (pp. 113-121). New York: Raven Press.

Thilmann, AF (1988). Alternation: The tendency for reflex EMG bursts to appear only in every second cycle during imposed sinusoidal movements of the human ankle joint. *J Physiol* 399: 60P.

Thilmann, AF, Fellows, SJ, & Garms, E (1991a). The mechanism of spastic muscle hypertonus. Variation in reflex gain over the time course of spasticity. *Brain* 114: 233-244.

Thilmann, AF, Fellows, SJ, & Ross, HF (1991b). Biomechanical changes at the ankle joint after stroke. *J Neurol Neurosurg Psychiat* 54: 134-139.

Tracey, DJ, Walmsey, B, & Brinkman, J (1980). "Long-loop" reflexes can be obtained in spinal monkeys. *Neurosci Lett* 18: 59-65.

Traub, MM, Rothwell, JC, & Marsden, CD (1980a). Anticipatory postural reflexes in Parkinson's disease and other akinetic-rigid syndromes and in cerebellar ataxia. *Brain* 103: 393-412.

Traub, MM, Rothwell, JC, & Marsden, CD (1980b). A grab reflex in the human hand. *Brain* 103: 869-884.

Turvey, MT (1990a). Coordination. *Amer Psychol* 45: 938-953.

Turvey, MT (1990b). The challenge of a physical account of action: A personal view. In HTA Whiting, OG Meijer, & PCW van Wieringen (Eds.) *The natural-physical approach to movement control* (pp. 57-92). Amsterdam: VU University Press.

Turvey, MT, Rosenblum, LD, Schmidt, RC, & Kugler, PN (1986). Fluctuations and phase symmetry in coordinated rhythmic movements. *J Exp Psychol: Hum Percept Perform* 12: 564-583.

Turvey, MT, Schmidt, RC, & Rosenblum, LD (1989). "Clock" and "motor" components in absolute coordination of rhythmic movements. *Neurosci* 33: 1-10.

Turvey, ML, Schmidt, RC, Rosenblum, LD, & Kugler, PN (1988). On the time allometry of coordinated rhythmic movements. *J Theor Biol* 130: 285-325.

Uno, Y, Kawato, M, & Suzuki, R (1989). Formation and control of optimal trajectory in human multijoint arm movement. *Biol Cybern* 61: 89-101.

Vallbo, AB (1971). Muscle spindle response at the onset of isometric voluntary contractions. Time difference between fusimotor and skeletomotor effects. *J Physiol* 218: 405-431.

Vallbo, AB (1974). Human muscle spindle discharge during isometric voluntary contractions. Amplitude relations between spindle frequency and torque. *Acta Physiol Scand* 90: 319-336.

Vallbo, AB (1981). Basic patterns of muscle spindle discharge in man. In A Taylor & A Prochazka (Eds.) *Muscle receptors and movement* (pp. 263-275). London: MacMillan Press.

Vincken, MH, Gielen, CCAM, & Denier van der Gon, JJ (1983). Intrinsic and afferent components in apparent muscle stiffness in man. *Neuroscience* 9: 529-534.

Viviani, P (1985). Segmentation and coupling in complex movements. *J Exp Psychol* 11: 828-845.

Viviani, P, & McCollum, G (1983). The relation between linear extend and velocity in drawing movements. *Neurosci* 10: 211-218.

Viviani, P, & Schneider, R (1991). A developmental study of the relationship between geometry and kinematics in drawing movements. *J Exp Psychol: Hum Percept Perform* 17: 198-218.

Viviani, P, Soechting, JF, & Terzuolo, CA (1976). Influence of mechanical properties on the relation between EMG activity and torque. *J Physiol (Paris)* 72: 45-58.

Viviani, P, & Terzuolo, C (1980). Space-time invariance in learned motor skills. In GE Stelmach & J Requin (Eds.) *Tutorials in motor behavior* (pp. 525-533). Amsterdam: North-Holland.

Von Holst, E (1954). Relation between the central nervous system and the peripheral organs. *Brit J Animal Behav* 2: 89-94.

Von Holst, E (1937/1973). *The behavioral physiology of animal and man.* Coral Gables, FL: University of Miami Press.

Von Uexkull, J (1904). Die ersten ursachen des rhythmus in der tierreihe. *Ergebnisse d Physiologie* 3: 1. (cited after Magnus, 1924).

Von Uexkull, J (1909). Ein wort uber die schlangensterne. *Zentrabl fur Physiologie* 23: 1. (cited after Magnus, 1924).

Wadman, WJ, Denier van der Gon, JJ, Geuze, RH, & Mol, CR (1979). Control of fast goal directed arm movements. *J Hum Mov Studies* 5: 3-17.

Wallace, SA (1981). An impulse-timing theory for reciprocal control of muscular activity in rapid, discrete movements. *J Mot Behav* 13: 144-160.

Wallace, SA, & Weeks, DL (1989). Initiating voluntary movements: Wrong theories for the wrong behavior? *Behav Brain Sci* 12: 233-234.

Wallace, SA, & Wright, L (1982). Distance and movement time effects on the timing of agonist and antagonist muscles: A test of the impulse-timing theory. *J Mot Behav* 14: 341-352.

Walter, CB, & Swinnen, SP (1990). Kinetic attraction during bimanual coordination. *J Mot Behav* 22: 451-473.

Warren, WH, Young, DS, & Lee, DN (1986). Visual control of step length during running over irregular terrain. *J Exp Psychol: Hum Percept Perform* 12: 259-266.

Watt, DGD, Stauffer, EK, Taylor, A, Reinking, PH, & Stuart, DG (1976). Analysis of muscle receptors for connections by spike-triggering averaging. I. Spindle primary and tendon organ afferents. *J Neurophysiol* 39: 1375-1392.

Weber, F. (1846) *Wagners Handworted die Physiologic* 3: 81.

Weiss, C, Houk, JC, & Gibson, AR (1990). Inhibition of sensory responses of cat inferior olive neurons produced by stimulation of red nucleus. *J Neurophysiol* 64: 1170-1185.

Weiss, PL, Hunter, IW, & Kearney, RE (1988). Human ankle joint stiffness over the full range of muscle activation levels. *J Biomech* 21: 539-544.

Werner, W, Bauswein, E, & Fromm, C (1991). Static firing rates of premotor and primary motor cortical neurons associated with torque and joint position. *Exp Brain Res* 86: 293-302.

Welford, AT, Norris, AH, & Schock, NW (1969). Speed and accuracy of movement and their changes with age. In WG Koster (Ed.) *Attention and performance II* (pp. 3-15). Amsterdam: North-Holland.

Whiting, HTA (1984). *Human motor actions: Bernstein reassessed*. Amsterdam: Elsevier.

Whitney, DE (1969). Resolved motion rate control of manipulators and human prostheses. *IEEE Trans Man-Machine Syst* MMS-10: 47-53.

Wiener, N (1948). *Cybernetics*. New York: Wiley.

Wierzbicka, M, Wiegner, AW, Logigian, EL, & Young, RR (1991). Abnormal most-rapid isometric contractions in patients with Parkinson's disease. *J Neurol Neurosurg Psychiat* 54: 210-216.

Wierzbicka, M, Wiegner, AW, & Shahani, BT (1986). Role of agonist and antagonist muscles in fast arm movements in man. *Exp Brain Res* 62: 331-340.

Wiesendanger, M, Chapman, CE, Macpherson, J, Berger, W (1984). Neurophysiological investigations of tizanidine. In B Conrad, R Benecke, & HJ Bauer (Eds.): *Die klinische wertung der spastizitat* (pp. 39-55). Stuttgart: Schattauer.

Winer, BJ (1971). *Statistical principles in experimental design*. New York: McGraw-Hill.

Wing, AM (1988). A comparison of the rate of pinch grip force increases and decreases in Parkinsonian bradykinesia. *Neuropsychol* 26: 479-482.

Wing, AM, & Kristofferson, AB (1973). Response delays and the timing of discrete motor responses. *Percept Psychophys* 14: 5-12.

Wing, AM, & Miller, E (1984). Basal ganglia lesions and psychological analyses of the control of voluntary movement. In D Evered & M

O'Connor (Eds.) *Functions of the basal ganglia. Ciba foundation symposium N 107* (pp. 242-253). London: Pitman.

Winter, DA, Olney, SJ, Conrad, J, White, SC, Ounpuu, S, & Gage, JR (1990a). Adaptability of motor patterns in pathological gait. In JM Winters & SL-Y Woo (Eds.) *Multiple muscle systems. Biomechanics and movement organization* (pp. 680-693). New York: Springer-Verlag.

Winter, D, Ruder, GK, & MacKinnon, CD (1990b). Control of balance of upper body during gait. In JM Winters & SL-Y Woo (Eds.) *Multiple muscle systems. Biomechanics and movement organization* (pp. 534-541). New York: Springer-Verlag.

Wolpaw, JR (1983). Adaptive plasticity in the primate spinal stretch reflex: Reversal and redevelopment. *Brain Res* 278: 299-304.

Wolpaw, JR, Braitman, DJ, & Seegal, RF (1983). Adaptive plasticity in primate spinal stretch reflex: Initial development. *J. Neurophysiol* 50: 1296-1311.

Woollacott, MH, & Shumway-Cook, A (1986). The development of postural and voluntary motor control systems in Down's syndrome children. In MG Wade (Ed.) *Motor skill acquisition and the mentally handicapped: Issues in research and training* (pp. 45-71). Amsterdam: North-Holland.

Woollacott, MH, Von Hosten, C, & Rosblad, B (1988). Relation between muscle response onset and body segmental movements during postural perturbations in humans. *Exp Brain Res* 72: 593-604.

Worringham, CJ, & Stelmach, GE (1990). Practice effects on the preprogramming of discrete movements in Parkinson's disease. *J Neurol Neurosurg Psychiat* 53: 702-704.

Wright, CE, & Meyer, DE (1983). Conditions for a linear speed-accuracy trade-off in aimed movements. *Quart J Exp Psychol* 35A: 279-296.

Wrisberg, CA, & Winter, TP (1985). Reproducing the end location of a positioning movement: The long and short of it. *J Mot Behav* 17: 242-254.

Wylie, RM, & Tyner, CF (1981). Weight-lifting by normal and deafferented monkeys: Evidence for compensatory changes in ongoing movements. *Brain Res* 219: 172-177.

Yang, JF, Winter, DA, & Wells, RP (1990). Postural dynamics in the standing human. *Biol Cybern* 62: 309-320.

Yamanishi, J, Kawato, M, & Suzuki, R (1980). Two coupled oscillators as a model for the coordinated finger tapping by both hands. *Biol Cybern* 37: 219-225.

Zajac, FE (1989). Muscle and tendon: Properties, models, scaling, and application to biomechanics and motor control. *CRC Crit Rev Biomed Eng* 17: 359-411.

Zajac, FE, & Gordon, ME (1989). Determining muscle's force and action in multi-articular movements. *Exer Sport Sci Rev* 17: 187-230.

Zajac, FE, Gordon, ME, & Hoy, MG (1986). Physiological classification of muscles into agonist-antagonist muscle action groups: Theory and methodology based on mechanics. *Abstr Soc Neurosci* 12: 1424.

Zanone, PG, & Kelso, JAS (1992). Evolution of behavioral attractors with learning: Nonequilibrium phase transitions. *J Exp Psychol: Hum Percept Perform* 18: 403-421.

Zatsiorsky, VM (1986). Mechanical work and energy expenditure in human motion. In IV Knets (Ed.) *Contemporary problems of biomechanics, vol. 3, Optimization of the biomechanical movements* (pp. 14-32). Riga: Zinatne Publ, House.

Zelaznik, HN, Hawkins, B, & Kisselburgh, L (1983). Rapid visual feedback processing in single-aiming movements. *J Mot Behav* 15: 217-236.

Zelaznik, HN, Mone, S, McCabe, GP, & Thaman, C (1988). Role of temporal and spatial precision in determining the nature of the speed-accuracy trade-off in aimed-hand movement. *J Exp Psychol: Hum Percept Perform* 14: 221-230.

Zieglgansberger, W, Howe, JR, & Sutor, B (1988). The neuropharmacology of baclofen. In H Muller, J Zierski, & RD Penn (Eds.) *Local-spinal therapy of spasticity* (pp. 37-49). Berlin: Springer-Verlag.

Zuylen, EJ van, Gielen, CCAM, & Denier van der Gon, JJ (1988). Coordination and inhomogenous activation of human arm muscles during isometric torques. *J Neurophysiol* 60: 1523-1548.

Index

All entries beginning with symbols are located before all alphabetical entries.

NORTHERN MICHIGAN UNIVERSITY LIBRARY

3 1854 005 124 709